International Max Planck Research School (IMPRS)
for Maritime Affairs
at the University of Hamburg

Hamburg Studies on Maritime Affairs

Volume 30

More information about this series at http://www.springer.com/series/6888

Vasco Becker-Weinberg

Joint Development of Hydrocarbon Deposits in the Law of the Sea

Springer

Vasco Becker-Weinberg
International Max Planck Research School
for Maritime Affairs
Max Planck Institute for Comparative
and International Private Law
Hamburg
Germany

Erstgutachter: Prof. Dr. Rainer Lagoni, LL.M (Columbia)
Zweitgutachter: Prof. Dr. Markus Kotzur, LL.M. (Duke)
Tag der mündlichen Prüfung: 30.4.2014

ISSN 1614-2462 ISSN 1867-9587 (electronic)
ISBN 978-3-662-43569-4 ISBN 978-3-662-43570-0 (eBook)
DOI 10.1007/978-3-662-43570-0

Library of Congress Control Number: 2014944322

Springer Heidelberg New York Dordrecht London

Printed on acid-free paper

Springer is part of Springer Science+Business Media (www.springer.com)

To my family

Preface

This book corresponds to the doctoral dissertation submitted to the Faculty of Law of the University of Hamburg in 2013, under the supervision of Professor Dr. Rainer Lagoni.

The completion of this study was only possible due to the generous conditions provided by the International Max Planck Research School for Maritime Affairs at the University of Hamburg (IMPRS), where I was a fulltime scholar from 2008 to 2012. I am thankful to Professor Dr. Jürgen Basedow, Director of the IMPRS and the Max Planck Institute for Comparative and International Private Law (MPI-Hamburg), and Professor Dr. Anatol Dutta and Mrs. Barbara Schröder, Coordinators of the IMPRS. I am particularly grateful to Professor Dr. Peter Ehlers and Professor Dr. Marian Pascke for their support and also for their dedication to the IMPRS, and all my colleagues for the passionate and wide-ranging debates. Appreciation is also owed to the administrative and library staff at the MPI-Hamburg, the Max Planck Institute for Public International Law (Heidelberg) and the Institute of the Law of the Sea and of Maritime Law at the University of Hamburg. I am moreover thankful to the Faculty of Law of the University of Hamburg and especially Professor Dr. Peter Wetzels, as well as Professor Dr. Markus Kotzur for accepting to act as second reviewer (*Zweitgutachter*) and Professor Dr. Stefan Oeter for taking part in the doctoral examination.

Writing the doctoral dissertation was an individual undertaking, but not a solitary one. During the years that I have had the pleasure and privilege of working with Professor Dr. Rainer Lagoni, I have vastly benefited from his immeasurable knowledge of the law of the sea and his wisdom in dealing with many of its complexities. In addition to being a leading authority on joint development agreements of offshore hydrocarbon deposits and having written some of the most significant and landmark works on the subject, Prof. Lagoni is an exceptional person and mentor, always available to listen and genuinely interested in debating new ideas and to question established ones. I am extremely fortunate that Prof. Lagoni accepted to be my supervisor and shall remain forever indebted to him. This work was brought to completion greatly because of his encouragement and enthusiasm.

July 2014 Vasco Becker-Weinberg

Contents

Abbreviations

ASEAN	Association of Southeast Asian Nations
Authority	International Seabed Authority
BIICL	British Institute of International and Comparative Law
CLCS	Commission on the Limits of the Continental Shelf
Convention on Biological Diversity	Convention on Biological Diversity, 1992
EEZ	Exclusive Economic Zone
FPSOs	Floating Production, Storage and Offloading Facilities
FSU	Floating Storage Units
ICJ	International Court of Justice
ILC	International Law Commission
IMO	International Maritime Organization
ITLOS	International Tribunal for the Law of the Sea
L.N.T.S.	League of Nations Treaty Series
MARPOL 73/78	International Convention for the Preservation of Pollution from Ships, 1973, as amended by the Protocol of 1978 relating to the International Convention for the Prevention of Pollution from Ships, 1973
MODU	Mobile Offshore Drilling Units
MoU	Memorandum of Understanding
Nm	Nautical miles
OPRC 1990	International Convention on Oil Pollution Preparedness, Response and Co-operation, 1990
OSPAR	Convention for the Protection of the Marine Environment of the North–East Atlantic, 1992
P.C.A.	Permanent Court of Arbitration
P.C.I.J.	Permanent Court of International Justice
Seabed Disputes Chambers	Seabed Disputed Chambers of the International Tribunal for the Law of the Sea
U.N.T.S.	United Nations Treaties Series

UN Charter	Charter of the United Nations, 1945
UNCLOS	United Nations Convention on the Law of the Sea, 1982
UNESC	United Nations Economic and Social Council
UNGA	United Nations General Assembly
UNSC	United Nations Security Council
UNTAET	United Nations Transitional Administration in East Timor
Vienna Convention	Vienna Convention on the Law of Treaties, 1969
Vienna Convention on Succession of States	Vienna Convention on Succession of States in respect of Treaties, 1978
1958 Convention on the Continental Shelf	Geneva Convention on the Continental Shelf, 1958
1958 Convention on the High Seas	Geneva Convention on the High Seas, 1958
1958 Convention on the Territorial Sea and Contiguous Zone	Geneva Convention on the Territorial Sea and Contiguous Zone, 1958
1972 London Convention	Convention on the Prevention of Marine Pollution by Dumping of Waste and Other Matter, 1972
1989 MODU Code	Code for the Construction and Equipment of Mobile Offshore Drilling Units, 1989
1995 Fish Stocks Agreement	Agreement for the Implementation of the Provisions of the United Nations Convention on the Law of the Sea of 10 December 1982 relating to the Conservation and Management of Straddling Fish Stocks and Highly Migratory Fish Stocks, 1995

Chapter 1
Introduction

The increase of claims by coastal States to extend jurisdiction over adjacent maritime areas may be partly attributed to the advancement and availability of the technology that allows for the exploration and exploitation of resources at depths that until a few years ago were unreachable to mankind. While most of the earth is covered by water, States' aspiration to secure and control access to mineral resources has resulted, on many occasions, in conflict between different coastal States regarding the delimitation of maritime boundaries, as well as the sovereignty over islands and other features and their respective adjacent maritime areas. This background is particularly prevailing in emerging economies with a growing population rate and consequential increase of urban centres. Living resources are likewise valuable for these regions as an important source of protein and, therefore, aggravate the complexity of the overall context. Many of these intricate and potentially irreconcilable situations have contributed towards several of the existing conflicts regarding the delimitation of maritime boundaries and are amongst the main reasons for enduring disputes in certain parts of the world.

When mineral resources straddle a maritime boundary line or are found in areas of overlapping claims, States are subject to certain obligations and to act in accordance with the principles of cooperation and to act in good faith. In the case of maritime areas that are claimed by two or more States, the legal regime requires that States make every effort to enter into interim measures, lacking, however, clarification as to the extent or content of these efforts. Also, in the case of mineral resources that straddle a maritime boundary, coastal States are not required by the *United Nations Convention on the Law of the Sea* (UNCLOS)[1] to enter into an agreement regarding the exploration and exploitation of these resources, similarly as they are not bound to share with the International Seabed Authority (Authority)

[1] *United Nations Convention on the Law of the Sea*, done at Montego Bay on 10 December 1982 and entered into force on 16 November 1994, published at 1833 U.N.T.S. 3 [UNCLOS].

V. Becker-Weinberg, *Joint Development of Hydrocarbon Deposits in the Law of the Sea*,
Hamburg Studies on Maritime Affairs 30, DOI 10.1007/978-3-662-43570-0_1

the mineral resources that lie across the Area and areas subject to national jurisdiction.

The law of the sea merely requires the delimitation of maritime boundaries by agreement and does not address issues pertaining to territorial disputes. Additionally, States are often reluctant to submit disputes to International Courts and Tribunals, partly due to the uncertainty of its outcome, particularly as the presence of mineral resources is not given specific relevance by international jurisprudence when drawing a maritime boundary. States further complicate matters by adopting a nationalistic rhetoric, which may be difficult to retract from in order to achieve a resolution of the dispute, and also because of their ambition to guarantee exclusive control over maritime areas with great mineral resource-potential.

The concept of joint development is not regulated in UNCLOS or in related international legal instruments, or for that matter in the *Geneva Conventions on the Law of the Sea*.[2] UNCLOS also does not include any rules regarding seabed activities in disputed maritime areas. Notwithstanding, several States have considered joint development a valuable alternative that allows for the development of mineral resources in areas of overlapping claims pending the delimitation of maritime boundaries without hindering their claims, including in areas of the continental shelf beyond 200 nautical miles (nm). Likewise, States have resorted to joint development agreements after the delimitation of maritime boundaries regarding transboundary mineral resources, even though mineral resources clauses included in delimitation treaties have, in most cases, a mere programmatic character, without predicting the sharing of resources. Similar joint development agreements could eventually be considered for the purpose of developing mineral resources in the Area that lie across limits of national jurisdiction.

It remains, therefore, to be seen whether the law of the sea has developed a rule requiring States to cooperate in the exploration and exploitation of mineral resources in the situations mentioned before, or to adopt conservation measures ensuring the efficient development of mineral resources and the avoidance of waste.

This dissertation analyses the joint development of hydrocarbon deposits in the law of the sea as a particular example of cooperation between States. It will begin, in Chap. 2, by examining the origin and concept of joint development and the relevant distinctions between agreements signed before and after the delimitation

[2] *Convention on the High Seas*, done at Geneva on 29 April 1958 and entered into force on 30 September 1962, published at 450 U.N.T.S. 11 [1958 Convention on the High Seas]; *Convention on the Continental Shelf*, done at Geneva on 29 April 1958 and entered into force on 10 June 1964, published at 499 U.N.T.S. 311 [1958 Convention of the Continental Shelf]; *Convention on the Territorial Sea and Contiguous Zone*, done at Geneva on 29 April 1958 and entered into force 10 September 1964, published at 516 U.N.T.S. 205 [1958 Convention on the Territorial Sea and Contiguous Zone]; *Convention on Fishing and Conservation of the Living Resources of the High Seas*, done at Geneva on 29 April 1958 and entered into force on 20 March 1966, published at 559 U.N.T.S. 285; *Optional Protocol of Signature concerning the Compulsory Settlement of Disputes*, done at Geneva on 29 April 1958 and entered into force on 30 September 1962, published at 450 U.N.T.S. 169.

of maritime boundaries. This Chapter will then refer to the purpose of joint development, taking into consideration the characteristics of hydrocarbon deposits, particularly in what concerns the efficient development of these and the adoption of conservation measures, such as unitization. The dissertation will subsequently make in Chap. 3 an assessment of the legal regime applicable to the different maritime areas considering, foremost, the rights of coastal States to explore and exploit mineral resources and the rights of other States in these areas. Since the practice of joint development is conventional in its legal nature and, therefore, is subject to international law, Chap. 3 will also consider the power of coastal States to negotiate and enter into joint development agreements before and after the delimitation of maritime boundaries. It will also take into account the efforts made to harmonize State practice regarding joint development of mineral resources. The relevance and legal nature of mineral resources included in delimitation treaties and joint development agreements will be studied in Chap. 4, together with States' obligation regarding transboundary mineral resources in the absence of such clauses. Additionally, the dissertation will also examine in Chap. 7 the role that international jurisprudence has given to mineral resources in maritime delimitation and what it has considered as being States' obligations in situations where two or more States share common mineral resources.

Chapters 5–7 will examine the legal regime and nature of State practice on joint development before and after the delimitation maritime boundaries, while Chap. 9 will consider the possibility of joint development being applicable to mineral resources in the Area that lie across limits of national jurisdiction. In particular, Chap. 5 will look into State practice regarding joint development agreements signed after the delimitation of maritime boundaries and the relevant applicable legal regime, mostly concerning coastal States' jurisdiction and enforcement over installations and structures, including their removal and decommissioning, as well as their liability for pollution caused from seabed activities.

The legal regime applicable to joint development in areas of overlapping claims is examined in Chaps. 6 and 7. Whilst Chap. 7 looks specifically into the Asia--Pacific region as a valuable case-study of maritime areas of overlapping claims and where the (potential) presence of mineral resources has a fundamental influence on current disputes, Chap. 6 examines the essential legal and functional aspects of joint development before the delimitation of maritime boundaries. Chapter 6 also studies the relevant aspects of States' entitlement to establish a joint development regime in disputed maritime areas and the identification of the joint development area, as well as the rights of coastal States to carry out seabed activities and their obligation to protect and preserve the marine environment in disputed maritime areas.

The dissertation will end with its conclusions in Chap. 10, being the ensuing legal question whether the treaty practice produced by joint development agreements thus far is sufficiently consistent and uniform and what legal assumptions can be drawn from it. It will also make an assessment of the adequacy of the law of the sea regarding the exploration and exploitation of mineral resources found in maritime areas of overlapping claims or that straddle maritime boundaries.

The overall purpose of the dissertation is to examine the legal regime applicable to the different situations in which the joint development of offshore hydrocarbon deposits is possible, taking into consideration the rules provided in international law and in particular the law of the sea.

Chapter 2
The Concept and Purpose of Joint Development of Hydrocarbon Deposits

2.1 The Concept of Joint Development

In a broad sense, joint development agreements can be defined as cooperative efforts between two or more States for the exploration and exploitation of mineral resources that straddle a maritime boundary or are found in areas of overlapping claims.[1]

Under UNCLOS and customary international law, non-living marine natural resources in the seabed and subsoil of the territorial sea[2] or of archipelagic water[3] are subject to the sovereignty of the coastal State. The coastal State exercises sovereign rights for the purpose of exploring or exploiting such resources in its exclusive economic zone (EEZ)[4] or continental shelf, including beyond 200 nm.[5] Despite these rules being clear and simple, legal and practical problems occur in two particular situations: first, if an offshore hydrocarbon deposit straddles a boundary line and, second, if such resources are found in a maritime area that is claimed by two or more States. These problems arise mainly in the case of oil and gas because of the migratory nature of these resources. In order to find adequate solutions for these problems, States have developed joint development agreements

---

[1] Vasco Becker-Weinberg, "Theory and practice of joint development in international law", in: *Cooperation and Development in the South China Sea*, edited by Zhiguo Gao, Yu Jia, Haiwen Zhang and Jilu Wu (China Democracy and Legal System Publishing House: Beijing, 2013), p. 85. Vasco Becker-Weinberg defines joint development agreements as self-regulating conventional instruments subject to international law, signed between two or more States holders of a legal title, although independent of such rights as claimed by the intervening States, concerning the maritime areas where natural resources are found in the seabed and marine subsoil, as well as undertaking of all activities deemed necessary without foregoing the rights and freedoms of third States granted under international law. Becker-Weinberg notes that any definition of joint development must take into consideration the continuous evolving State practice, as well as the discretionary powers of States to negotiate and enter into international agreements.

[2] Article 2(2) of UNCLOS.

[3] Article 49(2) of UNCLOS.

[4] Article 56(1) lit. (a) of UNCLOS.

[5] Article 77(1) of UNCLOS.

V. Becker-Weinberg, *Joint Development of Hydrocarbon Deposits in the Law of the Sea*, Hamburg Studies on Maritime Affairs 30, DOI 10.1007/978-3-662-43570-0_2

States are nonetheless bound to specific international obligations pending and during the implementation of a joint development agreement. These include, inter alia, the obligations of mutual-restraint and to share information, and of due regard for the sovereignty and sovereign rights of other States. Yet, irrespective of these obligations, the exercise of States' discretionary powers to define the structure and legal regime of joint development regimes of offshore hydrocarbon deposits has resulted in significant disparity between the agreements implemented thus far.

As the subsequent analysis demonstrates, there is a difference in the legal nature of joint development agreements implemented before or after the delimitation of maritime boundaries or as part of a boundary settlement. However, they do not alter the spatial organization of the seas and oceans or the rights and obligations of States in the different maritime areas, nor do they affect their claims to disputed maritime areas. Indeed, when implemented in disputed maritime areas, joint development regimes are by their nature interim arrangements that are dependent, amongst others, on the final delimitation of maritime boundaries, the economic viability of joint exploration and exploitation activities, the exhaustion of resources or the alteration of their characteristics.

Essentially, joint development agreements of offshore hydrocarbon deposits have a functional character that aims to create a legal regime applicable to the exploration and exploitation of hydrocarbon deposits in any of the particular situations mentioned above.

2.2 The Purpose of Joint Development

2.2.1 Characteristics of Hydrocarbon Deposits

Similarly to the 1958 Convention on the Continental Shelf, UNCLOS states that the natural resources of the continental shelf include the mineral and other non-living resources of the seabed and subsoil.[17] Although the Convention includes the seabed and subsoil in the EEZ, it refers to the legal regime of the continental shelf concerning the rights of coastal States for the exploration and exploitation of hydrocarbon deposits.[18]

Hydrocarbon deposits are a concentration of organic sediments composed of hydrogen and carbon atoms found in basins trapped in pore spaces of rock in the subsoil that may be liquid or solid. They may also be in the form of gas, depending on its molecules, pressure conditions and temperature.

In an article published in 1938, Northcutt Ely metaphorically described hydrocarbon deposits as engines that are put into operation by releasing the pressure trapped inside. This rock pressure is caused by the compressed gas or

[17] Article 2(4) of the 1958 Convention on the Continental Shelf and Article 77(4) of UNCLOS.

[18] Article 56(3) of UNCLOS.

water within the deposit. When a source-rock where petroleum is trapped is drilled, this operation causes the release of the pressure inside and consequently its content is forced to the surface. There are different types of oil displacement mechanisms and each requires different extraction procedures in order for production to be efficient. As drilling operations are carried out, the subterraneous pressure inside the deposit diminishes and its underground characteristics are altered.[19] Pressure of deposits is also influenced by the used drilling mechanism, the times a certain deposit has been drilled, the point or points where it has been drilled and the drive of existing pressure. It may occur that operators have to introduce additional pressure, such as the reinjection of gas, in order to recover the remaining content.

These characteristics make it very difficult for simultaneous drilling operations to take place, as the mobility of petroleum through geological layers of the subsoil caused by the release of pressure allows these resources to move within the permeable source-rock.[20] The existence of two or more points of release of pressure may hinder the chances of any of the operators recovering a significant part of the petroleum trapped. When competing drilling is in progress, operators would not easily accept that part of the respective production should be put back into the ground merely to provide another operator's production with the pressure required to drive the latter's oil to the surface, at least, not without sharing some of the benefits. This rush to retrieve as much oil and gas as possible in a short period of time would lead to an uncontrolled drilling and to the lost and waste of resources. Thus, M.W. Mouton's adage: "*never two straws in one glass*".[21]

2.2.2 Exploration and Exploitation Activities of Offshore Mineral Resources

The 1958 Conventions on the Continental Shelf and on the High Seas[22] and UNCLOS refer in several provisions to the rights of coastal States to exploit the EEZ, the continental shelf and the Area, and to explore their natural resources,

---

[19] Northcutt Ely, "The conservation of oil", in: 51 *Harvard Law Review* (1937–1938), pp. 1219–1220.

[20] On petroleum migration *see* Knut Bjørlykke, "Introduction to petroleum geology", in: *Petroleum Geoscience: From Sedimentary Environments to Rock Physics* (Springer Verlag: Berlin, 2011), pp. 10–11; and "Petroleum migration", *ibid*, pp. 349–360.

[21] M.W. Mouton, "The Continental Shelf", in: 85 *Recueil des Cours de l'Académie de Droit International* (1954-I), p. 421.

[22] Articles 1, 2(1) and (2), 4, and 5(1) and (2) of the 1958 Convention on the Continental Shelf, and Articles 24 and 26(2) of the 1958 Convention on the High Seas.

even though without providing a definition or identifying the content of either one of these concepts.[23]

The product-chain of petroleum begins with the discovery and extraction of resources and ends with making them available to consumers after being refined. Exploration and exploitation activities are included in the upstream part of the product-chain of petroleum, while the downstream part of this chain traditionally includes the transportation (also often referred autonomously as midstream), storage, refining, logistics, marketing and distribution. More specifically, exploration includes the prospecting and surveying operations that may eventually lead to the discovery and identification of the location and characteristics of hydrocarbon deposits and collecting other relevant data for determining the viability of initiating production, while exploitation activities consist of the extraction or production operations of these resources.

UNCLOS establishes that coastal States exercise sovereign and exclusive rights for the purpose of exploring and exploiting the natural resources of the continental shelf and that, consequently, only the coastal State has the exclusive right to authorize and regulate drilling on the continental shelf.[24] As a result, in disputed maritime areas and pending agreement between the relevant States, only operations that do not involve drilling or any physical change to the marine environment may be carried out.[25] In these cases and for the purpose of this work, such operations should not be considered as included in exploitation activities.

2.2.3 From the Rule of Capture to Efficient Development

The modern narrative of oil and gas development begins with the discovery of oil in Pennsylvania in the nineteenth century. Early development of onshore hydrocarbon deposits was built upon the analogy of the legal principle known as the rule of capture, according to which the owner of a certain tract of land had the right to make his the game or animals ferae naturae that came through or could be found in his respective property, as part of his right of property to the land.[26] This

[23] *Preamble*, Articles 1(3), 56(1), 65, 69 to 73, 77, 79, 82(1) and (2), 85, 118, 123, 142(2), 151(8), 153, 155(1), (2) and (4), 160(2) lit ((f.) ii), 162(2) lit. ((o) ii) and lit. (x), 163(8), 165(1) and (2) lit. (l), 168(2), 193, 194(3), 246(5) lit. (a), 249(2), 266(2), 269 lit. (a), 277 lit. (a) and 314(2), and Annexes III and IV and Article 7(1) of Annex VI of UNCLOS.

[24] Articles 77(1) and (2) and 81 of UNCLOS.

[25] *Guyana/Suriname, Award of the Arbitral Tribunal, 17 September 2007*, paras. 466–467 and 480–481 [online: http://www.pca-cpa.org/upload/files/Guyana-Suriname%20Award.pdf (accessed on August 2013)].

[26] Bruce M. Kramer and Owen L. Anderson, “The rule of capture—an oil and gas perspective”, in: 35 *Environmental Law* (2005), p. 949. Also *see* George W. Hardy, “Drainage of oil and gas from adjoining tracts—a further development”, in: 6 *Natural Resources Journal* (1966), pp. 47–51; James A. Veasey, “The law of oil and gas—part I”, in: 18 *Michigan Law Review* 445 (1919–1920), pp. 454, 464–465.

perception laid the path for the development of the oil and gas industry based on the principle of *first-come, first-serve* or "*drill-and-produce-as-you-please method*"[27] and provided the stimulus for an uncontrolled drilling activity in the United States of America.

During the early years of oil production, the rule of capture became a core principle of this activity under the umbrella of the Fourteenth Amendment to the Constitution of the United States of America, which protects in almost absolute terms owners' property rights from States' attempts to deprive any person of property.[28] The protection provided by the Fourteenth Amendment consolidated the perception that the owner of a tract of land was entitled to the oil and gas found under the surface for the purpose of exploration and exploitation.[29] Shielded by their rights as owners or lessees and in the absence of a compulsory conservation regime, operators had the right to explore and develop the natural resources of the soil and subsoil without due regard for neighbouring oil exploration and exploitation activities.[30]

The outcome of applying the rule of capture to onshore oil production in the United States of America was the launch of competitive drilling amongst owners of a common hydrocarbon deposit. In most cases, this lead to the *black gold* rush of the 1920s due to the excess of low-cost oil and the rapid exhaustion of resources, as well as to the mismanagement of production and the difficulty in providing adequate storage for the amount of oil extracted.[31]

The increase of demand for oil and gas in the United States of America and the fact that States in this country found themselves on many occasions powerless to regulate and supervise the significant number of wells being drilled on a regular basis, resulted in a crisis of production that ultimately lead to the need for the

---

[27] Robert E. Hardwicke, "The rule of capture and its implications as applied to oil and gas", in: 13-4 *Texas Law Review* (June 1935), p. 395.

[28] The text of the Fourteenth Amendment to the Constitution of the United States of America reads: "*All persons born or naturalized in the United States, and subject to the jurisdiction thereof, are citizens of the United States and of the State wherein they reside. No State shall make or enforce any law which shall abridge the privileges or immunities of citizens of the United States; nor shall any State deprive any person of life, liberty, or property, without due process of law; nor deny to any person within its jurisdiction the equal protection of the laws*".

[29] *Ryan Consolidated Petroleum Corporation v. Pickens, Supreme Court of Texas*, 1955, 155 Tex. 221, 285 S.W.2d 201; *Westmoreland and C. Nat Gas Co. v. De Witt*, 1899, 130 Pa.235. *See* Laura H. Burney, "A pragmatic approach to decision making in the next era of oil and gas jurisprudence", in: 16 *Journal of Energy, Natural Resources and Environmental Law* (1996), pp. 19–32.

[30] *Brown v. Spilman*, 155 U.S. Supreme Court 665, 14 January 1895, p. 92.

[31] Jacqueline Lang Weaver and David F. Asmus, "Unitizing oil and gas fields around the world: a comparative analysis of national laws and private contracts", in: 28 *Houston Journal of International Law* 3 (2006), pp. 7, 35.

implementation of conservation measures by some oil producing States.[32] However, although courts in the United States of America became aware of the need for the adoption of conservation measures, some of its jurisprudence considered that the protection provided by the Fourteenth Amendment did not allow for the curtailment of owners' property rights and therefore considered the rule of capture to be vested in the same.[33] This was the case, for example, of Texas which considered that under the Fourteenth Amendment the owner of a tract of land was entitled to the oil and gas produced from wells drilled thereon, even though part of the oil or gas may have migrated from neighbouring tracts of land.[34]

The fact remained that some courts in the United States of America considered that in the absence of a statute that empowered administrative bodies to direct the adoption of conservation measures of an entire field, the goal of field-wide unitization could only be reached by voluntary cooperation.[35] The compromise between, on the one hand, an absolute concept of property rights included in the rule of capture and the concern for the protection of resources, and on the other hand, the safeguard of a collective interest manifested in the avoidance of waste of resources and the protection of neighbouring operators' rights, had in most cases resulted in the sacrifice of the latter. It would, therefore, fall upon an "*ordinarily prudent operator*" to adopt the necessary precautionary actions.[36] In fact, the Supreme Court of the United States of America recognized that "*the owner of the soil has the exclusive right to reduce to possession the deposits of natural gas and*

[32] Charles Haglund, "The new conservation movement with respect to petroleum and natural gas", in: 22 *Kentucky Law Journal* (1933–1934), p. 551; Robert E. Hardwicke, "Unitization statutes: voluntary action or compulsion", in: 24 *Rocky Mountain Law Review* (1951–1952), pp. 38–43; J. Howard Marshall and Norman L. Meyers, "Legal planning of petroleum production", in: 41 *Yale Law Journal* (1931), p. 33; Weaver/Asmus, *supra* note 31, p. 34.

[33] *See Ryan Consolidated Petroleum Corporation v. Pickens*, *supra* note 29. The Supreme Court of Texas concluded that the Railroad Commission of Texas was an administrative body that did not have, under the law of Texas and the Constitution, the power to render ineffective the rule of capture, since it would have an impact on property rights. Also *see Carter Oil Co. v. Dees*, Appellate Court of Illinois, Fourth District, 1950, 340Ill.App.449, 92 N.E.2d 519. On conservation measures in Illinois *see* Howard W. Campbell, "Oil and gas conservation in Illinois", in: 1959 *University of Illinois Law Forum* (Summer 1959), pp. 570–583.

[34] On the applicability of the rule of capture in Texas *see* Dylan O. Drummond, Lynn Ray Sherman and Edmond R. McCarthy, Jr., "The rule of capture in Texas—still so misunderstood after all these years", in: 37 *Texas Tech Law Review* (2004–2005), pp. 1–97; Proctor L. Thomas, "Prospects for compulsory fieldwide unitization in Texas", in: 44 *Texas Law Review* (1965–1966), pp. 510–532.

[35] *Dobson v. Arkansas Oil and Gas Commission*, Supreme Court of Arkansas, 1951, 218 Ark. 160, 235 S.W. 2d 33. The statute of Arkansas was eventually amended to empower the Oil and Gas Commission to compel the unitization of tracts under different ownership. *See* Susan Webber Wright, "The Arkansas law of oil and gas", in: 9-2 *University of Arkansas at Little Rock Law Journal* (1986–1987), pp. 234–235.

[36] *Millette v. Phillips Petroleum Co.*, Supreme Court of Mississippi, in Banc, 1950, 209 Miss. 687, 48 So.2d 344. On the conservation legislation of Mississippi *see* Robert G. Rogers and E. Spivey Gault, "Mississippi compulsory field-wide unitization", in: 44 *Mississippi Law Journal* (1973), pp. 185–207.

oil found beneath the surface of his land", although adding that considering the fugitive character of oil and gas "*the owner may not follow the natural gas when it shifts from beneath his own to the property of someone else within the gas field*".[37] Evidence of the legacy of the rule of capture being applicable to game or animals ferae naturae.

Since there was no legal presumption that when the oil of one well was drained that part of that oil would be from neighbouring wells, even if technology could eventually allow such assessment,[38] oil was considered as a part of the realty until brought to the surface and susceptible of being sold in place. As a result, the operator had the right to take full charge of the land so far as necessary to get the oil from under it and as long as entitled to the oil for that purpose. Furthermore, considering the "*power of self-transmission*" of oil, the rule of capture was viewed as the sole possibility that could offer assurance as to its ownership, based on the principle that the property belonged to the owner or lessee of the land upon which the well is located and that manages to drill this resource to the surface.[39] Therefore, the means available under the rule of capture for neighbouring operators that witnessed oil and gas being drained from their respective tracts of land was to proceed likewise in order to protect the resources found under their property, but without granting the title to drain oil and gas from neighbouring land. This implied drilling of a well on property under which oil is being drained away by a well on adjacent property to make up for the loss of oil from the first property.

Only with the protection of property lines and the development of offset wells drilled on their own tracts of land could operators avoid the complete depletion of hydrocarbon deposits susceptible of being drilled from either side of a boundary line.[40] The practical solution would seem to be to allow free competition in the development of a common field as a means to achieve just distribution of the resources of a deposit.[41]

As early as the rule of capture was being applied to oil production, there was a growing plea for the adoption of legal conservation measures as an alternative and

---

[37] *Ohio Oil Co. v. Indiana*, Supreme Court of the United States of America, 1900, 177 U.S. 190,20 S.Ct.576, 44 L.Ed. 729. On correlative rights *see* Peter D. Junger, "The Wyoming oil and gas conservation act—private rights and public policy", in: 13 *Wyoming Law Journal* (1958–1959), pp. 32–37; Willmer D. Masterson, Jr., "A 1952 survey of gas and oil law", in: 6 *Southwestern Law Journal* (1952), pp. 4–8; David Edward Pierce, "Coordinated reservoir development—an alternative to the rule of capture for the ownership and development of oil and gas—part I", in: 4 *Journal of Energy Law and Policy* (1983), pp. 50–55; Maurice H. Merrill, "Recent unitization cases", in: 6 *Oklahoma Law Review* (1953), pp. 169–170.

[38] *Japhet v. McRae*, Commission of Appeals of Texas, Section B, 1925, 276 S.W. 669.

[39] Veasey, *supra* note 26, p. 453.

[40] Maurice H. Merrill, "The evolution of oil and gas law", in: 13 *Mississippi Law Journal* (1940–1941), p. 290; Hardwicke, *supra* note 27, p. 393.

[41] *Barnard v. Monongahela Natural Gas Co.*, Supreme Court of Pennsylvania, 1906, 216 Pa. 362, 65 A. 801; A.W. Walker, Jr., "Property rights in oil and gas and their effect upon police regulation of production", in: 16 *Texas Law Review* (1937–1938), p. 374; Haglund, *supra* note 32, p. 544.

as a replacement of this rule. Unitization of hydrocarbon deposits provided for the conservation of these resources and the prevention of disputes between different operators that could jeopardize production. It also allowed for the exchange of technical information and allocation of means and equipment that could lead to the reduction of costs, as well as the rational exploration and exploitation of a deposit as an integrated unit and consequently allowing for better recoveries, low operating costs and an equitable distribution of production.[42]

Among the leading advocates for the adoption of conservation measures, Charles Haglund made the case for proration[43] and eventually compulsory unitization as the most adequate means to stabilize the oil industry. In part because it would limit and control the amount of production and consequently reduce the quantity that was being wasted.[44] Charles Haglund believed that a statute for compulsory proration would be preferable, since only through the latter would it be possible to determine the acreage content of a deposit and allow several owners to become common tenants.[45] Similarly, William P. Z. German considered that proration and unit operations allowed operators to retrieve as much oil as it was initially found, although concluding that "[*p*]*roration is a helpful remedy; but unit operation is a cure, for it can more nearly attain a fair distribution as well as prevent underground waste*".[46]

Bruce M. Kramer and Owen L. Anderson expressed in their critical opinion of the rule of capture and its "*evil twin*" the offset drilling rule that the owner of a land drained by neighbouring wells could not take part in such development, let alone claim damages from the respective neighbouring operator.[47] In a different

---

42 Bernard Taverne, *Co-operative Agreements in the Extractive Petroleum Industry* (Kluwer Law International: London, The Hague, Boston, 1996), pp. 95–100. Also *see* Keith Blinn, Claude Duval, Honoré Le Leuch and André Pertuzio, *International Petroleum Exploration and Exploitation Agreements. Legal, Economic and Policy Aspects* (Euromoney Publications: Bristol, 1986), p. 212; Barbara Kwiatkowska, "Economic and environmental considerations in maritime boundary delimitations", in: *International Maritime Boundaries*, vol. 1, edited by Jonathan I. Charney and Lewis M. Alexander (Martinus Nijhoff Publishers: Dordrecht, Boston, London, 1993), p. 80. On the relevance of the technological advancement for the development of oil and gas in the North Sea *see* Peter R. Odell, "Oil and gas exploration and exploitation in the North Sea", in: 1 *Ocean Yearbook* (1978), pp. 139–159.

43 Proration is a method of waste prevention through production control and allocating production on a per-well basis, while recognizing the correlative rights of neighbouring owners providing them with an equitable share of the oil as it existed originally in the pool and which constituted the proportionate share of the reservoir's energy for the production of oil. *See* Haglund, *supra* note 32, pp. 548, 552.

44 *See* W. P. Z. German, "Compulsory unit operation of oil pools", in: 20 *California Law Review* (1932), p. 116; David W. Eckman, "Statutory fieldwide oil and gas units: a review for future agreements", in: 6 *Natural Resources Law* (1973), p. 381.

45 Haglund, *supra* note 32, p. 566.

46 German, *supra* note 44, pp. 115, 130. Also *see* Howard R. Williams, "Conservation of oil and gas", in: 65 *Harvard Law Review* (1951–1952), p. 1168.

47 Kramer/Anderson, *supra* note 26, pp. 900, 911, 953–954. Differently *see* Hardwicke, *supra* note 27, pp. 397–398, 403–408.

view, Robert E. Hardwicke considered that the tendency was to make the rule of capture an "*alibi which is relied upon to prove innocence of voluntary wrong-doing*"[48] or as a cause of the offset drilling rule. Accordingly, with the development of the technology that could allow determining and perhaps control the direction of migrating oil and gas, the rule of capture would not be wholly inappropriate. Instead, it would be quite logical and necessary, although recognizing that it could also be an incentive along with the offset drilling rule for unhealthy and wasteful competitiveness between neighbouring operators.[49]

Eventually, some jurisprudence in the United States of America came to accept that legislation on drilling of neighbouring oil fields should be enacted, despite the fact that, on its own, it would not be completely effective.[50] The adoption of compulsory unitization and pooling by some States in the United States of America permitted that "[*t*]*he history of oil and gas production* (…) *ranged from chaos to an enviable standard of efficiency*".[51] The difference between the two concepts is that pooling implies joining land to create a single-well drilling unit and maintain spacing of wells, while unitization consists of two or more operators exploring a hydrocarbon deposit as a single unit, sharing its costs and revenues, with or without merging the respective legal titles.[52]

In the early 1920s, Henry Doherty was among the first to call for a federal unitization law. Although not successful at first, he would set the path for the discussion of compulsory unitization of onshore production and later for the adoption of statutes, despite some States resisting to implement conservation measures.[53] The proposal perceived that only through conservation statutes or a federal compulsory unitization law as a means of replacing States that did not

---

[48] Robert E. Hardwicke, "Oil conservation: statutes, administration, and court review", in: 13 *Mississippi Law Journal* (1941), p. 384.

[49] Hardwicke, *supra* note 27, pp. 393–394, 420.

[50] A. Allen King, "Pooling and unitization of oil and gas leases", in: 46 *Michigan Law Review* (1947–1948), pp. 312, 317.

[51] Bill Prewett, "Act 134 compulsory unitization of oil and gas pools", in: 5 *Arkansas Law Review* (1950–1951), pp. 392–394. Also *see* Blakely M. Murphy, "The legislative and administrative concept of oil and gas conservation in Arkansas, 1917–1947", in: 1 *Arkansas Law Review* 236 (1946–1947), pp. 236–247; Richard J. Pierce, "State regulation of natural gas in a federally deregulated market: the tragedy of the commons revisited", in: 73 *Cornell Law Review* (1987–1988), p. 25; Richard W. Hemingway, *The Law of Oil and Gas*, 3rd edition (West Publishing Co.: St. Paul, Minnesota, 1991), pp. 427–428. On compulsory pooling *see* H. P. Pressler, Jr., "Legal problems involved in cycling gas in gas fields", in: 24 *Texas Law Review* (1945–1946), pp. 29–32; Weaver/Asmus, *supra* note 31, pp. 12, 34; King, *ibid*, pp. 321–325.

[52] See Jacqueline Lang Weaver, *Unitization of Oil and Gas Fields in Texas* (Resources for the Future, Inc.: Washington, D.C., 1986), p. 1. Also *see* Weaver/Asmus, *supra* note 31, p. 6; Hemingway, *ibid*, pp. 427–428; German, *supra* note 44, p. 111; King, *supra* note 50, 313; Hardwicke, *supra* note 32, pp. 29–30. Differently, David W. Eckman considers the two terms as synonyms *see* Eckman, *supra* note 44, p. 340.

[53] On the proposals of Henry Doherty for the adoption of conservation measures *see* Weaver, *ibid*, pp. 37–38.

approve conservation statutes, as it had happen with Texas,[54] would States achieve the increase production, while simultaneously avoiding waste and protecting property rights. Some authors tried to make the case for compulsory unitization that determined the legal regime that could, in certain aspects, facilitate operators to enter into unitization agreements. Yet, in practice, voluntary unitization proved to be more successful, since operators' participation remained essential to establish the relevant contractual framework.[55] The difficulties in implementing a compulsory unitization regime without owners' intervention would lead to the conclusion that waste prevention would be better achieved through voluntary unit operations.[56]

The debate on the need to adopt conservation measures could be divided in two different sets of arguments: between those that upheld the use of compulsory unitization as the most effective conservation measure and those that, while not necessarily defending the rule of capture, understood that compulsory unitization was not in conformity with the existing legal regulations. Only voluntary unitization would be a legally and constitutionally acceptable solution, since the implementation of compulsory unitization would constitute a direct State interference in production that could harm more than benefit production, and which resulted in a de facto act of confiscation as no limit in production ought to be imposed.[57]

The implementation of compulsory unitization would also breach antitrust laws and therefore, statutes should necessarily have to include measures to avoid the possibility of unit operations being deemed an infringement of these laws.[58] Furthermore, reliability between operators would also suffer, given that operators

---

[54] *See* Acts 1977, 65th Leg. p. 2517, ch. 871, art. I, para. 1, eff. Sept. 1, 1977, amended by Acts 1979, 66th Leg., p. 673, ch. 300, para. 1, eff. May 29, 1979; Acts 1981, 67th Leg., p. 3166, ch. 830, para. 2, eff. June 17, 1981; Acts 1995, 74th Leg., ch. 870, para. 1, eff. Sept. 1, 1995; Acts 1995, 74th Leg., ch. 871, para. 1, eff. Sept. 1, 1995. Despite House Bill 25, 12 August 1931, Texas did pass a compulsory pooling bill in 1965. On the adoption of compulsory unitization legislation in Texas *see* Ernest Smith, "The Texas compulsory pooling act", in: 43 *Texas Law Review* (1965), pp. 1003–1034; *ibid*, 44 *Texas Law Review* (1966), pp. 387–423; Thomas, *supra* note 34, pp. 510–532.

[55] Weaver, *supra* note 52, p. 258.

[56] Joe S. Bain, "Rostow's proposals for petroleum policy", in: 57-1 *Journal of Political Economy* (1949), pp. 55–60; Eugene V. Rostow, "A reply", in: 57-1 *Journal of Political Economy* (1949), pp. 60–68; Joe S. Bain, "A rejoinder", in: 57-1 *Journal of Political Economy* (1949), pp. 68–69; Leslie Moses, "Some legal and economic aspects of unit operations of oil fields", in: 21 *Texas Law Review* (1942–1943), pp. 750–751, 771.

[57] Hardwicke, *supra* note 27, pp. 408, 412; Veasey, *supra* note 26, p. 456.

[58] Allen L. Handlan and Kevin L. Sykes, "Pooling and unitization: legal and ethical considerations", in: 19 *Tulsa Law Journal* (1983–1984), pp. 333–334; Stephen L. McDonald, "Unit operation of oil reservoirs as an instrument of conservation", in: 49 *Notre Dame Law* (1973–1974), pp. 314–315; Wendell J. Doggett, "Practical legal problems encountered in the formation, operation and dissolution of fieldwide oil and gas units", in: *16 Oklahoma Law Review* (1963), pp. 25–29; Eckman, *supra* note 44, pp. 362–363, 384–385; King, *supra* note 50, p. 326.

would prefer to willingly adopt measures that improved production, rather than being compelled to do so.[59]

Similar to some States in the United States of America, other countries have passed legislation that envisage the unitization of hydrocarbon deposits as a means of preventing competitive drilling.[60] Accordingly, States or the relevant national authorities would intervene and establish the applicable conditions for the efficient development of common hydrocarbon deposits in the absence of agreement between operators. States that have not passed legislation on unitization may still establish the appropriate obligations for operators before granting the right to undertake exploration and exploitation activities.[61]

Despite domestic laws enforcing compulsory unitization, under international law the problem remains concerning the development of hydrocarbon deposits that straddle a maritime boundary or that are found in maritime areas where States' claims overlap. In fact, despite international law having ruled-out the rule of capture,[62] it has not set forth a general rule applicable to the exploration and exploitation of common offshore hydrocarbon deposits, or the adoption of conservation measures such as unitization or, for that matter, joint development.[63]

When executing a unitization regime, States or the entities responsible for the implementation of these regimes will necessarily enter into agreements with operators, which in most cases are oil companies. These agreements will also include provisions ensuring that operators will cooperate should their rights fall on

---

59 German, *supra* note 44, p. 113; Hardwicke, *supra* note 27, 384.

60 *See* Getting the Deal Through, *Oil Regulation 2012*, edited by Bob Palmer (Law Business Research Ltd.: London, 2012).

61 On unitization agreements *see* Bernard Taverne, *An Introduction to the Regulation of the Petroleum Industry. Laws, Contracts and Conventions, International Energy and Resources Law and Policy Series* (Graham and Trotman/Martinus Nijhoff: London, Dordrecht, Boston, 1994), pp. 147–170; *Ibid, supra* note 42, 95–100. Also *see* Scott Styles, "Joint Operating Agreements", in: *Oil and Gas Law. Current Practice and Emerging Trends*, edited by Greg Gordon and John Paterson (Dundee University Press: Dundee, 2007), pp. 265–301; Nicola MacLeod, "Unitisation", *ibid*, pp. 313–324; Weaver/Asmus, *supra* note 31, pp. 70–98.

62 John C. Jacobs, "Unit operation of oil and gas fields", in: 57 *Yale Law Journal* (1947–1948), p. 1208; Jacques Vuilleimier, "Vers un régime juridique propre à la recherché et à l'exploitation des gisements de pétrole", in: 76 *Zeitschrift für Schweizerisches Recht/Revue de Droit Suisse/Revista di Diritto Svizzero* (1957-2), pp. 286a-293a; Charles Robson, "Transboundary petroleum reservoirs: legal issues and solutions", in: *The Peaceful Management of Transboundary Resources*, edited by Gerald H. Blake, William J. Hildesley, Martin A. Pratt, Rebecca J. Ridley and Clive H. Schofield (Graham and Trotman/Martinus Nijhoff: London, Dordrecht, Boston: 1995), p. 6; Philippe Manin, "Le Traité de Frigg", in: 24 *Annuaire Français de Droit International* (1978), pp. 796–798; Albert E. Utton and Paul D. McHugh, "On an institutional arrangement for developing oil and gas in the Gulf of Mexico", in: 26 *Natural Resource Journal* (1986), pp. 722–732.

63 Rainer Lagoni, "Oil and gas deposits across national frontiers", in: 73-1 *American Journal of International Law* (January 1979), p. 237. Also *see* Julio A. Barberis, "Los recursos minerales compartidos entre Estados y el derecho internacional", in: 18–19(3) *Derecho de la Integración* (March-July 1975), p. 49.

a shared deposit. Such cooperation shall be in accordance with the terms and conditions determined by the unitization regime and that ought to be reflected in the applicable legislation and the respective concession or licensing regime, or in the production sharing contract signed with operators, amongst other.[64]

Similarly to the courts in the United States of America in the early twentieth century, International Courts and Tribunals have long understood the problems represented by common hydrocarbon deposits and the need for the efficient development of mineral resources. However, as it will be analysed in Chap. 8, international jurisprudence has not acknowledge the obligation for neighbouring States to engage in the cooperative development of common mineral resources, nor has it recognized the existence of a rule requiring States to adopt conservation measures in these cases.

From the comparison with conservation measures adopted under domestic law it can be concluded that it is not possible to construct a similar obligation in international law, especially when the rights that States may exert under domestic law do not have a parallel in international law. While under domestic law States may impose the adoption of conservation measures, in international law and particularly in the law of the sea there is no rule requiring States to adopt similar measures regarding offshore hydrocarbon deposits.[65]

2.2.4 Distinction Between Unitization and Joint Development

There are conceptual differences between unitization and joint development agreements of offshore hydrocarbon deposits, despite the fact that both may include the development of mineral resources when they straddle a maritime boundary line or are found in a disputed maritime area.

Unitization agreements essentially envisage the preservation and development of an identified hydrocarbon deposit as a single unit in order to ensure its efficient production and to avoid waste. These agreements when signed between States provide the legal framework for the development of identified offshore hydrocarbon deposits and often constitute an umbrella under which other intricate legal relations will be established.

---

[64] Taverne, *supra* note 42, pp. 79, 85; MacLeod, *supra* note 61, p. 325.

[65] Differently *see* Alberto Székely, "The international law of submarine transboundary hydrocarbon resources: legal limits to behaviour and experiences for the Gulf of Mexico", in: 26 *Natural Resource Journal* (1986), p. 766. Also *see* Zhiguo Gao, "Legal aspects of joint development in international law", in: *Sustainable Development and Preservation of the Oceans: The Challenges of UNCLOS and Agenda 21*, edited by Mochtar Kusuma-Atmadja, Thomas A. Mensah and Bernard H. Oxman (The Law of the Sea Institute, University of Hawaii: Hawaii, 1983), p. 643; *Ibid*, "The legal concept and aspects of joint development in international law", in: 13 *Ocean Yearbook* (1998), p. 115.

States may also agree that access to offshore development is only reserved to national oil companies and consequently establish in the unitization regime the legal framework applicable to the exploration and exploitation of hydrocarbon deposits by these entities as one single unit. However, when private entities also take part in offshore development, States often provide in the unitization regime the terms and conditions that will be included in the licenses or concessions to be granted, or in the agreements to be signed with operators by the relevant States or State-owned entities. These terms and conditions may also include that different operators shall either voluntarily or compulsorily enter into unitization agreements with other operators for the purpose of achieving an efficient production of resources.[66] In some situations, States may create a specific body to oversee the implementation and technical aspects regarding the unitization of offshore hydrocarbon deposits, such as, for example, the definition of blocks or license areas, or requiring operators to appoint a unit operator in the event that two or more operators establish a unitization regime in the terms previously described. In this respect, unitization agreements can be more or less complex legal regimes and include, inter alia, the identification of the participation of different operators in the exploration of a hydrocarbon deposit to be developed as a single unit, the definition of unit percentages, their redetermination and adjustments, unit operator's and non-operators' responsibilities, the establishment of a unit management committee or joint operating committee as applicable, creation of a unit development plan including budget and working plan, as well as sole risks operations for the purpose of redetermination, setting transportation and storage conditions, defining non-unit operations and expenditures, such as administrative costs, payment of consultants and experts. Unitization agreements may also provide an applicable tax and accounting regime, as well as provisions on withdrawal, default, forfeiture, assignment and pre-emption rights and penalty clauses, and other relevant conditions, such as confidentiality, notices, information, authorization procedures, applicable law and dispute resolution mechanisms.[67]

---

[66] Taverne, *supra* note 42, pp. 18–19. Also *see* Keyuan Zou, *China's Marine Legal System and the Law of the Sea*, edited by Vaughan Lowe (Martinus Nijhoff Publishers: Leiden, Boston, 2005), p. 155; Thomas Wälde, "Investment policies in the international petroleum industry—responses to the current crisis", in: *Petroleum Investment Policies in Developing Countries*, edited by Nicky Beredjick and Thomas Wälde (Graham and Trotman: London, 1988), pp. 21–22; Richard Young, "Offshore claims and problems in the North Sea", in: 59-3 *American Journal of International Law* (July 1965), p. 517; International Law Association, *supra* note 15, p. 4; Gao, "Legal aspects..." *ibid*, pp. 632–633 and "The legal..." *ibid*, p. 112.

[67] Taverne, *supra* note 42, pp. 95–100. For an example *see* Article 6(2) of the *Agreement between the United States of America and the United Mexican States concerning Transboundary Hydrocarbon Reservoirs in the Gulf of Mexico*, done at Los Cabos on 20 February 2012 [online: http://www.state.gov/p/wha/rls/2012/185259.htm (accessed August 2013)]. This provision determines that unitization agreements shall include the identification of the limits of the unit area and of any transboundary deposit, the identity of the operators and their respective participating interests, the methodology used to calculate the allocation of production, a development plan for the exploration and exploitation of the unit area, including the estimated number and timing of wells, and a mechanism for delivery and approval of subsequent changes to such plan, the

Differently, joint development agreements of offshore hydrocarbon deposits are signed between two or more States and establish the conditions for the undertaking of exploration and exploitation activities of offshore hydrocarbon deposits that may be found in the seabed and marine subsoil of a designated maritime area, before or after the delimitation of maritime boundaries. As it will be analysed in Chaps. 5 and 6, these regimes may also include provisions applicable to other aspects that States may consider relevant in accordance with their discretionary powers and within the limits of international law and in particular of the law of the sea. These include, for example, the protection and preservation of the marine environment or the safety of navigation. In comparison with unitization agreements, joint development regimes generally have a broader scope and regulate matters regarding existing or potentially existing hydrocarbon resources, amongst other relevant legal and functional aspects.

In the event that a common hydrocarbon deposit is discovered during the implementation of a unitization agreement, States could either decide to extend the existing legal framework to include the newly discovered deposit, or enter into a new unitization regime regarding the latter. States could further agree to disagree and do nothing,[68] or create a buffer zone or a "*protective perimeter*" in order to avoid developing deposits close to the boundary line and susceptible of interfering with other production activities.[69]

Similarly, should a hydrocarbon deposit extend between a joint development area and any maritime area of a third State, the States-parties and the third States must seek an understanding on the development of these resources before any exploration and exploitation activities can take place. The competitive drilling of wells in these situations could cause damages as a result of the depletion of mineral resources and eventually lead to their waste. It would also constitute a serious obstacle for the delimitation of maritime boundaries, particularly if there is no longer an economic interest as a result of the exhaustion of resources given their non-renewable character.[70]

(Footnote 67 continued)
effective date and term of the unitization agreement, the identity and appointment of the unit operator, the process for resignation and removal of the unit operator, and the process for appointment of a successor unit operator, provisions regarding the transfer of interests, and for an accurate measurement of production, procedures for ensuring accurate payments of royalties and other proceeds, safety and environmental measures to be taken under the national law of each State, provisions for appropriate information sharing between the unit operator and each State, and procedures for the redetermination of the allocation of production, including a timetable or the events that rigger such redetermination.

68 Robin R. Churchill and Alan V. Lowe, *The Law of the Sea*, 3rd edition (Manchester University Press: Manchester, 1999), p. 200.

69 Mouton, *supra* note 21, p. 421.

70 Rainer Lagoni, "Interim measures pending maritime delimitation agreements", 78-2 *American Journal of International Law* (April 1984), p. 366; Thomas Wälde, "Financial and contractual perspectives in negotiating joint petroleum development agreements", in: *Joint Development of Offshore Oil and Gas. Volume II. The Institute's Revised Model Agreement.*

The diverse State practice under domestic and international law regarding the use of conservation measures is essentially the result of the problematic regarding their enforcement. While States may determine in their domestic law or under the umbrella of a joint development regime the implementation of unitization regimes between operators, they have been overwhelmingly reluctant to do so in respect of agreements applicable to the delimitation of maritime boundaries or the implementation of cooperative efforts. States have generally considered that the inclusion of unitization or natural resources clauses in these arrangements does not correspond to the corroboration of their duty to cooperate in the production of shared or transboundary resources. These matters will be examined in Chap. 4.

2.3 The Role of Private Entities

Joint development agreements are international legal instruments signed exclusively between States that establish, amongst other, the legal framework for granting rights to operators for the exploration and exploitation of offshore hydrocarbon deposits straddling a boundary line or that are found in disputed maritime areas. Although operators are not parties to these agreements, oil companies are essential for their successful completion. They are frequently the driving force behind their implementation given their knowledge and capabilities to carry out offshore development of hydrocarbon deposits.

Despite the fact that in most cases States have given operators the rights to explore and exploit resources, certain joint development agreements create joint entities for the purpose of granting such rights to operators and overall managing the joint development area. Depending on the relation between States or joint entities and operators, and considering their rights and obligations associated with the undertaking of exploration and exploitation activities, operators may have different levels of participation in the implementation of the joint development agreement.[71] For example, if a joint development agreement determines that operators shall enter into product sharing agreements, operators' participation is

(Footnote 70 continued)
Conference Papers. The Australia/Indonesia Zone of Co-operation Treaty 1989, edited by Hazel Fox (The British Institute of International and Comparative Law: London, 1990), p. 156; Mark J. Valencia, "Joint jurisdiction and development in southeast Asia seas: factors and candidate areas", in: *Geology and Hydrocarbon Potential of the South China Sea and Possibilities of Joint Development*, edited by Mark J. Valencia (Pergamon Press: New York, Oxford, Toronto, Sydney, Paris, Frankfurt, 1985), p. 576; Juraj Andrassy, "Les relations internationales de voisinage", in: 79 *Recueil des Cours de l'Académie de Droit International* (1951-II), p. 110; Albert E. Utton, "Institutional arrangements for developing North Sea oil and gas", in: 9-1 *Virginia Journal of International Law* (December 1968), p. 69.

71 Ibrahim F.I. Shihata and William T. Onorato, "Joint development of international petroleum resources in undefined and disputed areas", in: *Boundaries and Energy: Problems and Prospects*, edited by Gerald Blake, Martin Pratt, Clive Schofield and Janet Allison Brown (London/The

far greater when compared with the traditional concession or licensing regimes.[72] In the case of exploration and exploitation activities in disputed maritime areas, consideration should also be given to the protection of the rights granted to oil companies in the event of termination of the joint development regime.

Exploration and exploitation activities are extremely costly and risky endeavours that require vast investments in research and development of the technology that allows for reaching hydrocarbon deposits in remote places and at greater depths and often in extremely difficult circumstances. Oil companies are also required to make large and long-term capital investments to collect detailed information on the resource-potential of different parts of the world without the certainty as to its outcome. In fact, it will take several years and considerable amount of investment before initiating production of any site or even reaching the conclusion that production is not viable. Moreover, downstream investment is also extremely high, particularly as the demand for fossil fuels continues to grow due to the increase of population and urban centres, especially in developing countries with emerging economies.[73] These States often do not have the knowledge or the financial capability to carry out the necessary research or to create a national oil industry for the purpose of exploration and exploitation of offshore hydrocarbon deposits. In these cases, the investment of multinational oil companies is indispensable for the exploration and exploitation of these resources, as well as for the creation of refining and transportation infrastructures where they are lacking. The result of oil companies redirecting their investments to emerging economies that have a great resource-potential and where demand is very high and construction costs are comparatively lower[74] is the shortening of the product-chain and consequently the reduction of costs connected with the importation and transportation of crude oil. On the other hand, such redirection of investments creates an excess of refining capacity in developed countries where demand is smaller and national reserves are declining.[75]

(Footnote 71 continued)
Hague/Boston 1998), pp. 445–447; D.H. Anderson, "Strategies for dispute resolution: negotiating joint agreements", in: *ibid*, pp. 480–481.

[72] On product sharing agreements *see* Taverne, *supra* note 42, pp. 158–219.

[73] *See* Organization of the Petroleum Exporting Countries, *World Oil Outlook 2011*, edited by James Griffin (OPEC Secretariat: 2011), pp. 223–227 [online: http://www.opec.org/opec_web/static_files_project/media/downloads/publications/WOO_2011.pdf (accessed on August 2013)].

[74] *Ibid*, p. 229.

[75] *Ibid*, pp. 7, 13. The report refers that the commercial primary energy demand is expected to increase by 51 % over the period 2010–2035, of which fossil fuels will continue to account for over 80 %.

Chapter 3
Joint Development Agreements in International Law in General

3.1 Sovereignty, Sovereign Rights and Jurisdiction of Coastal States

3.1.1 Maritime Areas Within and Beyond National Jurisdiction

The legal order of the seas and oceans is structured, on the one hand, in maritime areas that are subject to the jurisdiction of coastal States and, on the other hand, those areas that are beyond national jurisdiction and that consequently cannot be claimed or allocated to any State. The first include the territorial sea,[1] internal waters,[2] international straits,[3] archipelagic waters of an archipelagic State,[4] the contiguous zone,[5] the EEZ[6] and the continental shelf;[7] and the second consist of the high seas[8] and the Area.[9]

The maritime areas that are subject to the sovereignty or jurisdiction of a costal State can be distinguished between those that are subject to the territorial sovereignty of a costal State, i.e. the territorial sea, internal waters, international straits and archipelagic waters, and those where States exercise functionally limited sovereign rights and jurisdiction, namely, the continental shelf, including beyond 200 nm, and the EEZ that, when established, includes the contiguous zone.

---

[1] Article 2 of UNCLOS.
[2] Article 8 of UNCLOS.
[3] Article 34 of UNCLOS.
[4] Article 46 of UNCLOS.
[5] Article 33 of UNCLOS.
[6] Article 55 of UNCLOS.
[7] Article 76 of UNCLOS.
[8] Article 86 of UNCLOS.
[9] Article 133 of UNCLOS.

V. Becker-Weinberg, *Joint Development of Hydrocarbon Deposits in the Law of the Sea*, Hamburg Studies on Maritime Affairs 30, DOI 10.1007/978-3-662-43570-0_3

The difference between maritime areas that are under the territorial sovereignty of the coastal States and those that are not, is related with the rights and obligations that coastal States may exercise in each maritime area, including for the purpose of the development of non-living marine natural resources.[10] If one moves from the internal waters and territorial sea through the EEZ and continental shelf onto the high seas and in the Area, the strength and scope of the coastal State's jurisdiction decreases. Accordingly, diverse jurisdictional problems may arise with regard to joint development of hydrocarbon deposits in different maritime areas.

Beyond and adjacent to the territorial sea extends ipso facto the continental shelf, which comprises the seabed and subsoil of the submarine areas up to the outer edge of the continental margin or to a distance of 200 nm from the baselines that are drawn along the coast, or 350 nm in the case of the extended continental shelf or 100 nm from the 2,500 m isobath.[11] The coastal States exercise over the continental shelf sovereign rights for the purpose of exploring its natural resources.[12] In addition, the coastal State may establish, beyond and adjacent to its territorial sea, an EEZ with a breadth of up to 200 nm[13] where it has sovereign rights for the purpose of exploring and exploiting, conserving and managing the living and non-living natural resources of the water column and of the seabed and its subsoil.[14] It also has exclusive rights with regard to the construction, operation and use of artificial islands, installations and structures in the EEZ[15] and on its continental shelf,[16] over which it enjoys jurisdiction.[17]

The aforementioned rights and obligations are, therefore, connected with the organization of the seas and oceans and with the legal regime applicable to the maritime areas subject to the sovereignty or jurisdiction of coastal States. In the maritime areas that are subject to the jurisdiction of a coastal State but beyond its territorial sovereignty, namely the EEZ and the continental shelf, the exercise of rights is subject to those matters identified in UNCLOS. These rights, often referred to as sovereign rights, are limited to a particular maritime area and to certain matters such as the development of resources. Accordingly, the exercise of

[10] *Continental Shelf case (Tunisia/Libyan Arab Jamahiriya), Judgment, I.C.J. Reports 1982*, para. 104: "*The fact that a given area is territorial sea or internal waters does not mean that the coastal State does not enjoy "sovereign rights for the purpose of exploring it and exploiting its natural resources"; it enjoys those rights and more, by virtue of its full sovereignty over that area*".

[11] Article 76(5) of UNCLOS.

[12] Article 77(1) of UNCLOS.

[13] Article 57 of UNCLOS.

[14] Article 56(1) lit. (a) of UNCLOS.

[15] Article 60 of UNCLOS.

[16] Article 80 of UNCLOS.

[17] Article 56(1) lit. (b) of UNCLOS.

the right of coastal States to develop resources in the EEZ and the continental shelf should not be considered as being part of their territorial sovereignty, since the EEZ and the continental shelf are not part of the coastal State's territory.

3.1.2 *Mineral Resources in the EEZ, the Continental Shelf and the Area*

The reference to mineral resources can be found in the legal rules applicable to the EEZ,[18] the continental shelf[19] and the Area.[20] Considering, however, that the resources of the Area are the 'common heritage of mankind' and that all activities in the Area shall be carried out for the 'benefit of mankind', no State may claim or exercise sovereignty or sovereign rights over any part of the Area or its resources. UNCLOS further establishes that no such claim will be recognised under international law.[21] Indeed, the principle of the common heritage of mankind applicable to the Area, together with the freedom of the high seas, prevents coastal States from making sovereignty claims over any of these maritime areas and their resources.[22] Consequently, excluding the possibility of developing the resources that extend from the continental shelf to the Area, States may only enter into joint development agreements regarding mineral resources that straddle a boundary line or that are found in areas of overlapping claims based on a valid legal title, on in other words, that would potentially be subject to national jurisdiction if States would agree on the delimitation of maritime boundaries. As such, it is necessary to determine coastal States' rights and obligations under the relevant legal regimes applicable in the different maritime areas. Likewise, the study of the legal rules on maritime delimitation of the EEZ and the continental shelf and the understanding of the importance given to mineral resources in the settlement of maritime boundaries are necessary to identify the rights and obligations of coastal States with respect to the development of mineral resources that are found in areas of overlapping claims.[23]

---

18 Article 56(1) lit. (a) of UNCLOS.

19 Article 77(4) of UNCLOS.

20 Article 133 lit. (a) of UNCLOS.

21 Articles 136, 137 and 140(1) of UNCLOS.

22 Articles 89 and 137(1) of UNCLOS.

23 Yoshifumi Tanaka, *The International Law of the Sea* (Cambridge University Press: Cambridge, 2012), pp. 7, 186. Also *see* Francisco Orrego Vicuña, *Los Fundos Marinos y Oceanicos. Jurisdicción nacional y Régimen internacional* (Editorial Andres Bello: Santiago, 1976), pp. 176–180; Jorge A. Vargas, "The legal nature of the patrimonial sea: a first step towards the definition of the exclusive economic zone", in: 22 *German Yearbook of International Law* (1979), pp. 142–177.

3.1.3 In the EEZ

3.1.3.1 Legal Status

The EEZ is a maritime area extending up to 200 nm from the baseline. It is characterized by the fact that it exists *ipso jure*, meaning that it depends on prior proclamation by the respective coastal State, although there is no obligation for a coastal State to claim an EEZ.[24] The concept of the EEZ emerged as a result of State practice and was later included and its regime developed in UNCLOS.[25] Considering the generalized State practice demonstrated through maritime delimitation treaties and national legislation, it has been often considered that the concept of the EEZ is now part of customary international law.[26]

The EEZ has also been referred to as a "*separate functional zone of a sui generis character*",[27] particularly when compared with the territorial sea and the high seas. In fact, while the continental shelf regime provides that the rights of coastal States in this maritime area do not affect the legal status of the superjacent waters given the underlining principle of the freedom of the high seas,[28] the EEZ does not maintain a high seas residual character. Indeed, the legal status of the EEZ was one of the most controversial issues during the negotiation and drafting of UNCLOS.[29] The Convention defines that the EEZ shall be "*subject to the specific legal regime*

---

[24] Article 57 of UNCLOS. *See* Alexander Proelβ, "Ausschlieβliche Wirtschaftszone", in: *Handbuch des Seerechts*, edited by Wolfgang Graf Vitzthum (C.H. Beck: Munich, 2006), pp. 222–230; David Joseph Attard, *The Exclusive Economic Zone in International Law* (Claredon Press: Oxford, 1987), pp. 212–221, 235, 271; International Law Association, *supra* note 15 in Chap. 2, pp. 1–32.

[25] This is the case, for example, of the *Declaration of Santiago de Chile*, done in 18 August 1952 and signed by Chile, Ecuador and Peru. Accordingly, coastal States had sovereign and exclusive rights over the adjacent maritime waters to a minimum distance of 200 nm, which corresponded to the characteristics of the coasts of these countries and the area where the most relevant living resources were concentrated. *See* Pilar Armanet, "The economic interest underlying the first declaration on a maritime zone", in: *The Exclusive Economic Zone. A Latin American Perspective*, edited by Francisco Orrego Vicuña (Westview Press: Boulder Colorado, 1984), pp. 27–30.

[26] Edward Duncan Brown, *The International Law of the Sea, vol. 1, Introductory Manual* (Dartmouth: Aldershot, Broofield USA, Singapore, Sydney, 1994), pp. 224–225, 217–218, 245. Also *see* Donald R. Rothwell and Tim Stephens, *The International Law of the Sea* (Hart Publishing: Oxford and Portland, Oregon, 2010), pp. 84–85; Tanaka, *supra* note 23, p. 124.

[27] Churchill/Lowe, *supra* note 68 in Chap. 2, p. 166. Also *see* Brown, *ibid*, pp. 218–220; Tanaka, *ibid*, p. 128; Rothwell/Stephens, *ibid*, pp. 83–85.

[28] The text of Article 78 of UNCLOS reads: "*1. The rights of the coastal State over the continental shelf do not affect the legal status of the superjacent waters or of the air space above those waters. 2. The exercise of the rights of the coastal State over the continental shelf must not infringe or result in any unjustifiable interference with navigation and other rights and freedoms of other States as provided for in this Convention.*"

[29] *Memorandum by the President of the Conference on document A/CONF.62/WP.10* (Document A/CONF.62/WP.10/Add.1), 22 July 1977, p. 68.

established in this Part, under which the rights and jurisdiction of the coastal State and the rights and freedoms of other States are governed by the relevant provisions of this Convention."[30] Moreover, although the EEZ includes certain aspects of the regime of the territorial sea, it is not, however, included in the territorial sovereignty of the coastal State. The functional character of the EEZ which is reflected in the nature of the rights of coastal States and those of other States in this maritime area, makes the use of the term 'exclusive' exceptionally unimpressive.

3.1.3.2 Rights and Duties of Coastal States

Considering the scope of this work, the analysis of States' rights in the EEZ shall be restricted only to include the legal rules applicable to the exploitation of mineral resources, consequently, excluding marine living resources, as well as, for that matter, the production of energy from the water, currents and winds, to which the Convention also refers to.[31]

The EEZ regime provides that the coastal State has sovereign rights for the purpose of exploring and exploiting mineral resources in the water column and the seabed and subsoil. States also have jurisdiction with regard to establishing installations and structures, including offshore platforms.[32] This reference to sovereign rights should not be viewed in the same light as the rights exercised in the territorial sea or the sovereignty rights referred in the continental shelf regime. The rights of coastal States in the EEZ should be considered as preferential rights for the purpose of the development of marine natural resources, in the sense that, in particular situations, the coastal States must grant States access to these resources.[33]

In what concerns the exploration and exploitation of mineral resources, UNCLOS does not provide similar rules for sharing of resources. In this case, the rights of the coastal State regarding mineral resources could be rightfully considered as exclusive, even though, as outlined above, these are sovereign rights but that do not establish the coastal State' sovereignty over mineral resources as provided for in the continental shelf regime.[34] UNCLOS ensures the compatibility between both regimes referring that in the seabed and subsoil coastal States shall act in conformity with the continental shelf regime, however, without implying a

[30] Article 55 of UNCLOS. *See* Churchill/Lowe, *supra* note 68 in Chap. 2, p. 84: "*The EEZ combines characteristics of the territorial sea and the high seas, but cannot be assimilated to either. It is a sui generis zone with its own distinctive regime. (...) It is an amalgam, or 'multifunctional' zone, in which coastal states enjoy sovereign rights in relation to economic resources, and also for certain other matters including environmental protection.*" Also *see* Brown, *supra* note 26, p. 219.

[31] Article 56(1) lit. (a) of UNCLOS.

[32] Articles 56(1) lit. (a), ((b)i), 60(1) lit. (b) and 208(1) of UNCLOS.

[33] Articles 56(2), 69 and 70, of UNCLOS.

[34] Article 77(1) of UNCLOS.

merger of the two.[35] All the same, there is a significant difference concerning the rights of third States in the continental shelf and the EEZ, independently of the aforementioned substantive integration of both regimes in what concerns the rights of coastal States in the seabed and subsoil.[36]

3.1.3.3 Rights and Duties of Other States

The introduction of the EEZ regime by UNCLOS was viewed by some States as imposing limitations on the traditional freedoms of the high seas, and consequently a possible curtailment of the rights of other States in this maritime area, or what has been considered having a "*creeping jurisdiction*" effect.[37] Even before the EEZ regime had been developed, Cecil J. B. Hurst had already referred that it would be necessary to secure the compatibility between coastal States' sovereignty claims of the seabed and the rights of third States regarding the exploitation of other resources.[38]

UNCLOS includes several provisions that safeguard the rights and duties of third States in the EEZ. Article 58(1) provides that all States enjoy in the EEZ the freedoms of navigation, overflight and of laying submarine cables and pipelines.[39] In paragraph 3 of this provision, the Convention also specifies that States shall have due regard for the rights and obligations of coastal States and shall comply with its laws and regulations in the EEZ. The due regard obligation had already been included in Article 2 of the 1958 Convention on the High Seas referring that the freedoms of the high seas should be exercised with "*reasonable regard to the interests of other States*". In this respect, the *Fisheries Jurisdiction* case considered that the concept of 'reasonable regard' implied that the rights of coastal States were not absolute and that, consequently, the different interests of the United Kingdom and Iceland in the high seas should be reconciled.[40] This decision notwithstanding, the freedom of fishing in the high seas referred in Article 87(1) lit.

---

[35] Article 56(3) of UNCLOS.

[36] Francisco Orrego Vicuña, *The Exclusive Economic Zone. Regime and Legal Nature under International Law* (Cambridge University Press: Cambridge, New York, Port Chester, Melbourne, Sydney, 1989), p. 71.

[37] Brown, *supra* note 26, p. 236. On creeping jurisdiction and its influence on the creation of customary law *see* John A. Knauss, "Creeping jurisdiction and customary international law", in: 15-2 *Ocean Development and International Law* (1985), p. 210; Barbara Kwiatkowska, "Creeping jurisdiction beyond 200 miles in the light of the 1982 Law of the Sea Convention and state practice", in: 22-2 *Ocean Development and International Law* (1991), pp. 159–160; Erik Franckx, "The 200-mile: between creeping jurisdiction and creeping common heritage?", in: 39 *George Washington International Law Review* (2007), pp. 476–485.

[38] Cecil J. B. Hurst, "Whose is the bed of the Sea?", in: 4 *British Yearbook of International Law* (1923–1924), p. 43.

[39] Articles 55, 56(2) and 87 of UNCLOS.

[40] *Fisheries Jurisdiction (United Kingdom v. Iceland), Merits, Judgment, I.C.J. Reports 1974*, paras. 52, 55, 57 to 60, 66, 68 to 70.

(e) of UNCLOS is evidently not applicable in the EEZ, seeing that, in the EEZ, the coastal State has the sovereign right to develop these resources.[41] Likewise, the freedom of scientific research is dependent on the permission of the relevant coastal States.[42] Nonetheless, the content or extent of the due regard obligation remains to be considered in light of the circumstances of each case when different uses are in conflict. This is the same *ratio* included in Article 59 of UNCLOS[43] that is applicable to situations where the Convention does not determine the rights and jurisdiction of coastal States and other States in the EEZ, despite the fact that UNCLOS did not include these situations within the scope of the compulsory settlement mechanism in Part XV.[44]

3.1.4 On the Continental Shelf

3.1.4.1 Legal Status

The continental shelf regime is traditionally considered having its origin in the well-known proclamation of President of the United States of America, Harry S. Truman,[45] although there had been previous examples of coastal States' claims regarding the seabed and subsoil adjacent to their land territory.[46]

---

41 Article 56(1) lit. (a) of UNCLOS.

42 Articles 56(1) lit. ((b)ii), 87(1) lit. (f) and 246(2) of UNCLOS.

43 The text of Article 59 of UNCLOS reads: "*In cases where this Convention does not attribute rights or jurisdiction to the coastal State or to other States within the exclusive economic zone, and a conflict arises between the interests of the coastal State and any other State or States, the conflict should be resolved on the basis of equity and in the light of all the relevant circumstances, taking into account the respective importance of the interests involved to the parties as well as to the international community as a whole.*"

44 Article 297(1) lit. (a) of UNCLOS, *a contrario.*

45 *Policy of the United States with respect to the Natural Resources of the Subsoil and Sea Bed of the Continental Shelf,* Proclamation n. 2667, 28 September 1945, 10 Fed. Reg. 12.303 (1945).

46 *North Sea Continental Shelf cases (Federal Republic of Germany/Denmark; Federal Republic of Germany/Netherlands), Judgment, I.C.J. Reports 1969,* para. 47. One of these examples is the *Treaty between Great Britain and Northern Ireland and Venezuela relating to the submarine areas of the Gulf of Paria,* done at Caracas on 26 February 1942, published at 205 L.N.T.S. 122 and reproduced in: *International Maritime Boundaries,* vol. 1, *supra* note 42 in Chap. 2, pp. 651–654. On the legal concept of the continental shelf *see* Rainer Lagoni, "Festlandsockel", in: *Handbuch des Seerechts,* edited by Wolfgang Graf Vitzthum (C.H. Beck: Munich, 2006), pp. 176–180; Suzette Suarez, *The Outer Limits of the Continental Shelf. Legal Aspects of their Establishment* (Springer: Berlin, Heidelberg, New York, 2008), pp 21–74. Also *see* Juraj Andrassy, *International Law and the Resources of the Sea* (Columbia University Press: New York, London, 1970), pp. 49–69; H. Lauterpracht, "Sovereignty over submarine areas", in: 27 *British Yearbook of International Law* (1950), pp. 376–433; José Luis de Azcárraga y Bustamante, "Los derechos sobre la plataforma continental", in: 2 *Revista Española de Derecho Internacional* (1949), pp. 47–99; Richard Young, "Recent developments with respect to the continental shelf", in: 42 *American Journal of International Law* (1948), pp. 849–857.

The Truman Proclamation adopted the criteria of contiguity of the continental shelf, thus recognizing the right of every coastal State to exercise jurisdiction and control over this maritime space and the natural resources found therein, as well as preventing States from consolidating their sovereignty claims based on the appropriation of the seabed and subsoil and the development of its natural resources.[47]

The Truman Proclamation did not determine the width of the sea over which the continental shelf extended, nor did it provide a method of delimitation regarding maritime boundaries between neighbouring States, merely referring to delimitation in accordance with equitable principles. Nonetheless, the Truman Proclamation had a surprising effect.[48] With the generalization of similar claims by coastal States,[49] the consuetudinary nature of the continental shelf regime rapidly became consolidated in international law,[50] namely the recognition of the inherent and

---

[47] Kriangsak Kittichaisaree, *The Law of the Sea and Maritime Boundary Delimitation in South-East Asia* (Oxford University Press: Oxford, New York, 1987), pp. 57–63.

[48] Gilbert Appolis, *L'Emprise Maritime de L'Etat Côtier* (Editions A. Pedone: Paris, 1981), p. 135; United Nations, "La doctrine du plateau continental et la pratique", in: *Yearbook of the International Law Commission, 1950, vol. 2*, pp. 108–113.

[49] David Colson, "The delimitation of the outer continental shelf between neighboring status", in: 97 *American Journal of International Law* (2003), pp. 91–107; Nguyen Quoc Dinh, Patrick Dailler and Alain Pellet, *Droit International Public*, 7th edition (Librairie Générale de Doit et de Jurisprudence/E.J.A.: Paris, 2002), pp. 1185–1187; J.-A. Carrillo-Salcedo, *El Derecho Internacional en un Mundo en Cambio* (Tecnos: Madrid, 1985), p. 102; Mohamed Bennouna, "Les droits d'exploitation des ressources minérales des océans", in: *Le Nouveau Droit international de la Mer*, edited by Bardonnet and Virally (A. Pedone: Paris, 1983), pp. 121–127; Charles Rosseau, *Droit International Public. Les Relations Internationales*, vol. 4 (Sirey: Paris, 1980), p. 433; René-Jean Dupuy, *L'Océan Partagé* (A. Pedone: Paris, 1979), pp. 105–116, 104; Laurent Lucchini and Michel Voelckel, *Droit de la Mer: La Mer et son Droit, Les espaces maritimes*, vol. 1 (A. Pedone: Paris, 1990), pp. 54–56, 230–266, 258.

[50] Jean Combacau, *Le Droit de la Mer* (Presses Universitaires de France: Paris, 1985), pp. 47–57; F.V. Garcia Amador, *Continental Shelf and the Exclusive Economic Zone/Le Plateau Continental et la Zone Économique Exclusive*, edited by Donat Pharand and Umberto Leanza (Martinus Nijhoff Publishers: Dordrecht, Boston, London, 1993), pp. 13–18; Shabtai Rosenne, "Geography in international maritime boundary-making", in: *Law of the Sea*, edited by Hugo Caminos (Ashgate: Dartmouth, Aldershot, Burlington US, Singapore, Sydney, 2001), pp. 234–237; Tulio Treves, "La limite extérieure du plateau continental: évolution récente de la pratique", in: 35 *Annuaire Français de Droit International* (1989), pp. 724–732; Daniel Bardonnet, "Frontières terrestres et frontières maritimes", in: 35 *Annuaire Français de Droit International* (1989), pp. 23–38; Victor Prescott, "National rights to hydrocarbon resources of the continental margin beyond 200 nautical miles", in: *Boundaries and Energy: Problems and Prospects. International Boundary Studies Series*, edited by Gerald Blake, Martin Pratt, Clive Schofield and Janet Allison Brown (Kluwer Law International: London, The Hague, Boston: 1998), pp. 84, 51–82; Nuno Marques Antunes, "O novo regime jus-internacionalista do mar: a consagração ex vi pacti de um mare nostrum", in: *Estudos em Direito Internacional Público* (Almedina: Coimbra, 2004), pp. 18–31; David Ong, "The legal status of the 1989 Australia-Indonesia Timor Gap Treaty following the end of Indonesian rule in East Timor", in: 31 *Netherlands Yearbook of International Law* (2000), p. 113; Philippe Cahier, "Les sources du droit relatif à la délimitation du plateau continental", in: *Le Droit International au service de la*

exclusive right of every coastal State to exploit the natural resources of the seabed and subsoil of the territory contiguous to the land mass and beyond the territorial sea.[51]

The Truman Proclamation acknowledged the predominance of the legal or juridical notion of continental shelf by defending that this maritime space should be considered an extension of the land mass of the territory of the coastal State, despite the fact that States did not agree on the criteria according to which the notion of continental shelf should be defined.[52] The ICJ would also later establish a clear association between the sovereign right of a State to explore and exploit the resources of its continental shelf as that of a State with regard to its land, under the principle that "*the land dominates the sea.*"[53] The 1958 Convention on the Continental Shelf acknowledged the sovereign rights of each coastal State over the respective continental shelf,[54] even though many difficulties arose regarding the interpretation of the relevant rules[55] given that the notion of continental shelf

(Footnote 50 continued)
Paix, de la Justice et du Développement, mélanges Michel Virally (Pedone: Paris, 1991), pp. 175–179; Umberto Leanza, "La délimitation du plateau continental et de la zone économique exclusive: une introduction", in: *The Continental Shelf and the Exclusive Economic Zone/Le Plateau Continental et la Zone Économique Exclusive*, edited by Donat Pharand and Umberto Leanza (Martinus Nijhoff Publishers: Dordrecht, Boston, London, 1993), pp. 37–54; Jonathan I. Charney, "Progress in international maritime boundary delimitation law", in: 90-1 *American Journal of International Law* (1996), pp. 227–229, 230; Prosper Weil, "Des espaces maritimes aux territoires maritimes: vers une conception territorialiste de la délimitation maritime", in: *Le Droit International au service de la Paix, de la Justice et du Développement.* Mélanges Michel Virally (Pedone: Paris, 1991), pp. 501–511; Rosseau, *supra* note 49, pp. 334–337, 436–440; Brown, *supra* note 26, pp. 19, 161, 165; Tanaka, *supra* note 23, pp. 21–22, 132–133; Rothwell/Stephens, *supra* note 26, pp. 100–101; Carrillo-Salcedo, *supra* note 49, p. 211; Churchill/Lowe, *supra* note 68 in Chap. 2, pp. 7–12, 142–145; Lucchini/Voelckel, *supra* note 49, pp. 250–256.

[51] *North Sea Continental Shelf cases*, *supra* note 46, para. 19.

[52] Gilbert Gidel, *Le Plateau Continental* (International Bar Association: 1952), p. 4.

[53] *North Sea Continental Shelf cases*, *supra* note 46, paras. 43, 96; *Aegean Sea Continental Shelf, Judgment, I.C.J. Reports 1978*, para. 86; *Continental Shelf* (Tunisia/Libya), *supra* note 10, para. 73; *Case Concerning Maritime Delimitation in the Black Sea (Romania v. Ukraine), Judgment, I.C.J. Reports 2009,* para. 77; *Territorial and Maritime Dispute* (*Nicaragua v. Colombia*), *Judgement, I.C.J. Reports 2012,* para. 140.

[54] Article 2 of the 1958 Convention on the Continental Shelf. *See* Dominique Carreau, *Droit International*, 8th edition (Pedone: Paris, 2004), pp. 160–161; Jean-Pierre Lévy, "Les Nations Unies et la Convention de 1982 sur le Droit de la Mer", in: *Revue Belge de Droit International* (1995-1), p. 27; Treves, *supra* note 50, 732–733; Churchill/Lowe, *supra* note 68 in Chap. 2, p. 144; Dinh/Dailler/Pellet, *supra* note 49, p. 1185; Lucchini/Voelckel, *supra* note 126, pp. 52, 94–98.

[55] On the definition of the continental shelf in the 1958 Convention on the Continental Shelf *see* Francisco Orrego Vicuña, *The Changing International Law of High Seas Fisheries* (Cambridge University Press: Cambridge, 1999), pp. 200–226; Shigeru Oda, *International Control of Sea Resources* (Martinus Nijohff Publishers: Dordrecht, Boston, London, 1989), pp. 147–172; Jean Devaux-Charbonnel, "Le régime juridique de la recherche et de l'exploitation du pétrole dans le plateau continental", in: *Annuaire Français de Droit International* (1956-2), pp. 320–333.

neither adopted the geomorphologic and geophysical notion, nor the legal notion of the Truman Proclamation.[56]

Article 1 of the 1958 Convention on the Continental Shelf provides that the delimitation of the outer limits of the continental shelf would correspond to the seabed and subsoil of the adjacent submarine areas of the coastal States until the depth of 200 m,[57] or beyond which the development of natural resources could be carried out.[58] Although the general repetition of this practice allowed States to claim a continental shelf of 200 nm, the assessment of the method applicable to the delimitation could not be completed based on the general reference to equitable principles made in Article 6 of the same Convention.[59]

UNCLOS adopted a more ample definition of continental shelf, granting States an entitlement of 200 nm and the possibility for its extension.[60] The rights and

---

[56] Article 1 of the 1958 Convention on the Continental Shelf. *See Yearbook of the International Law Commission, 1953, vol. 2, Documents of the Fifth session including the report of the Commission to the General Assembly*, paras. 65, 66 and 73, pp. 213–215. Also *see* Resolution (UNGA) n. 2574 (XXIV), 15 December 1969, *Question of the reservation exclusively for peaceful purposes of the sea-bed and the ocean floor, and the subsoil thereof, underlying the high seas beyond the limits of present national jurisdiction, and the use of their resources in the interests of mankind.* This Resolution recognized that the definition of the continental shelf provided by the 1958 Convention on the Continental Shelf did not precisely define the limits of the maritime area over which the State exercises sovereign rights for the purpose of developing natural resources and that customary international law was inconclusive on this matter.

[57] This was the limit usually marked on nautical charts and also where it was generally considered that the continental shelf came to an end. *See* John C. Colombos, *The International Law of the Sea*, 6th edition (David McKay Company Inc.: New York, 1967), p. 76.

[58] Article 1 lit. (a) of the 1958 Convention on the Continental Shelf.

[59] *North Sea Continental Shelf cases, supra* note 46, paras. 71, 81. Also *see* Lewis M. Alexander, "Baseline delimitations and maritime boundaries", in: *Law of the Sea* (Ashgate: Dartmouth, Aldershot, Burlington US, Singapore, Sydney: 2001), pp. 22–24; Barbara Kwiatkowska, "Equitable maritime boundary delimitation: a legal perspective", *ibid*, pp. 243–247, 256–257; Jonathan I. Charney, "Ocean boundaries between nations: a theory for progress", in: 78-3 *American Journal of International Law* (1984), pp. 583, 587–594; M. D. Blecher, "Equitable delimitation of continental shelf", in: 73 *American Journal of International Law* (1979), pp. 61–65.

[60] Article 76 of UNCLOS. *See Delimitation of the Maritime Boundary in the Gulf of Maine Area, Judgment, I.C.J. Reports 1984*, paras. 157–163, 230; *Continental Shelf* (*Libyan Arab Jamahiriya/Malta*), *Judgment, I.C.J. Reports 1985*, para. 61; *Continental Shelf* (Tunisia/Libya), *supra* note 10, para. 127. Also *see* Gerard J. Tanja, *The Legal Determination of International Maritime Boundaries. The Progressive Development of Continental Shelf, EFZ and EEZ Law* (Kluwer Law and Taxation Publishers: Deventer, Boston, 1990), pp. 207–253; Alex G. Oude Elferink, *The Law of Maritime Boundary Delimitation: A Case Study of the Russian Federation*, edited by Shigeru Oda (Martinus Nijhoff Publishers: Dordrecht, Boston, London, 1994), pp. 73–84; Edward Duncan Brown, *Sea-bed Energy and Minerals: the International Legal Regime. The Continental Shelf*, vol. 1 (Martinus Nijhoff Publishers: Dordrecht, Boston, London, 1992), pp. 19–23; L.D.M Nelson, "The roles of equity in the delimitation of maritime boundaries", in: 84-4 *American Journal of International Law* (October 1990), p. 846; Shigeru Oda, *The Law of the Sea in our Time - I. New Developments 1966–1975* (Sijthoff: Leyden, 1977), p. 254.

duties established by UNCLOS in the continental shelf do not introduce substantial differences to those already existing under the 1958 Convention on the Continental Shelf, except for those aspects specifically concerning the continental shelf beyond 200 nm. These include the revenue-sharing system referred in Article 82 of UNCLOS in respect of the exploitation of non-living resources of the continental shelf beyond 200 nm and which will be discussed in Chap. 6.

3.1.4.2 Rights and Duties of Coastal States

States have the exclusive and inalienable sovereign right to develop the resources found in the continental shelf,[61] regardless of its prior proclamation or occupation. In this respect, the ICJ in the *North Sea Continental Shelf* cases considered that the continental shelf existed ipso facto and ab initio.[62]

Although both the EEZ and the continental shelf regimes include the seabed and subsoil and regard sovereign rights for the functional purpose of developing natural resources,[63] there is only exclusivity with regard to the continental shelf.[64] In the continental shelf, only coastal States are entitled to exercise the inherent and exclusive sovereign right of exploring and exploiting the non-living resources and sedentary species,[65] safe for coastal States' explicit consent that other States may undertake similar operations. This means that if a coastal State chooses not to explore and exploit the non-living resources and the sedentary species found in the

[61] David Ong, "Joint development of common offshore oil and gas deposits: "mere" state practice or customary International Law?", in: 93-4 *American Journal of International Law* (1999), p. 777; Philip Allott, "Mare Nostrum: a new international Law of the Sea", in: 86 *American Journal of International Law* 4 (1992), pp. 767–768; Rosseau, *supra* note 49, pp. 358–359; Dinh/Dailler/Pellet, *supra* note 49, p. 1192; Carrillo-Salcedo, *supra* note 49, p. 214; Combacau, *supra* note 50, pp. 58–67; Lucchini/Voelckel, *supra* note 49, pp. 164–169; Bennouna, *supra* note 49, p. 125; Bardonnet, *supra* note 50, pp. 22–23, 42; Prescott, V., *supra* note 50, pp. 51–52; Lagoni, *supra* note 63 in Chap. 2, pp. 216–218.

[62] Article 77 (2) *in fine* (3) of UNCLOS. *North Sea Continental Shelf cases*, *supra* note 46, para. 19.

[63] Articles 56(1), 57 and 76(1) of UNCLOS.

[64] Francisco Orrego Vicuña, "La zone économique exclusive dans la législation et la pratique des États", in: *Droit de La Mer*, vol. 2, coordination by Jean Combacau and Pierre-Marie Dupuy (Pedone: Paris, 1990), pp. 44–45, *ibid*, *supra* note 36, p. 71; Julio César Lupinacci, "The legal status of the exclusive economic zone in the 1982 Convention on the Law of the Sea", in: *The Exclusive Economic Zone. A Latin American Perspective*, edited by Francisco Orrego Vicuña (Westview Press: Boulder Colorado, 1984), pp. 105–111; Barbara Kwiatkowska, *The 200 Mile Exclusive Economic Zone in the New Law of the Sea* (Martinus Nijhoff Publishers: Dordrecht, Boston, London, 1989), pp. 91–92; Hugo Caminos, "The regime of fisheries on the exclusive economic zone", in: *The Exclusive Economic Zone. A Latin American Perspective*, edited by Francisco Orrego Vicuña (Westview Press: Boulder Colorado, 1984), pp. 151–155; Attard, *supra* note 24, pp. 192–210.

[65] Article 77(2) and (4) of UNCLOS and Article 2(2) of the 1958 Convention on the Continental Shelf. *See Yearbook of the International Law Commission*, *supra* note 56, para. 71, p. 214.

continental shelf, no other State may do so without the latter's consent.[66] This is also similar to coastal States' exclusive right to drill in the continental shelf, which was not included in the 1958 Convention on the Continental Shelf.[67]

Furthermore, contrary to the living resources of the EEZ, coastal States on the continental shelf are not subject to the obligation of conservation and optimum utilization and do not have to share access to mineral resources or the benefits resulting from their exploitation.[68] Additionally, the applicability of the continental shelf regime to sedentary species through Article 68 of UNCLOS excludes these living resources from the freedom of fishing and from the obligation of conservation of coastal States. The main reason for the particular regime of sedentary species would appear still to be related with the historical understanding that these resources have special features and perhaps ought to best be considered as belonging to the soil or bed of the sea, instead of to the water column.[69]

3.1.4.3 Rights and Duties of Other States

Similar to the provisions already included in the 1958 Convention on the Continental Shelf, coastal States' rights over the continental shelf do not have an absolute character, in the sense that they do not know any limits.[70] The legal regime of the continental shelf coexists with that of the EEZ and, in its absence, the high seas, the general principle being that coastal States shall exercise their rights and perform their duties with due regard for the rights and duties of other States.[71] In this respect, UNCLOS clearly establishes that the rights of the coastal States over the continental shelf do not affect the legal status of the superjacent waters and air space above those waters or the rights and freedoms of other States,[72] such as the right to lay submarine cables and pipelines,[73] the freedom of navigation or the activities regarding international cooperation and promotion of marine scientific research.[74]

---

66 Article 77(2)(4) of UNCLOS.

67 Article 81 of UNCLOS.

68 Articles 61, 62 and 63 of UNCLOS.

69 Colombos, *supra* note 57, p. 159.

70 Articles 3, 4 and 5(1)(6) of the 1958 Convention on the Continental Shelf.

71 I.T.L.O.S., *Dispute concerning Delimitation of the Maritime Boundary between Bangladesh and Myanmar in the Bay of Bengal, Judgment of 14 March 2012, Case n. 16*, para. 475.

72 Article 78 of UNCLOS.

73 Article 79 of UNCLOS.

74 Part XIII of UNCLOS.

3.2 Power to Negotiate and Enter into Joint Development Agreements

3.2.1 Agreements Signed Before or After Maritime Delimitation

States' power to negotiate and enter into joint development agreements is an essential aspect of the characterization of these regimes as international binding agreements subject to international law. The making, entering into force, application, interpretation, modification and termination of joint development agreements are matters that are regulated by the principles of the law of treaties[75] established in the *Vienna Convention on the Law of Treaties* (Vienna Convention)[76] and which, for the most part, were declaratory of already existing international law and are regarded as customary international law.[77] These refer in particular to the assessment that joint development agreements are treaties creating legally binding obligations for the relevant States-parties. However, the power to negotiate and enter into joint development agreements is different from the preceding question of States' legitimacy, seeing that States only have the ability to negotiate and determine the legal regime applicable to the exploration and exploitation of non-living marine natural resources while holders of a valid legal title under international law. For example, in the *East Timor case*, Portugal claimed that when the *Timor Gap Treaty* was signed following Indonesia's occupation of East Timor, Australia had violated the right of self-determination of the Timorese people and the right of permanent sovereignty over their natural resources, as well as the rights and duties of Portugal regarding the territory of East Timor. Because the ICJ considered that it lacked jurisdiction due to the absence of Indonesia's consent upon which the necessary assessment of the lawfulness of the

---

[75] The law of treaties deals with the law governing how treaties are made and operated, whereas treaty law addresses the substance of treaties, namely the rights and obligations they create. *See* Anthony Aust, *Modern Treaty Law and Practice* (Cambridge University Press: Cambridge, 2007), pp. 2, 6. Also *see* Malcom Shaw, *International Law* (Cambridge University Press: Cambridge, 2003), pp. 810–860.

[76] *Vienna Convention on the Law of Treaties*, done at Vienna on 23 May 1969 and entered into force on 27 January 1982, published at 1155 U.N.T.S. 331 [Vienna Convention].

[77] I.T.L.O.S., *Responsibilities and obligations of States sponsoring persons and entities with respect to activities in the Area, Advisory Opinion* (*Request for Advisory Opinion submitted to the Seabed Disputes Chamber*), *1 February 2011, Case n. 17*, para. 57: "*Among the rules of international law that the Chamber is bound to apply, those concerning the interpretation of treaties play a particularly important role. The applicable rules are set out in Part III, Section 3 entitled "Interpretation of Treaties" and comprising Articles 31 to 33 of the 1969 Vienna Convention on the Law of Treaties (hereinafter "the Vienna Convention"). These rules are to be considered as reflecting customary international law. Although the Tribunal has never stated this view explicitly, it has done so implicitly by borrowing the terminology and approach of the Vienna Convention's articles on interpretation (...). The ICJ and other international courts and tribunals have stated this view on a number of occasions (...)*".

latter's conduct would depend on, the Court did not take into consideration the arguments presented by Portugal.[78] Notwithstanding, it demonstrates the importance and centrality of legal title in respect of the acquisition of territorial soveringty and consequentialy the right to explore non-living marine natural resources.

Joint development agreements include the creation of joint development areas or, at least, the identification of the location of or the resources that are to be jointly explored and exploited. Regarding agreements signed after the delimitation of maritime boundaries, the right to undertake exploration and exploitation activities of non-living resources derives *ipso jure* from the coastal States' rights over the continental shelf and its resources. Differently, before the delimitation of maritime boundaries, such activities depend on the coastal States' legal title to the respective maritime area, which may overlap with that of other States. In these cases, the power to negotiate and to entre into joint development agreements depends on States' entitlement to the disputed maritime area. This matter will be subsequently examined in Chap. 6.

3.2.2 Relevant Principles of the Law of Treaties

Often joint development regimes are referred to as agreements, treaties, memorandum of understanding, exchange of notes, letters of intent, joint declarations, protocol, or other equivalents. However, the use of different arrangements and terminology to refer to joint development regimes does not determine their legal nature.

The Vienna Convention defines a treaty as "*an international agreement concluded between States in written form and governed by international law*".[79] This definition does not adopt any specific classification and is sufficiently broad to include a wide range of regimes, provided that they fulfil this definition's relevant elements, i.e. an international agreement governed by international law[80] between

[78] *East Timor (Portugal v. Australia), Judgment, I.C.J. Reports 1995*, para. 35. Also *see* Pierre-Marie Dupuy, "A general stocktaking of the connections between the multilateral dimension of obligations and codification of the law of responsibility", in: 13-5 *European Journal of International Law* (2002), p. 1056; Christine M. Chinkin, "East Timor moves into the World Court", in: 4-2 *European Journal of International Law* (1993), pp. 208–218; Maria Clara Maffei, "The case of East Timor before the International Court of Justice - some tentative comments", in: 4-2 *European Journal of International Law* (1993), pp. 225, 227, 231.

[79] Article 2(1) lit. (a) of the Vienna Convention.

[80] Richard K. Gardiner, *Treaty Interpretation* (Oxford University Press: Oxford, 2010), p. 20; Aust, *supra* note 75, p. 17.

States or with international organizations.[81] It rules out all arrangements signed by States, or other legal entities, particularly those created under a joint development agreement, and foreign operators regarding the development of offshore hydrocarbon deposits that only envisage the regulation of the relations between States or joint-entities and operators regarding exploration and exploitation activities.[82]

The use of a more informal composition to establish a joint development regime, such as memoranda of understanding, may not necessarily have repercussion in its legal standing, seeing that there are no substantive requirements regarding the form of treaties.[83] The wording used in an agreement is, nonetheless, essential for verifying States' determination to be bound by it and, for that matter, that States have given their express consent, either by signature, exchange of instruments, ratification, acceptance, approval or accession.[84] For example, a memorandum of understanding may include legally binding obligations, or merely constitute a political commitment.[85] As such, joint development regimes may be interpreted differently depending on the expressions used, since not all agreements between States carry legally binding obligations and may instead only envisage the adoption of best efforts.[86]

---

[81] Articles 1, 2(1) and 5 of the *Vienna Convention on the Law of Treaties between States and International Organizations or between International Organizations*, done at Vienna on 21 March 1986, not yet in force. Also *see* James Crawford, *Brownlie's Principles of Public International Law* (Oxford University Press: Oxford, 2012), pp. 369, 179.

[82] *Anglo-Iranian Oil Co. case (jurisdiction), Judgment of July 22nd, 1952: I.C.J. Reports 1952*, p. 112.

[83] Crawford, *supra* note 81, p. 371; Aust, *supra* note 75, pp. 25–27, 32–57. In this respect the ICJ concluded in the *Aegean Sea Continental Shelf* case that: "*On the question of form, the Court need only observe that it knows of no rule of international law which might preclude a joint communiqué from constituting an international agreement to submit a dispute to arbitration or judicial settlement. (...)* [T]*he Court must have regard above all to its actual terms and to the particular circumstances in which it was drawn up.*" (*supra* note 53, para. 96). Also *see Land and Maritime Boundary between Cameroon and Nigeria (Cameroon* v. *Nigeria: Equatorial Guinea intervening), Judgment, I.C.J. Reports 2002*, paras. 263–264.

[84] Articles 12 to 17 and 31(1) of the Vienna Convention. The ICJ noted that: "(...) *the first duty of a tribunal which is called upon to interpret and apply the provisions of a treaty, is to endeavour to give effect to them in their natural and ordinary meaning in the context in which they occur. If the relevant words in their natural and ordinary meaning make sense in their context, that is and end of the matter. If, on the other hand, the words in their natural and ordinary meaning are ambiguous or lead to an unreasonable result, then, and then only, must the Court, by resort to other methods of interpretation, seek to ascertain what the parties really did mean when they used these words.*" *Competence of Assembly regarding admission of the United Nations, Advisory Opinion: I.C.J. Reports 1950*, p. 8. Also *see Arbitral Award of 31 July 1989, Judgment, I.C.J. Reports 1991*, para. 48.

[85] Anthony Aust, "Alternatives to treaty-making: MOUs as political commitments", in: *The Oxford Guide to Treaties*, edited by Duncan B. Hollis (Oxford University Press: Oxford, 2012), p. 49.

[86] On the interpretation of treaties *see* Gardiner, *supra* note 80, pp. 141–350; "The Vienna Convention rules on treaty interpretation", in: *The Oxford Guide to Treaties*, edited by Duncan B. Hollis (Oxford University Press: Oxford, 2012), pp. 475–506.

The use of less formal and non-binding arrangements that do not constitute a treaty and, therefore, are not subject to international law, provides for swift negotiation and adoption procedures. They also give States the possibility of keeping such arrangements confidential and, whenever possible, avoiding publication, even though the lack of international publication, in itself, is not a defining element of a treaty.[87] This informality also contributes towards making amending and terminating arrangements easier, seeing that they frequently do not include dispute settlement mechanisms with recourse to a third party. The main disadvantage arises, however, when States fail to agree on the status of these arrangements.[88] In these situations, an International Court or Tribunal will have to decide whether States intended to conclude a treaty or not.[89]

3.2.3 Joint Development Agreements and Third States

The general rule included in the Vienna Convention is that no treaty may constitute rights or obligations for a third State without its consent.[90] It has been long established by international jurisprudence that:

> It cannot be lightly presumed that stipulations favourable to a third State have been adopted with the object of creating an actual right in its favour. There is however nothing to prevent the will of sovereign States from having this object and this effect. The question of the existence of a right acquired under an instrument drawn between other States is therefore one to be decided in each particular case: it must be ascertained whether the States which have stipulated in favour of a third State meant to create for that State and actual right which the latter has accepted as such.[91]

In what concerns treaties providing rights for third States, the Vienna Convention refers that States' assent is required but is not dependent on any specific act of acceptance. It also addresses situations where third States' consent may be presumed.[92] An important exception to this rule is when certain provisions of a treaty

[87] Article 102 of the *Charter of the United Nations*, done at San Francisco on 26 June 1945 and entered into force on 24 October 1945 [online: http://treaties.un.org/doc/Publication/CTC/uncharter.pdf (accessed on August 2013) [UN Charter].

[88] Aust, *supra* note 85, pp. 60–64.

[89] *Case concerning the Land, Island and Maritime Dispute (El Salvador/Honduras: Nicaragua intervening), Judgment, I.C.J. Reports 1992*, paras. 373–380.

[90] Articles 2(1) lit. (h) and 34 of the Vienna Convention.

[91] *Free Zones of the Upper Savoy and the District of Gex* (*France v. Switzerland*)*, 1932 P.C.I.J., Series A/B*, No. 46, p. 147.

[92] Article 36(1) of the Vienna Convention. *See* Crawford, *supra* note 81, pp. 384–386.

become part of customary law.[93] In these cases, third States would be subject to these rules regardless of their consent,[94] even tough, in the case of joint development agreements, so far they have not established rights or obligations for third States.

As a result, in the case of joint development regimes in disputed maritime areas, only if all relevant States take part in this common effort, can these be adopted. In other words, two States may implement a joint development agreement in a disputed maritime area on condition that it does not include areas claimed by third States and that it does not hinder these States' right of exploration and exploitation of the resources found therein. Likewise, after the delimitation of maritime boundaries, joint development of transboundary non-living marine natural resources can only be carried out if all States that have the right to explore and exploit these resources take part in such a cooperative effort.

3.2.4 State Succession in Joint Development Agreements

The principle of *pacta sunt servanda* is a cornerstone of international public law and in particular of the law of treaties, whereby every treaty in force is binding upon the relevant States and must be implemented in good faith.[95] One of the most complex issues is the impact that State succession has in respect of treaties.[96] In

---

[93] On customary international law *see* International Law Association, *Report of the International Committee on Formation of Customary (General) International Law,* by Rein Mullerson (London Conference 2000), p. 751; International Law Commission, *Fragmentation of International Law: Difficulties arising from the Diversification and Expansion of International Law*, by Martti Koskenniemi, Doc. A/CN.4/L.682, 13 April 2006 (Fifty-eighth session: Geneva, 1 May-9 June and 3 July-11 August 2006) [online: http://untreaty.un.org/ilc/documentation/english/a_cn4_l682.pdf (accessed on August 2013]. Also *see* Michael Byers, *Custom, power and the power of rules: international relations and customary International Law* (Cambridge University Press: Cambridge, 1999), pp. 130–165; Maurice Mendelson, "The formation of customary international law", in: 272 *Recueil des Cours de l'Académie de Droit International* (1998), pp. 245–293, *ibid*, "The subjective element in customary international law", in: 66 *British Yearbook of International Law* (1995), pp. 177–208; Michael Akehurst, "Custom as a source of international law", in: 47 *British Yearbook of International Law* 1974–1975 (1977), pp. 31–42; Maarten Bos, "The recognized manifestations of international law. A new theory of "sources" ", in: 20 *German Yearbook of International Law* (1977), pp. 25–42; Anthony A. d'Amato, "Trashing customary international law", in: 81 *American Journal of International Law* (1971), pp. 102–103; *Ibid*, "Treaties as a source of general rules of international law", in: 3 *Harvard International Law Club Bulletin* (1961–1962), pp. 23–28; Crawford, *supra* note 81, pp. 33–34.

[94] Article 2(6) UN Charter, *supra* note 87, and Article 75 of the Vienna Convention. *See* Crawford, *ibid*, p. 33; Shaw, *supra* note 75, pp. 835–836.

[95] Article 26 of the Vienna Convention.

[96] Article 2(1) lit. (b) of the *Vienna Convention on Succession of States in respect of Treaties*, done at Vienna on 22 August 1978 and entered into force on 6 November 1996, published at 1946 U.N.T.S. 3 [Vienna Convention on Succession of States]. This Convention defines 'succession of States' as the replacement of one State by another in the responsibility for the international relations of territory.

the case of joint development agreements this is particularly relevant concerning the safeguard of the interests of other States-parties, as well as the rights acquired by operators. The concern lays in determining whether international law protects such interests and rights and what differences, if any, may result from agreements implemented before or after the delimitation of maritime boundaries, or as an integral part of a boundary treaty.

The Vienna Convention provides that States may not invoke the fundamental change of circumstances as grounds for terminating or withdrawing from a treaty that establishes a boundary.[97] This is an important exception to the 'clean slate' or tabula rasa rule.[98] Other relevant rules are also found in customary international law and in the *Vienna Convention on Succession of States in respect of Treaties* (Vienna Convention on Succession of States),[99] even though the latter is only applicable to State succession after its entry into force.[100]

The Vienna Convention on Succession of States provides that succession of States shall not affect the principles of international law affirming the permanent sovereignty of every people and every State over its natural wealth and resources, ensuring that they fully benefit from the advantages provided by these resources.[101] This notion is also understood in the nature of coastal States' rights regarding the exploration and exploitation of non-living marine natural resources of the seabed and subsoil.[102] Accordingly, in addition to the said principle of permanent sovereignty over natural resources, joint development regimes implemented before or after the delimitation of maritime boundaries or when included in a boundary treaty, are not applicable to the successor State in light of the principle of stability or "*intangibility*" of boundaries, which is enshrined in the *uti possidetis* doctrine.[103] There is no difference, in this respect, between joint development agreements implemented before or after the delimitation of maritime

---

[97] Article 62(2) lit. (a) of the Vienna Convention. Also *see* Article 11 of the Vienna Convention on Succession of States, *supra* note 96.

[98] Crawford, *supra* note 81, pp. 438–439.

[99] *Supra* note 96.

[100] Article 7(1) of the Vienna Convention on Succession of States.

[101] Article 13 of the Vienna Convention on Succession of States. Also *see* Nico Schrijver, *Sovereignty over Natural Resources: Balancing Rights and Duties* (Cambridge University Press: Cambridge, 1997), pp. 21–22, 260–274.

[102] Article 2(1) and (3) of the 1958 Convention on the Continental Shelf and Articles 56(1) lit. (a) and 77(1) of UNCLOS.

[103] *Territorial Dispute Case (Libyan Arab Jamahiriya v Chad), Judgment, I.C.J. Reports 1994*, para. 73; *Frontier Dispute, Judgment, I.C.J. Reports 1986*, paras. 20–26, 29, 30, 33; *Case concerning the Temple of Preah Vihear (Cambodia* v. *Thailand), Merits, Judgment of 15 June 1962, I.C.J. Reports 1962*, p. 34; *The Eritrea – Yemen Arbitration, Award of the Arbitral Tribunal in the First Stage of the Proceedings (Territorial Sovereignty and Scope of the Dispute), 3 October 1996*, para. 153 [online: http://www.pca-cpa.org/upload/files/EY%20Phase%20I.PDF (accessed on August 2013)]. *See* Wolff Heintschel von Heinegg, "Treaties, fundamental change of circumstances", in: *Max Planck Encyclopedia of Public International Law* [online: http://www.mpepil.com (accessed on August 2013)].

boundaries or as part of a boundary treaty. In the case of joint development agreements signed before the delimitation of maritime boundaries, the State succeeding to any of the States that implemented the agreement, be that, for example, as a result of a process of decolonisation, would not be subject to keeping the agreement in force. Likewise, in the case of joint development agreements implemented after the delimitation of maritime boundaries, or as part of a boundary treaty, the State succeeding to any of the States that implemented the joint development regime is not bound by the latter, without prejudice of maintaining the boundaries established before of, or parallel to, the joint development agreement.

State succession and the principle of stability do not hinder the consideration of the principles of international law of self-determination and the peoples' sovereignty of their natural resources. There are, nonetheless, implications regarding the rights of operators as a result of State succession. The State succeeding to any of the States that implemented the agreement must respect the rights acquired under the legal order of its predecessor, even though it has the right to curtail such rights, subject to the relevant international rules, for example, on the protection of foreign investments, nationalization or expropriation, namely that these actions are carried out for a public purpose, are not discriminatory and compensation is paid.[104]

3.3 Attempts to Propose Rules on Common Mineral Resources and to Standardize State Practice

3.3.1 The Work of the International Law Commission on Shared Natural Resources

The notion of shared natural resources was initially considered by the International Law Commission (ILC) with respect to international watercourses.[105] Only later was the reference to mineral resources and shared natural resources included in a syllabus prepared during the ILC's fifty-second session, in the year 2000, for the purpose of selecting the topics for its long-term work programme.[106]

---

[104] Vasco Becker-Weinberg, "A nacionalização do petróleo e o princípio do aproveitamento conjunto entre Estados", edited by M. Oliveira Martins (AAFDL: Lisbon, 2008), pp. 376–377, 387–396. Also *see* Andreas Zimmermann, "State Succession in other matters than treaties", in: *Max Planck Encyclopedia of Public International Law* [online: http://www.mpepil.com (accessed on August 2013)]; André von Walter, "Arbitration on oil concession disputes", *ibid.*

[105] International Law Commission, *Shared natural resources: first report on outlines*, by Chusei Yamada, Doc. A/CN.4/533 (Fifty-fifth session: Geneva, 5 May-6 June and 7 July-8 August 2003) [online: http://untreaty.un.org/ilc/documentation/english/a_cn4_533.pdf (accessed on August 2013)].

[106] *Official Records of the General Assembly, Fifty-fifth Session, Supplement No. 10* (A/55/10), extract from *Yearbook of the International Law Commission, 2000, vol. 2*, p. 131, para. 729(3).

The ILC decided that the scope of its effort should be limited to the natural resources within the jurisdiction of two or more States considering certain applicable principles, such as the duty to cooperate, the equitable and reasonable utilization and participation, which in turn included unitization and examples of regimes for shared resources, the prevention and abatement of significant harm and procedure for situations in which harm is caused, exchange of data and information, management through a joint management mechanism, and non-discrimination.[107]

In a later report on shared natural resources, the ILC referred to the question of the relationship between its work on transboundary aquifers and that on oil and natural gas. The Commission noted that an overwhelming majority of States found that the law on transboundary aquifers should be treated independently of any future work on oil and natural gas. It concluded that the differences between aquifers and oil and gas and their economic and social relevance were very significant. There was also a great disparity facing the problems related to their management, as well as regarding the environmental impacts and effects and the commercial aspects of the two. The ILC also found that exploration and exploitation of energy resources were a complex endeavour and that the gathering and assessing of State practice on oil and natural gas would take a relatively long time. The ILC, therefore, concluded that the proposed articles on aquifers would not necessarily apply to oil and natural gas, but that the work on aquifers would be helpful in determining the potential direction, substance and value of any work that might be carried out on oil and natural gas.[108]

The law of transboundary aquifers presented by the ILC recognized the sovereignty of each State over the portion of a transboundary aquifer or aquifer system located within its territory. It also mentioned that States should utilize these resources according to the principle of equitable and reasonable utilization. The law of transboundary aquifers presented by the ILC also provided that States should establish joint mechanisms of cooperation on the basis of sovereign equality, territorial integrity, sustainable development, mutual benefit and good faith.[109]

Similar rules could not, however, be envisaged for the development of hydrocarbon deposits that straddled a boundary line or, for that matter, which are found in maritime areas of overlapping claims. On the one hand, the continental shelf regime establishes the exclusive sovereign right of the coastal State to develop its resources, without referring to an obligation of cooperation. On the other hand, the development of non-living marine resources, particularly of oil and

---

[107] *Ibid*, Annex, Section 3, p. 141.

[108] International Law Commission, *Fifth report on shared natural resources*, by Chusei Yamada, Doc. A/CN.4/591 (Sixtieth session: 5 May–6 June and 7 July–August 2008), pp. 2–3 [online: http://daccess-dds-ny.un.org/doc/UNDOC/GEN/N08/249/11/PDF/N0824911.pdf?OpenElement (accessed on August 2013)].

[109] *Ibid*, pp. 17–24.

gas, takes into consideration the fact that these are non-renewable resources and their development should be guided by its intrinsic economic worth.

The ILC eventually considered not pursuing the work on shared natural resources regarding oil and gas based on the recommendation of its own working group. Firstly, because it considered that the specific and complex issues relating to transboundary reserves had been adequately addressed by States for a number of years through bilateral arrangements and cooperation. Secondly, in areas pending the delimitation of maritime boundaries and unless States agreed to circumvent the problem of delimitation, offshore development of mineral resources was connected with the resolution of these disputes and that, as a result, the ILC should refrain from considering matters relating to maritime boundary delimitation, seeing that these should be considered on a case-by-case basis by the relevant States. Furthermore, the proposal made for the gathering and analysing information on State practice concerning oil and gas or even the preparation of a model agreement was considered not being very productive, in particular due to the specificities of each case. Thirdly, the ILC recognized that State practice of including in maritime delimitation agreements provisions for the joint exploitation of oil and gas deposits straddling a boundary line was a better solution by allowing States to negotiate an outcome regarding common mineral resources instead of pursing a process of progressive development and codification of international law, particularly when considering that the development of offshore mineral resources was a highly technical matter.[110]

The attempt of the ILC to prepare rules on shared mineral resources ultimately resulted in the recognition that the development of these resources would be more efficiently dealt with by the Sates due to the unique nature of these resources and of the circumstances in which the respective exploration and exploitation activities took place.

3.3.2 A Need for a Model Agreement?

The British Institute of International and Comparative Law (BIICL) published a model agreement of joint development of offshore hydrocarbon deposits with an explanatory commentary for the purpose of being used by academics, government lawyers and legal practitioners. It was meant to offer a "*practical, flexible and comprehensive solution to the development of offshore oil and gas resources where no general maritime boundary has been agreed between two coastal States.*"[111]

[110] International Law Commission, *Report of the International Law Commission* (Sixty-second session: 3 May–4 June and 5 July–6 August 2010), Doc. A/65/10 (Chapter VII: Shared natural resources), pp. 342–344, paras. 374–384 [online: http://untreaty.un.org/ilc/reports/2010/english/chp12.pdf (accessed on August 2013)].

[111] Hazel Fox et al., *Joint Development of Offshore Oil and Gas. A Model Agreement for States for Joint Development with Explanatory Commentary*, directed by Maurice Bathurst (BIICL: London, 1989), p. 1.

The model agreement was intended to be neutral regarding the interests of developed and developing countries, while at the same time sufficiently flexible to adjust to their needs.[112] Its aim was to offer a practical solution of an agreement which was hoped that States would make it the basis for negotiations and drafting of agreements on joint development.[113]

The model agreement and commentary were the result of considerable efforts made over several years. Its preparation was a multidisciplinary exercise based on principles of international law, as well as the examination of existing documentation on joint development and a survey and analysis of the respective practice, in order to identify the problems and possible solutions.[114]

The model agreement took into consideration three possible compositions: a compulsory joint operating venture between the States or their nationals; the creation of a joint authority for the purpose of developing the joint development area; and the possibility of allowing a single State to manage the joint development area while paying a share of the revenue to the other State.[115] These alternatives allowed for the analysis of the merits and disadvantages of each one and the identification of the elements that are common to all and that should be included in any joint development agreement, as well as of the mechanisms for cooperation that had best worked in practice in joint development agreements implemented to that date.[116]

The model presented by BIICL provided a structure to guide States during negotiations and when choosing the relevant joint development regime. It took into consideration the framework used in most comprehensive joint development regimes in force at the time of its preparation, while allowing for different variations to suit the specificities of each case.[117] The model agreement was also prepared considering the potential parts of the world that would most likely adopt a joint development regime. These were mainly regions considered to have significant hydrocarbon-potential and where the relevant States did not agree on the delimitation of maritime boundaries.[118]

Other studies have also tried to distinguish between different models of joint development agreements. This is the case of the study of Anette Flormann-Pfaff that identified five different models for cooperation between States.[119] The first model included the development of the joint development area by only one State, being that the other State or States would participate on the revenues resulting from the

---

112 *Ibid.*

113 *Ibid*, p. 14.

114 *Ibid*, pp. 10, 115–153.

115 *Ibid*, p. 1.

116 *Ibid*, pp. 10–11.

117 *Ibid*, pp. 388–416.

118 *Ibid*, p. 155.

119 Annette Flormann-Pfaff, *Lagerstätten im Völkerrecht: Joint Development: Zusammenarbeit bei anerkannten und streitigen Grenzen* (Duncker und Humblot: Berlin, 1994), pp. 201–209, 376–377.

respective exploration and exploitation activities undertaken by that State and which would be subject to its legal rules. A second model and less straightforward than the previous one provided that all States would engage in such activities but separately and on the respective side of the maritime boundary or joint development area. In this case, States would exchange information regarding relevant aspects of the exploration and exploitation activities, such as production plans and apportionment of resources, either directly or through a commission created for that purpose. A third model considered that different operators would be granted rights by the participating States and having the possibility of together undertaking exploration and exploitation activities through a *"joint" or "single" "operator"* in accordance with a joint venture or joint operating agreement. According to the fourth and more complex model, States would create an enterprise with legal personality for the purpose of exploring and exploiting the joint development area, equally sharing the respective costs and benefits.[120] And, finally, a fifth model that consisted of States creating an entity with legal personality and with a wide range of powers and autonomy regarding the management of resources of the joint development area that would allow entering into agreements with operators for the purpose of undertaking exploration and exploitation activities.

Despite of the value of the models presented by BIICL and Anette Flormann-Pfaff, there has been a significant disparity between the different joint development agreements that have been implemented by States. As Anette Flormann-Pfaff also recognized, it is not possible to limit the possibility of States adjusting or creating an agreement that best serves their interests. The existing disparity is mostly the result of States' discretionary powers and the specific circumstances in which each agreement is negotiated and implemented and which often vary considerably from agreement to agreement. The final agreement corresponds, in most cases, to an achievable outcome, rather than the initial aspired commitment.

Joint development agreements do depend on States reaching a consensus on certain essential legal elements, namely the identification of the designated area and of the natural resources to be exploited, the establishment of a jurisdictional and legal framework, as well as of the terms and conditions under which the joint operations are to take place.[121] BIICL considered that these elements would perhaps be to imprecise, adding that joint development agreements should include

---

[120] On the possibility of joint development with the use a corporate entity "*possessing all of the necessary functional responsibilities of supervision, administration, political oversight and State-to- State consultation, while, equally, allowing (...) the potential and power to conduct the actual physical development of the resource*" *see* William T. Onorato, "Joint development in the international petroleum sector: the Yemeni variant", in: 39 *International and Comparative Law Quarterly* (1990), pp. 660–662.

[121] "*Als wesentliche Elemente muss eine Joint Development-Übereinkunft bestimmte Regelungen enthalten. Erstens muss sie die entsprechende Zone definieren. Zweitens sind die Ressourcen zu bestimmen, auf die sie Anwendung findet. Drittens müssen die Hoheitsbefugnisse und das anwendbare Recht ausgewählt werden. Viertens müssen sich die Parteien über die Umstände und Einzelheiten der Erforschung und Ausbeuteung der Ressourcen einschließlich der Verteilung der Kosten und Gewinne einigen*" Lagoni, *supra* note 46, p. 281.

the designation of the joint development area, method of appointment, choice of operator, financial provisions, the regulatory authority and the identification of the laws to be applied, safety and health provisions, requirements prohibiting pollution and for the protection of the marine environment, and a procedure for dispute settlement.[122]

State practice has indeed demonstrated that joint development regimes have included different structures and a wide variety of applicable terms and conditions. Depending on the negotiation process, some agreements have been established straightforwardly and others have taken quite a number of years to be implemented due to the significant disparities between States' proposals and views regarding the envisaged joint development regime.[123] The differences in the practice of States underline the unique context under which each regime is negotiated and the influence that certain factors have on States' position before and during negotiations. From political rhetoric deeply embedded in national views regarding a dispute on the delimitation of maritime boundaries, to the incompatibilities between the different domestic legal regimes and the diverse functional concepts of the common development of offshore hydrocarbon deposits, such as who should have access to operations and in what manner should access be granted, or what should be the scope and organization applicable to the management of the joint development area. For example, the BIICL model-agreement was based on a two-State composition,[124] despite the fact that there have been attempts to establish trilateral joint development regimes, including within the frame of regional efforts towards the settlement of disputes on the delimitation of maritime boundaries.[125]

[122] Fox et al., *supra* note 111, p. 46.

[123] Also *see* Mark J. Valencia, *South-East Asian Seas: Oil Under Trouble Waters. Hydrocarbon Potential, Jurisdictional Issues, and International Relations* (Oxford University Press: Oxford, New York, 1985), pp. 118–119. In the same sense *see* G. H. Blake and R. E. Swarbrick, "Hydrocarbon and international boundaries: a global overview", in: *Boundaries and Energy: Problems and Prospects*, edited by Gerald Blake, Martin Pratt, Clive Schofield and Janet Allison Brown (Kluwer Law International: London, The Hague, Boston, 1998), p. 14.

[124] "[J]*oint development as an agreement between two States to develop so as to share jointly in agreed proportions by inter-States cooperation and national measures the offshore oil and gas in a designated zone of seabed and subsoil of the continental shelf to which both or either of the participating States are entitled in international law.*" Fox et al., *supra* note 111, p. 45. Robin R. Churchill defines joint development zone as "*an area where two or more States have, under international law, sovereign rights to explore and exploit the natural resources of the area and where the States concerned have agreed to engage in such exploration and exploitation under some form of common or joint arrangement.*" Robin R. Churchill, "Joint development zones: international legal issues", in: *Joint Development*, vol. 2, *supra* note 70 in Chap. 2, p. 55.

[125] On the French proposal for a joint development regime with Spain and Italy *see* Umberto Leanza, "The delimitation of the continental shelf of the Mediterranean Sea", in: 8-3 *International Journal of Marine and Coastal Law* (1993), p. 388. Also *see* the *Agreement between the Government of the Republic of Indonesia, the Government of Malaysia and the Government of the Kingdom of Thailand Relating to the Delimitation of the Continental Shelf Boundaries in the Northern Part of the Strait of Malacca*, done on 21 December 1971,

The preparation of a model-agreement of joint development of offshore hydrocarbon deposits or, for that matter, any attempt at systematization or codification of joint development agreement,[126] would not necessarily favour the implementation of these regimes or their use whenever these resources would be found in disputed maritime areas or straddle a maritime boundary, nor would they contribute towards the consolidation of State practice in that respect. They are, nonetheless, extremely valuable for the study of the concept of joint development agreements and the analysis of their legal intricacy.

(Footnote 125 continued)
reproduced in: *International Maritime Boundaries*, vol. 2, edited by Jonathan. I. Charney and Lewis M. Alexander (Martinus Nijhoff Publishers: Dordrecht, Boston, London, 1993), pp. 1452–1454. This agreement provided that a hydrocarbon deposit that should straddle a boundary line will only be developed after consultation between the three States. On the efficiency of regional instruments in regulating the common resource use of the oceans *see* Judo Umarto Kusumowidagdo, *Consultations Clauses as Means of Providing for Treaty Obedience. A Study in the Law of Treaties* (Almqvist & Wiksell International: Stockholm, 1981), pp. 131–132. Also *see* Ian Brownlie, "Legal status of natural resources in international law (some aspects)", in: 162 *Recueil des Cours de l'Académie de Droit International* (1979-I), p. 289.

[126] William T. Onorato made a proposal for international codification of procedures to be implemented whenever States intended to develop a common offshore hydrocarbon deposit, as an important step towards achieving cooperation regarding these resources: "[a]*s a beginning, there is ample reason for an international technical convention establishing uniform exploitation procedures reflecting accepted standards of good oil practice which should apply where several countries are found tapping the same oil deposit. An international codification of such proved practices and procedures would be a useful first step towards providing a basis for even greater co-operation. An alternative step, however, would be to reach an international accord providing for unitised operations between countries drawing from a common source* (…)." William T. Onorato, "Apportionment of an international common petroleum deposit", in: 17 *International and Comparative Law Quarterly* (1968), pp. 101–102; Also *see* "Apportionment of an international common petroleum deposits", in: 26 *International and Comparative Law Quarterly* (1977), p. 336.

The preparation of a model agreement of joint development of offshore hydrocarbon deposits or, for that matter, any attempt at systematization or codification of joint development agreements [illegible] would not necessarily favour the implementation of these regimes or their use. Whenever these regimes could be found in disputed maritime areas or [illegible] [illegible] [illegible] [illegible] agreements [illegible] joint [illegible].

Chapter 4
Mineral Resources Clauses in Delimitation Treaties and Joint Development Agreements

4.1 Outline of State Practice

4.1.1 Maritime Delimitation Agreements

Most mineral resources clause only include the obligation of the States-parties to negotiate towards settling the terms applicable to the development of an offshore hydrocarbon deposit that straddles a maritime boundary line. This is the case, for example, of the maritime delimitation agreement between Indonesia and Malaysia:

> If any single geological petroleum or natural gas structure extends across the straight lines referred to in Article I and the part of such structure which is situated on one side of the said lines is exploitable, wholly or in part, from the other side of the said lines, the two [States-parties] will seek to reach agreement as to the manner in which the structure shall be most effectively exploited.[1]

A slightly more elaborated example of a mineral resources clause can be found in the agreement between France and Spain regarding the delimitation of the continental shelf in the Bay of Biscay, although the two States did not foresee the unitization of deposits within the joint development area created by this agreement, or between the latter and the continental shelf areas of the States-parties:

> 1. If a deposit of a natural resource lies across the continental shelves' dividing line, and if the portion of the deposit line is wholly or partially exploitable by means of installations located on the other side of the line, the [States-parties], acting in conjunction with the holders of the exploitation rights, if such exist, shall seek an

[1] Article 4 of the *Agreement between the Government of the Republic of Indonesia and the Government of Malaysia relating to the delimitation of the continental shelves between the two Countries*, done at Kuala Lumpur on 27 October 1969 and entered into force on 7 November 1969 [online: http://www.un.org/depts/los/LEGISLATIONANDTREATIES/PDFFILES/TREATIES/MYS-IDN1969CS.PDF (accessed on August 2013)], reproduced in: Shigeru Oda, *The International Law of the Ocean Development: Basic Documents* (Sijthoff: Leiden, 1972), pp. 424–425.

V. Becker-Weinberg, *Joint Development of Hydrocarbon Deposits in the Law of the Sea*, Hamburg Studies on Maritime Affairs 30, DOI 10.1007/978-3-662-43570-0_4

agreement on terms for exploiting the deposit so that such exploitation may be as profitable as possible and in such manner that each of the [States-parties] shall preserve all its rights over the natural resources of its continental shelf. This is particularly applicable if the method for exploiting the portion of the deposit located on one side of the dividing line affects conditions for exploiting the other portion of the deposit.

2. In the event that the natural resources of a deposit lying across the diving line have already been exploited, the [States-parties], acting in conjunction with the holders of exploitation rights, if such exist, shall seek an agreement on appropriate compensation.[2]

The maritime delimitation agreement signed by the Netherlands and Germany also provides a good example of a more advanced mineral resources clause:

1. If the existence of a mineral deposit in or upon the continental shelf of one of the Contracting Parties is established and the other Contracting Party is of the opinion that the said mineral deposit extends into or onto its continental shelf, the latter Party may notify the former Party accordingly, submitting the data on which it bases its opinion. If the said former Party does not share the opinion of the other Party, the arbitral tribunal referred to in article 5 shall, at the request of either Party, make a ruling on the question.
2. If the Contracting Parties agree on the question or the arbitral tribunal rules that the mineral deposit extends into or onto the continental shelf of both Parties, the Governments of the Contracting Parties shall, for the purpose of exploitation, adopt regulations which, with due regard to the interests of both Parties, take into account the principle that each Party has title to mineral resources situated in or upon its continental shelf. If any mineral resources have previously been extracted from the deposit extending across the boundary, the regulations shall also include provisions for appropriate compensation.
3. Regulations as referred to in paragraph (2) above may also, with the consent of the Governments of the Contracting Parties, be adopted wholly or partly between the entitled parties. An entitled party is any person who has a right to extract the mineral resources in question.
4. If regulations as referred to in paragraph (2) or (3) above have not been drawn up within a reasonable time, either Contracting Party may bring the matter before the arbitral tribunal referred to in article 5. In such cases, the arbitral tribunal may also make a ruling ex aequo et bono. The arbitral tribunal shall be empowered, after hearing the Contracting Parties, to issue interim orders.[3]

[2] Article 4 of the *Convention between the Government of the French Republic and the Government of the Spanish State on the delimitation of the continental shelves of the two states in the Bay of Biscay*, done on 29 January 1974 [online: http://www.un.org/depts/los/LEGISLATIONANDTREATIES/PDFFILES/TREATIES/FRA-ESP1974CS.PDF (accessed on August 2013)], reproduced in: *International Maritime Boundaries*, vol. 2, *supra* note 125 in Chap. 3, pp. 1728–1734. Also *see* Masahiro Miyoshi, "The joint development of offshore oil and gas in relation to maritime boundary delimitation", in: 2–5 *Maritime Briefing/International Boundaries Research Unit* (1999), pp. 31–32.

[3] Article 2 of the *Treaty concerning the Delimitation of the Continental shelf Under the North Sea*, done at Copenhagen on 28 January 1971 and entered into force on 7 December 1965, published at 857 U.N.T.S 130, reproduced in: *International Maritime Boundaries*, vol. 2, *supra* note 125 in Chap. 3, pp. 1844–1850.

Despite some resemblance between the general content of natural resources clauses commonly included in maritime delimitation agreements, there are some agreements that resort to other alternatives. This is the case of the maritime delimitation agreement signed between the United States of America and Mexico, in which these States agreed on a moratorium together with the implementation of a protective perimeter. Accordingly, the agreement established a period of 10 years during which no exploration or exploitation activities could take place in an area of 2.8 nm of the continental shelf within 1.4 nm of the boundary line. These States also agreed that, in accordance with the applicable national laws and regulations, they would facilitate the requests from the other State regarding geological and geophysical studies to help determine the possible presence and distribution of transboundary reservoirs and to share such geological and geophysical information in order to identify the possible existence and location of such resources.

> 1. Due to the possible existence of petroleum or natural gas reservoirs that may extend across the boundary set forth in Article I (hereinafter referred to as "transboundary reservoirs"), the Parties, during a period that will end ten (10) years following the entry into force of this Treaty, shall not authorize or permit petroleum or natural gas drilling or exploitation of the continental shelf within one and four-tenths (1.4) nautical miles of the boundary set forth in Article I. (This two and eight-tenths (2.8) nautical mile area hereinafter shall be referred to as "the Area".) (…)
> 3. The parties, by mutual agreement through an exchange of diplomatic notes, may modify the period set forth in paragraph 1.
> 4. From the date of entry into force of this Treaty, with respect to the Area on its side of the boundary set forth in Article I, each Party, in accordance with its national laws and regulations, shall facilitate requests from the other Party to authorize geological and geophysical studies to help determine the possible presence and distribution of transboundary reservoirs.
> 5. From the date of entry into force of this Treaty, with respect to the Area in its entirety, each Party, in accordance with its national laws and regulations, shall share geological and geophysical information in its possession in order to determine the possible existence and location of transboundary reservoirs.
> 6. From the date of entry into force of this Treaty, if a Party has knowledge of the existence or possible existence of a transboundary reservoir, it shall notify the other Party.[4]

---

[4] Article 4 of the *Treaty between the Government of the United States of America and the Government of the United Mexican States on the delimitation of the Continental Shelf in the Western Gulf of Mexico beyond 200 nautical miles*, done at Washington D.C. on 9 June 2000 [online: http://www.un.org/depts/los/LEGISLATIONANDTREATIES/PDFFILES/TREATIES/USA-MEX2000CS.PDF (accessed on August 2013)], reproduced in: *New Directions in the Law of the Sea: Regional & National Developments*, edited by Roy S. Lee and Moritaka Hayashi (Oceana Publications Inc.: Dobbs Ferry, New York, 2001) 2001–2003, at X.10.a(5). The United States of America and Mexico have also signed the *Treaty to Resolve Pending Boundary Differences and Maintain the Rio Grande and Colorado River as International Boundary*, done in the City of Mexico on 23 November 1970 and entered into force on 18 April 1972 [online: http://www.un.org/depts/los/LEGISLATIONANDTREATIES/PDFFILES/TREATIES/mex-usa1970.pdf (accessed on August 2013)], and the *Treaty on Maritime Boundaries between the United States*

The United States of America and Mexico agreed that in the aforesaid area and during the moratorium period, they would meet for the purpose of identifying, locating and determining the geological and geophysical characteristics of transboundary reservoirs and attempt to reach an agreement for their efficient and equitable exploitation. They also agreed that, after the moratorium period, both countries would share information regarding any decision to lease, license, grant concessions, or otherwise make available portions of the protective perimeter for petroleum or natural gas exploration or development. The United States of America and Mexico further determined that each State would inform the other when production of petroleum or natural gas resources was to begin. Both States agreed to ensure that the entities authorized to undertake activities within the perimeter would also observe the terms agreed in the maritime delimitation agreement regarding the development of transboundary reservoirs. States could ultimately agree on the implementation of a unitization or a joint development agreement.

1. With respect to the Area in its entirety, during the period set forth in paragraph 1 of Article IV:

(a) as geological and geophysical information is generated that facilitates the Parties' knowledge about the possible existence of transboundary reservoirs, including notifications by Parties in accordance with paragraph 5 of Article IV, the Parties shall meet periodically for the purpose of identifying, locating and determining the geological and geophysical characteristics of such reservoirs;
(b) the Parties shall seek to reach agreement for the efficient and equitable exploitation of such transboundary reservoirs; and
(c) the Parties shall, within sixty days of receipt of a written request by a party through diplomatic channels, consult to discuss matters related to possible transboundary reservoirs.

2. With respect to the Area in its entirety, following the expiry of the period set forth in paragraph 1 of Article IV

(a) a Party shall inform the other Party of its decisions to lease, license, grant concessions, or otherwise make available, portions of the Area for petroleum or natural gas exploration or development and shall also inform the other Party when petroleum or natural gas resources are to commence production; and
(b) a Party shall ensure that entities it authorizes to undertake activities within the Area shall observe the terms of the Treaty.[5]

(Footnote 4 continued)
of America and the United Mexican States, done at Mexico on 4 May 1978 and entered into force on 13 November 1997 [online: http://www.un.org/depts/los/LEGISLATIONANDTREATIES/PDFFILES/TREATIES/MEX-USA1978MB.PDF (accessed on August 2013)], reproduced in: 17 *International Legal Materials* 1073 (1978).

[5] Article 5, *idem*. On joint development in the Gulf of Mexico *see* Utton/Mchugh, *supra note 62 in Chapter 2*, pp. 717–718, 722–732;, Richard J. McLaughlin, "Maritime boundary delimitation and cooperative management of transboundary hydrocarbons in the ultra-deepwaters of the Gulf of Mexico", in: *Maritime Boundary Disputes, Settlement Processes, and the Law of the Sea*, edited by Seoung-Yong Hong and Jon Van Dyke (Martinus Nijhoff Publishers: 2009), pp. 225–230.

Similar to the protective perimeter regime established by the United States of America and Mexico, the maritime delimitation agreement implemented by Iran and Qatar provided that, in the event of a transboundary deposit extending across the boundary line and allowing its exploitation, wholly or partly, from the other side of that line, no well should be drilled on either side of which the "*producing section*" would be less than 125 m from the boundary line. Except if these States agreed otherwise regarding the manner in which the operations of both sides of the boundary line could be coordinated or unitized.[6] Similar use of a protective perimeter can also be found in the mineral resources clause of the maritime delimitation agreement signed by Iran and Saudi Arabia, whereby the two States agreed that there would not be any drilling operations within an area extending 500 m in width in the submarine areas on both sides of the boundary line.[7]

However, despite the fact that most mineral resources clauses included in maritime delimitation agreements do not envisage an obligation to enter into a unitization regime, there have been exceptions. This is the case of the agreement between Denmark and Norway that established that both countries would agree on the implementation of an arrangement regarding the development of straddling hydrocarbon deposits.[8] Likewise, the agreement between the United Kingdom and Norway implemented cross-boundary cooperation of petroleum activities between both States and established that transboundary hydrocarbon deposits would be considered for all terms and purposes as a unit, despite the fact that it did not include any reference to unitization or any specific form of a cooperative regime.[9]

---

[6] Article 2 of the *Agreement concerning the boundary line dividing the continental shelf between Iran and Qatar*, done at Doha on 20 September 1969 and entered into force on 10 May 1970 [online: http://www.un.org/depts/los/LEGISLATIONANDTREATIES/PDFFILES/TREATIES/IRN-QAT1969CS.PDF (accessed on August 2013)], reproduced in: Oda, *supra* note 1, pp. 417–418.

[7] Article 4 of the *Agreement concerning the Sovereignty over the Islands of Farsi and Al-Arabiyah and the Delimitation of the Boundary Line Separating the Submarine Areas between Iran and Saudi Arabia*, done at Teheran on 24 October 1968 [online: http://www.un.org/depts/los/LEGISLATIONANDTREATIES/PDFFILES/TREATIES/SAU-IRN1968SA.PDF (accessed on August 2013)], reproduced in: Oda, *supra* note 1, 415–416.

[8] Article 4 of the *Agreement between Denmark and Norway Relating to the Delimitation of the Continental Shelf*, done at Oslo on 8 December 1965 and entered into force on 22 June 1966 [online: http://www.un.org/depts/los/LEGISLATIONANDTREATIES/PDFFILES/TREATIES/DNK-NOR1965CS.PDF (accessed on August 2013)], published at 634 U.N.T.S. (1968), pp. 71–75.

[9] Articles 3.1, 3.2, 3.3 and 3.5 of the *Framework Agreement between the Government of the United Kingdom of Great Britain and Northern Ireland and the Government of the Kingdom of Norway Concerning Cross-boundary Petroleum Co-operation*, done at Oslo 4 April 2005 and entered into force on 10 July 2007 [online: http://www.official-documents.gov.uk/document/cm72/7206/7206.pdf (accessed on August 2013)], registered at U.N.T.S. n.º 44683, 24 January 2008. Also *see Agreement between the Government of the United Kingdom and Northern Ireland and the Government of the Kingdom of Norway relating to the Delimitation of the Continental Shelf between the Two Countries*, done at London on 10 March 1965 and entered into force on 29 June 1965, published at 551 U.N.T.S. 213; *Agreement between the Government of the United Kingdom of Great Britain and Northern Ireland and the Government of the Kingdom of Norway*

The agreement on the delimitation of maritime boundaries signed between Qatar and Abu Dhabi (UAE) also determined the equal sharing of the benefits and rights of the oil field referred to as *Al-Bunduq*, despite the fact that it would only be developed by an entity of Abu Dhabi (UAE) and subject to the terms of the respective concession.[10]

An example in which States have entered into unitization agreements in reference to natural resources clauses included in maritime delimitation agreements is that of the unitisation agreement signed by Trinidad and Tobago and Venezuela, following the maritime delimitation agreement signed by both States.[11] Accordingly, the natural resources clause included in this agreement provided that Trinidad and Tobago and Venezuela would seek to reach an understanding as to the manner in which hydrocarbon deposits should be most effectively exploited and on the sharing of costs and benefits. These States also signed a memorandum that envisaged establishing a procedure for the most effective development of offshore hydrocarbon deposits straddling a boundary line of the continental shelf

(Footnote 9 continued)
relating to the exploitation of the Murchison Field Reservoir and the offtake of petroleum therefrom, done at Oslo on 16 October 1979, published at 1249 U.N.T.S. 173; *Agreement between the Government of the United Kingdom of Great Britain and Northern Ireland and the Government of the Kingdom of Norway relating to the exploitation of the Frigg Field Reservoir and the transmission of gas therefrom to the United Kingdom*, done at London on 10 May 1976, published at 1098 U.N.T.S. 4; *Agreement between the United Kingdom of Great Britain and Northern Ireland and the Netherlands relating to the exploitation of the Markham Field Reservoirs and the offtake of petroleum therefrom*, done at The Hague on 26 May 1992, published at 1731 U.N.T.S. 156; *Agreement between the Government of the United Kingdom of Great Britain and Northern Ireland and the Government of Norway relating to the exploitation of the Statfjord Field Reservoirs and the offtake of petroleum therefrom*, done at Oslo on 16 October 1979, published at 1254 U.N.T.S. 379. On these agreements *see* MacLeod, *supra* note 61 in Chap. 2, pp. 328–332.

[10] Section 7 of the *Agreement concerning settlement of offshore boundaries and ownership of islands between Qatar and the United Arab Emirates (with declarations and map)*, done at Doha on 20 March 1969 [online: http://treaties.un.org/doc/Publication/UNTS/Volume%202402/v2402.pdf (accessed on August 2013)], published at 2402 U.N.T.S. 49.

[11] *Unitisation Agreement for the Exploitation and Development of Hydrocarbon Reservoirs of the Loran-Manatee Field that extends across the Delimitation Line between the Republic of Trinidad and Tobago and the Bolivarian Republic of Venezuela*, done at Caracas on 16 August 2010, [online: http://www.energy.gov.tt/content/Unitisation_Agreement_between_T&T_and_Venezuela-:_Loran-Manatee_Field.pdf (accessed on August 2013)], reproduced in: *International Maritime Boundaries*, vol. 1, *supra* note 42 in Chap. 2, pp. 685–689. *See* Article 7 of the *Treaty between the Republic of Trinidad and Tobago and the Republic of Venezuela on the Delimitation of Marine and Submarine Areas*, done at Caracas on 18 April 1990 and entered into force 23 July 1991 [online: http://www.un.org/depts/los/LEGISLATIONANDTREATIES/PDFFILES/TREATIES/TTO-VEN1990SA.PDF accessed on August 2013)], reproduced in: *International Maritime Boundaries*, vol. 1, *supra* note 42 in Chap. 2, pp. 685–689.

between the two countries, in addition to a letter of intent that sought the promotion of cooperation between the energy sectors.[12]

Nigeria also signed an umbrella unitization agreement with Equatorial Guinea following the signing of the maritime boundary delimitation agreement. The delimitation agreement provided that these States would try to establish a unitization agreement in the event that the maritime boundary should run through any field of hydrocarbon deposits. It also determined that States would establish unitization and other arrangements with the relevant concession holders to enable that the maritime area would be developed in a commercially feasible manner. An interesting aspect of this unitization regime is the fact that it established that the law of Equatorial Guinea would be applicable to the exploration and exploitation activities in the maritime area subject to the sovereignty of Nigeria, including matters such as employment, customs, tax, environment, amongst others, thus applying only one legal regime to operators and inevitably improving the undertaking of operations.[13]

4.1.1.1 Joint Development Agreements

Following the previously referred maritime delimitation agreements signed by the United States of America and Mexico, in 2012, these States entered a joint development agreement of transboundary resources that extend across the

---

[12] *Memorandum of Understanding Concerning the Procedure for Unitisation of Hydrocarbon Reservoirs that Extend Across the Delimitation Line between the Republic of Trinidad and Tobago and the Bolivarian Republic of Venezuela*, done at Caracas on 12 August 2003, and *Letter of Intent - Enhanced bilateral cooperation between the Republic of Trinidad and Tobago and the Bolivarian Republic of Venezuela in the Energy Sector*, done at Caracas on 12 August 2003 *see Increasing Local Content, Ministry of Energy and Energy Industries: Annual Administrative Report, 2003–2004*, p. 39 [online: http://www.energy.gov.tt/content/Annual_Administration_Report_2003-04.pdf (accessed on August 2013)].

[13] Article 6 of the *Treaty between the Federal Republic of Nigeria and the Republic of Equatorial Guinea Concerning their Maritime Boundary*, done at Malabo on 23 September 2000 and entered into force on 3 April 2002, published at 2205 U.N.T.S. 325 [online: http://www.un.org/depts/los/LEGISLATIONANDTREATIES/PDFFILES/TREATIES/NGA-GNQ2000MB.PDF (accessed on August 2013)], and *Protocol in Implementation of Article 6.2 of the Treaty between the Federal Republic of Nigeria and the Republic of Equatorial Guinea Concerning their Maritime Boundary*, done at Abuja on 2 April 2002, published at 2220 U.N.T.S. 410, reproduced in: *International Maritime Boundaries*, vol. 5, edited by David A. Colson and Robert W. Smith (Martinus Nijhoff Publishers: Leiden, Boston, 2005), pp. 3630–3637. Also *see* the *Treaty between Equatorial Guinea and Nigeria on Joint Exploration of Crude Oil, especially at the Zafiro-Ekanga Oil Field located at the maritime boundary of both countries*, done on 3 April 2002, not available [online: http://www.un.org/depts/los/LEGISLATIONANDTREATIES/PDFFILES/TREATIES/GNQ-NGA2002ZE.PDF (accessed on August 2013)]. Cameroon considered the Nigeria/Equatorial Guinea maritime delimitation agreement "*was concluded in utter disregard of its own rights*" *see Land and Maritime Boundary between Cameroon and Nigeria, supra note 83 in Chap. 3*, para. 243.

maritime boundaries established by the delimitation agreements.[14] According to the terms set forth in this agreement, the United States of America and Mexico must annually consult each other on exploration and exploitation activities, including the exchange of all relevant and available geological information, without prejudice, inter alia, of their obligation to inform and initiate consultations regarding the existence of a transboundary hydrocarbon deposit or of the detection of oil and natural gas during drilling operations on either side of the delimitation line.[15] This agreement also provides that any joint exploration and/or exploitation of a transboundary hydrocarbon deposit between operators must be approved by the States-parties.[16] Furthermore, the agreement establishes that States-parties shall not commence production regarding any transboundary deposit without reaching an agreement. In the event that production has commenced, the relevant State-party shall take the appropriate measures under national law to provide that production does not "*unduly prejudice the implementation of the agreement*".[17] Another key aspects of this agreement is that, should operators fail to agree on a unitization agreement or if States-parties do not succeed in reaching an understanding with the respective operators or through the joint commission established by the agreement, operators will be allowed to carry on with production of the relevant transboundary deposit and the United States of America and Mexico will exchange the relevant data on a monthly basis, being that the joint commission will endeavour to resolve issues related to the allocation of production.[18]

A further example of unitization regimes included in joint development agreements is that of the agreement between Iceland and Norway, which provides that if a deposit lies in its entirety in the "*economic zone*" of Norway but extends into the joint development area, the deposit shall be considered as being only within this area. Whereas, any hydrocarbon deposit located on both sides of the delimitation line between the two States' economic zones, or laying in its entirety on the economic zone of Iceland, but expanding into the joint development area, States-parties should agree on "*usual unitization principles for the distribution and exploitation of the deposit*".[19]

---

[14] *Agreement between the United States of America and the United Mexican States concerning Transboundary Hydrocarbon Reservoirs in the Gulf of Mexico*, *supra* note 67 in Chap. 2. This Agreement also provides in Article 24 that, upon its entry into force, the period of any moratorium on the authorization or permitting of petroleum or natural gas drilling or exploration of the continental shelf within the boundary area shall be terminated, as established in Article 4 of the Continental Shelf Agreement, *supra* note 4.

[15] Articles 4 and 5(1) *ibid.* This agreement defines geological information as the geological, geophysical or geochemical information and data resulting from exploration and exploitation, including, but not limited to, information from drilled wells and interpretations derived from such data, and which, subject to the respective national law, may be disclosed.

[16] Article 6(1) *ibid.*

[17] Article 7(1) *ibid.*

[18] Article 7(5) *ibid.*

[19] Article 8 of the *Agreement between Iceland and Norway on the Continental Shelf between Iceland and Jan Mayen*, done at Oslo on 22 October 1981, reproduced in: *International Maritime*

Differently, the agreement signed between Sudan and Saudi Arabia established that in the event that any accumulation or deposit of a natural resource extending across the boundary of the exclusive sovereign rights area of either State-party and the joint development area, the authority created by both countries will determine the manner in which it is to be exploited, provided that any decision taken shall guarantee for the State involved an equitable share of the revenues of the exploitation of such accumulation or deposit.[20]

In contrast, the agreement signed between Nigeria and Sao Tome and Principe provided that if a hydrocarbon deposit straddled a dividing line of the joint development area and any area of the continental shelf or the EEZ subject to the jurisdiction of one of these States-parties, both countries would endeavour to reach an agreement for the unitization of the deposit, upon notice by either State-party. Should an agreement not be reached within 9 months following such notice, the deposit would be fair and reasonably apportioned between Nigeria and Sao Tome and Principe in consideration of the moment when production of the said deposit began. A literal interpretation of the relevant provisions of this agreement would not create an obligation of information when such a situation has occurred because it uses the expression "*may give notice*" instead of *shall give notice*. Still, according to this agreement, should a hydrocarbon deposit straddle a dividing line of the joint development area and a maritime area subject to the jurisdiction of a third State, the entity created by both countries would consider seeking an agreement with the third State as to the manner in which the deposit could be most effectively developed and the respective revenues distributed, having due regard for the proportions of the resource located on either side of the dividing line.[21]

(Footnote 19 continued)
Boundaries, vol. 2, *supra note 125 in Chap. 3*, pp. 1762–1765; 20 *International Legal Materials* 787 (1981).

[20] Article 14 of the *Agreement between the Government of the Democratic Republic of the Sudan and the Government of the Kingdom of Saudi Arabia relating to the Joint Exploitation of the Natural Resources of the Sea-bed and Subsoil of the Red Sea in the Common Zone*, done at Khartoum on 16 May 1974 and entered into force 26 August 1974, reproduced in: *International and United States Documents on Oceans Law and Policy*, edited by John Norton Moore (William S. Hein: Buffalo New York, 1986), pp. 198–200.

[21] Article 31 of the *Treaty between the Federal Republic of Nigeria and the Democratic Republic of Sao Tome and Principe on the Joint Development of Petroleum and Other Resources, in Respect of Areas of the Exclusive Economic Zone of the Two States*, done at Abuja on 21 February 2001 and entered into force 16 January 2003, [online: http://www.un.org/Depts/los/LEGISLATIONANDTREATIES/PDFFILES/TREATIES/STP-NGA2001.PDF (accessed on August 2013)], reproduced in: *International Maritime Boundaries*, vol. 5, *supra* note 13, pp. 3649–3682.

4.1.2 Relevance of Mineral Resources Clauses

The analysis in Chap. 2 of the concept of unitization noted that, as a conservation measure, unitization ensured the efficient development of hydrocarbon deposits on account of their characteristics. Although unitization is a useful method to achieve optimal production of common offshore hydrocarbon deposits, States have not always determined its effective implementation during the negotiation of a boundary or joint development agreement. States seldom agree, beforehand, in acting together in order to equally share such resources and/or the revenues resulting from their exploitation as a single unit. Despite the general practice of including unitization or mineral resources clauses in maritime delimitation treaties and joint development agreements, these regimes only recognize the duty of the coastal State to inform the neighbouring coastal States on the existence of straddling offshore hydrocarbon deposits and also, on occasion, the obligation of mutual-restraint regarding their exploration and exploitation.

Mineral resources clauses have, in most cases, a programmatic character and generally envisage a reinforced obligation for States to cooperate in the event that a hydrocarbon deposit should extend across a boundary line or is found in an area of overlapping claims. This is the case of offshore hydrocarbon deposits that straddle across the boundaries of a joint development area controlled by different States or the boundary between a joint development area and the continental shelf of a coastal State. As it will be subsequently analysed, although most mineral resources clauses do not include an obligation for States to implement a unitization regime, there have been, nonetheless, exceptions and some agreements do in fact introduce alternative measures.[22]

4.1.3 Obligation to Include Mineral Resources Clauses

There is no obligation in international law to include mineral resources clauses in maritime delimitation agreements or in joint development agreements signed before or after the delimitation of maritime boundaries. Even though this has been the usage on many of these agreements, there is no consistent practice and even in those cases where similar clauses have been adopted, there is no uniformity as to their content or scope, particularly in joint development agreements. Furthermore, despite the fact that International Courts and Tribunals have long understood the problems represented by common hydrocarbon deposits and the need for the

---

[22] Laurent Lucchini and Michel Voelckel, *Droit de la Mer: Délimitation, Navigation et Pêche*, vol. 2 (A. Pédone: Paris, 1996), pp. 114–118. These authors identified the following alternatives should resources straddle a border line: (a) negotiate in order to enter into an agreement, (b) create a buffer zone, (c) enter into a unitization agreement, and (d) implement a cooperative regime such as a joint development regime. Also *see* Wälde, *supra* note 66 in Chap. 2, pp. 21–22.

efficient development of mineral resources, international jurisprudence has not acknowledge the existence of an obligation requiring neighbouring States to engage in the cooperative development of common mineral resources, nor has it recognized a rule demanding States to adopt conservation measures in these cases. In this respect, the Arbitral Tribunal in the arbitration between Eritrea and Yemen made the following assumption regarding the existing State practice and the legal obligations applicable to States in the case of transboundary resources:

> In paragraph 1 of its Prayer for Relief, Eritrea requests the Tribunal to determine that "The Eritrean people's historic use of resources in the mid-sea islands includes... mineral extraction". For reasons explained in paragraph 104 of this Award, the Tribunal is not in a position to accede to this request. However, it is of the view that, having regard to the maritime boundary established by this Award, the Parties are bound to inform one another and to consult one another on any oil and gas and other mineral resources that may be discovered that straddle the single maritime boundary between them or that lie in its immediate vicinity. Moreover, the historical connections between the peoples concerned, and the friendly relations of the Parties that have been restored since the Tribunal's rendering of its Award on Sovereignty, together with the body of State practice in the exploitation of resources that straddle maritime boundaries, import that Eritrea and Yemen should give every consideration to the shared or joint or unitised exploitation of any such resources.[23]

Therefore, despite the generalized use of mineral resources clauses in delimitation treaties, as well as in cooperative arrangements regarding the development of transboundary deposits or those found in areas of overlapping claims, it is not possible to conclude that States are legally required to include natural resources clauses in delimitation treaties.

4.1.4 Non-compliance with Mineral Resources Clauses

It may occur that a State does not act in accordance with its procedural duty to provide information, but does not, however, cause harm or damage to the neighbouring State by complying with its substantial obligation of self-restraint regarding the development of transboundary mineral resources.[24] In this case, the

---

[23] *The Eritrea - Yemen, Award of the Arbitral Tribunal in the Second Stage of the Proceedings (Maritime Delimitation), 17 December 1999*, para. 86. [online: http://www.pca-cpa.org/upload/files/EY%20Phase%20II.PDF (accessed on August 2013)].

[24] International Law Association, *Report of the International Committee on the Legal Aspects of the Conservation of the Environment*, by Dietrich Rauschning (Fifty-eighth Conference: Manila, 1978), pp. 390–411, and *Ibid* (Fifty-ninth Conference: Belgrade, 1980), pp. 548–550. *See* Brian D. Smith, *State Responsibility and the Marine Environment: the Rules of Decision* (Clarendon Press: Oxford, 1988), pp. 83–85: "*In summary, the principles of good neighbourliness and abuse of rights provide the underlying or animating logic of the preventive obligation defined in the preceding section. They do not, however, appear to have advanced from policy abstractions to concrete rules of decision which limit state prerogatives with respect to the environment*". Also *see* Andrassy, *supra* note 70 in Chap. 2, p. 80: "*Quoique les relations de voisinage existent pour*

first and foremost effect of withholding information on transboundary resources is evidently the hindering of the prospects for cooperation between the relevant States. Secondly, when pending the delimitation of maritime boundaries, it creates an advantage for the State withholding information on the said resources, particularly when negotiating the respective boundary line. This would also constitute a breach of States' obligation to negotiate in good faith. Thirdly, after the delimitation of maritime boundaries, failure by a State to inform a neighbouring State on the existence of common resources effectively results in preventing the latter from identifying and possibly having access to these resources, which are also subject to its sovereignty. The failure to inform could ultimately affect the economic and social progress and general welfare of a State, which is also a fundamental right of States.[25]

The interrelation between procedural and substantial obligations is not always evident, particularly when substantial obligations are complied with, even though procedural obligations have been breached. In the *Pulp Mills on the River Uruguay* case, the ICJ took into consideration the distinction between procedural and substantial duties in its evaluation of States' compliance with the duty of information, particularly regarding the obligation of one State to provide the other with the results of environmental impact assessments of activities that had transboundary effects and the adoption of measures to protect and preserve the environment:

> In the opinion of the Court, the obligation to notify is intended to create the conditions for successful co-operation between the parties, enabling them to assess the plan's impact on the river on the basis of the fullest possible information and, if necessary, to negotiate the adjustments needed to avoid the potential damage that it might cause; and that [t]he obligation to notify is therefore an essential part of the process leading the parties to consult in order to assess the risks of the plan and to negotiate possible changes which may eliminate those risks or minimize their effects.[26]

The Court concluded that the relevant State had breached its procedural obligations to inform the other State, seeing that the information on the environmental impact assessment on any plan that is liable to cause significant transboundary harm to another State would be required before any decision on the environmental viability of that plan could be reached.[27] Yet, despite the Court having recognized

(Footnote 24 continued)
la plupart entre Etats limitrophes, il peut arriver que certains phénomènes produisent leurs effets à une distance assez grande pour se faire sentir, non seulement dans l'Etat immédiatement contigu, mais aussi dans un Etat qui n'a pas de frontières communes avec celui où le phénomène a eu lieu. Nous pouvons imaginer d'autres cas pour cette hypothèse, mais le plus commun est celui d'un système fluvial comprenant les territoires de plus deux Etats. Il y a donc fait de voisinage, et droit de voisinage, sans qu'il y ait continuité territoriale".

[25] Articles 55 lit. (a) and 56 of the UN Charter. Also *see* Resolution (UNGA) n. 2625 (XXXV), 24 October 1970, *Declaration on principles of international law concerning friendly relations and co-operation among States in accordance with the Charter of the United Nations.*

[26] *Pulp Mills on the River Uruguay (Argentina* v. *Uruguay), Judgment, I.C.J. Reports 2010*, paras. 113, 115.

[27] *Ibid*, paras. 119, 120, 158.

a functional link between procedural and substantive obligations,[28] it did not give full consideration to their interdependence, particularly concerning the consequences of the breach of the procedural duties. The Court concluded that, provided substantive obligations have not been breached, the non-fulfilment of procedural obligations would not greatly matter and that, as a result, a declaration to that effect would constitute appropriate satisfaction. Indeed, no damage had resulted from the lack of information. The only obligations breached were the obligations of cooperation and acting in good faith and the non-compliance of procedural duties. The Court concluded that, as a result, there would be no claim for compensation.[29] Yet, it did not go unnoticed that the decision which provided that the non-compliance of procedural obligations had eventually no effect on the compliance of substantive obligations lacked the unanimous support of the judges.[30] Notwithstanding, non-compliance with the duty of information pursuant to a mineral resources clause constitutes, therefore, a breach of the respective delimitation treaty or cooperative arrangement, and in particular of the principle of *pacta sunt servanta*,[31] even if no consequence may be attributed to its non-compliance.

4.1.5 States' Obligation in the Absence of Mineral Resources Clauses

4.1.5.1 Obligation to Negotiate in Good Faith and of Mutual-Restraint

The principle of *bona fide* establishes the duty of States to act and negotiate in good faith, despite the fact that such a general obligation of conduct may lack some degree of specification.[32] This is the case, for example, of the general reference included in Article 83(1) of UNCLOS to international law and Article 38 of Statute of the ICJ, as well as to the achievement of an equitable solution regarding the delimitation of the continental shelf between States with opposite or adjacent coasts.[33]

The duty to act and negotiate in good faith implies that States must engage in reasonable and meaningful negotiations, while holders of a valid title under

---

[28] *Ibid*, para. 79.

[29] *Ibid*, para. 187.

[30] *See Pulp Mills on the River Uruguay*, *supra* note 26, *Joint Dissenting Opinion of Judges Al-Khasawneh and Simma*, paras. 26, 27.

[31] Article 26 of the Vienna Convention.

[32] *Fisheries Jurisdiction case, supra note 40 in Chap. 3*, para. 79(3)(c). *See* Article 2(2) of the UN Charter, *supra* note 87 in Chap. 3.

[33] *Maritime Delimitation in the Area between Greenland and Jan Mayen, Judgment, I.C.J. Reports 1993*, *Separate Opinion of Vice-President Oda*, paras. 67–68; *Ibid*, *Separate Opinion of Judge Schwebel*, paras. 127–128.

international law.[34] States must also enter negotiations in view of achieving an acceptable solution, although they are not required to continue with such negotiations if proven to be unproductive.[35] However, even when negotiations have reached a standstill States should adopt "*rules of engagement*" that essentially allow for the resuming of negotiations and the definition of norms of procedure or conduct, while pending the settlement on the delimitation of maritime boundaries or the implementation of a provisional arrangement. These norms would, at least, prevent States from acting uncoordinatedly.[36]

Should a deadlock situation be reached during negotiations, States ought to implement a *non-facere* regime or a "*joint non-development*" arrangement to reduce political tensions whilst protecting the environment, in addition to postponing the legal discussion regarding the sovereignty of the disputed maritime areas.[37] States may also agree to disagree and do nothing while keeping the ways of communication open or enter into arrangements regarding matters where reaching a consensus is possible, such as partial delimitation of maritime boundaries, joint development or implementing a regional cooperative effort.[38] States

[34] *See* International Law Association, *supra* note 15 in Chap. 2, p. 29; Lagoni, *supra* note 70 in Chap. 2 pp. 356–358; Brown, *supra* note 26 in Chap. 3, pp. 158–159.

[35] In respect to the fact that the negotiations between Cameroon and Nigeria did not lead to an agreement, the ICJ noted that Articles 74 and 83 of UNCLOS did not require that negotiations should be successful. The Court also observed that similar to other obligations to negotiate in international law, these have to be conducted in good faith. It further stated that Articles 74 and 83 do not require that judicial proceedings instituted be suspended while new negotiations are conducted. In this respect it concluded that: "*if, following unsuccessful negotiations, judicial proceedings are instituted and one of the parties then alters its claim, Articles 74 and 83 of the Law of the Sea Convention would not require that the proceedings be suspended while new negotiations were conducted. It is of course true that the Court is not a negotiating forum. In such a situation, however, the new claim would have to be dealt with exclusively by judicial means. Any other solution would lead to delays and complications in the process of delimitation of continental shelves and exclusive economic zones. The Law of the Sea Convention does not require such a suspension of the proceedings.*" *Land and Maritime Boundary between Cameroon and Nigeria, supra note 83 in Chap. 3*, para. 244.

[36] Peter D. Cameron, "The rules of engagement: developing cross-border petroleum deposits in the North Sea and the Caribbean", in: 55 *International and Comparative Law Quarterly* (2006), p. 570.

[37] Robert W. Smith, "Joint (development) zones: a review of past practice and thoughts on the future", in: *Sustainable Development and Preservation of the Oceans: The Challenges of UNCLOS and Agenda 21*, edited by Mochtar Kusuma-Atmadja, Thomas A. Mensah, Bernard H. Oxman (The Law of the Sea Institute/University of Hawaii: 1983), pp. 654–662.

[38] Douglas M. Johnston and Mark J. Valencia, *Pacific Ocean Boundary Problems. Status and Solutions*, edited by Shigeru Oda (Martinus Nijhoff Publishers: Dordrecht, Boston, London, 1991), pp. 24–29. Also *see* Sun Pyo Kim, *Maritime Delimitation and Interim Arrangements in North East Asia*, edited by Vaughn Lowe (Martinus Nijhoff Publishers: The Hague, London, New York, 2004), pp. 12, 94; D.H. Anderson, "Negotiating maritime boundary agreements: a personal view", in: *Maritime Delimitation*, edited by Rainer Lagoni and Daniel Vignes (Martinus Nijhoff Publishers: Leiden, Boston, 2006), pp. 137–138; Lagoni, *supra* note 70 in Chap. 2 pp. 360–361; Gao, "The legal concept..." *supra* note 65 in Chap. 2, p. 114.

must also refrain from unjustifiably delaying or breaking away from negotiations, as well as from disregarding agreed procedures or systematically refusing to take into consideration adverse proposals or interests, or also from not granting adequate time for reflection or reaction to proposals made during negotiations.[39] In the case concerning the *Gabčíkovo-Nagymaros Project*, the ICJ considered that the relevant States were required by the rule of *pacta sunt servanda* to find an agreed solution within the cooperative context of the legal regime in force between those States, even if finding a solution to a dispute should prevail over the literal application of the same, particularly if there is no obligation to reach an agreement.[40] The same could also be concluded regarding maritime delimitation agreements and provisional arrangements of a practical nature, particularly as UNCLOS does not foresee a *pactum de contrahendo*, but does, nevertheless, envisage a *pactum de negotiando*,[41] the scope of which may be summarized as States being bound to negotiate in good faith even when an agreement has not been reached.[42] This is also evident, for example, in the regime applicable to enclosed or semi-enclosed seas, whereby all bordering States, without exception, must cooperate and negotiate, even if with no obligation to enter into an agreement.[43] The wording of Article 123 of UNCLOS removes any doubts regarding the extent

---

[39] *Delimitation of the Maritime Boundary in the Gulf of Maine Area, supra note 60 in Chap. 3*, para. 112; *Fisheries Jurisdiction /United Kingdom v. Iceland), supra note 40 in Chap. 3*, paras. 73–76, 79(3); *North Sea Continental Shelf cases, supra note 46 in Chap. 3*, paras. 48, 85; *South West Africa Cases (Ethiopia* v. *South Africa; Liberia* v. *South Africa), Preliminary Objections, Judgment of 21 December 1962, I.C.J. Report 1962*, pp. 345–346; *Lac Lanoux Arbitration (France v. Spain), 24 International Law Reports 1957*, p. 128; *Case of the Mavrommatis Palestine Concessions, Judgment, 1924 P.C.I.J., Series A, No. 2*, p. 13. Also *see* Ana E. Bastida, Adaeze Ifesi-Okoye, Salim Mahmud, James Ross and Thomas Wälde, "Cross-border unitization and joint development agreements: an international law perspective", in: 29 *Houston Journal of International Law* (2006–2007), p. 375; José A. de Yturriaga, *The International Regime of Fisheries. From UNCLOS 1982 to the Presential Sea*, edited by Shigeru Oda (Martinus Nijhoff Publishers: The Hague, Boston, London, 1997), pp. 160–161; Martin A. Rogoff, "The obligation to negotiate in international law: rules and realities", in: 16 *Michigan Journal of International Law* (1994), pp. 153–171; Myers S. McDougal and William T. Burke, *The Public Order of the Oceans. A Contemporary International Law of the Sea* (Yale University Press: New Haven, London, 1962), pp. 87–88; Barberis, *supra* note 63 in Chap. 2, p. 56; Lagoni, *supra* note 70 in Chap. 2 pp. 355–356; Utton, *supra* note 70 in Chap. 2, p. 69; Cameron, *supra* note 36, p. 567.

[40] *Gabčíkovo-Nagymaros Project (Hungary/Slovakia), Judgment, I.C.J. Reports 1997*, paras. 141–142.

[41] International Law Association, *supra* note 15 in Chap. 2, p. 5.

[42] *See* Articles 63(1), 74(3), 83(3), 117, 118 and 123. Also *see* Jon M. Van Dyke, *Law of the Sea. The Common Heritage and Emerging Challenges*, edited by Harry N. Scheiber (Martinus Nijhoff Publishers: The Hague, London, Boston, 2000), pp. 26–35; *Continental Shelf (Tunisia/Libya), supra note 10 in Chap. 3, Judge Gros' Dissenting Opinion*, paras. 3, 4; and *North Sea Continental Shelf cases, supra note 46 in Chap. 3*, para. 48.

[43] Lewis M. Alexander, "Regionalism and the Law of the Sea: the case of semi-enclosed seas", in: 2-2 *Ocean Development and International Law Journal* (1974), pp. 151–186; Anastasia Strati, "Potential areas for joint development", in: *Joint Development*, vol. 2, *supra* note 70 in Chap. 2, pp. 109–122.

of the duty of cooperation of States bordering enclosed or semi-enclosed seas, or if the fulfilment of this obligation depends on a correlative duty to conclude international agreements, particularly against the interests and the will of any of the these States. This provision merely establishes the duty of cooperation regarding the coordination of activities concerning living resources, protection and preservation of the marine environment, and scientific research, excluding non-living resources.[44]

There are regions that having several and enduring examples of cooperation in enclosed or semi-enclosed seas, States have failed to adopt measures towards establishing an international legal regime for the joint development of non-living marine resources. This is the case of the Caribbean Sea where cooperation has mostly fallen upon the protection of the environment and development of marine natural living resources.[45] In this respect, Mark J. Valencia considered that UNCLOS gave "*two steps forward and one step back*" by promoting, on the one hand, cooperation between States bordering enclosed or semi-enclosed seas and, on the other hand, for not regulating the exercise of States' rights in these maritime areas, concluding that this provision would represent an obstacle to regional cooperation unless States agreed to grant such rights to a regional organization.[46]

The interpretation of Article 123 of UNCLOS in the light of the aforementioned principles of international law implies that States bordering enclosed or semi-enclosed seas cannot refuse to participate, in good faith, in meaningful negotiations in view of reaching an agreement. Furthermore, States may not practice any act or adopt any conduct detrimental to the legitimate interests of other States.

---

[44] Janusz Symonides, "Status of the enclosed and semi-enclosed seas", in: 27 *German Yearbook of International Law* (1984), pp. 327–329.

[45] Chris Carleton, "Maritime delimitation in complex island situations: a case study on the Caribbean Sea", in: *Maritime Delimitation*, edited by Rainer Lagoni and Daniel Vignes (Martinus Nijhoff Publishers: Leiden, Boston, 2006), pp. 165–188; L.D.M. Nelson, "The delimitation of maritime boundaries in the Caribbean", in: *Ocean Boundary Making: Regional Issues and Developments*, edited by Douglas M. Johnston and Phillip M. Saunders (Croom Helm: London, New York, Sydney, 1988), p. 142; Kaldone G. Nweihed, "EZ(uneasy) delimitation in the semi-enclosed Caribbean Sea: recent agreements between Venezuela and her neighbours", in: 8-1 *Ocean Development and International Law* (1980), pp. 1–33. On cooperative regimes regarding the protection of the marine environment in the Caribbean *see* Jill Sheppard and Bisessar Chakalall, "Regional planning to protect and develop the Caribbean marine environment", in: *A New Law of the Sea for the Caribbean. An Examination of Marine Law and Policy. Issues in the Lesser Antilles*, edited by Edgar Gold (Springer-Verlag: Berlin, Heidelberg, New York, London, Paris, Tokyo, 1988), pp. 163–174; Vaughn A. Lewis and Brian Challenger, "Regional cooperation and ocean development—the OECS experience", *ibid*, pp. 249–264; Edgar Gold and Carlyle L. Mitchell, "The new law of the sea in the Eastern Caribbean: towards a new ocean development action plan", *ibid*, pp. 265–276; Nelson Andrade Colmenares and J. Jairo Escobar, "Ocean and coastal issues and policy responses in the Caribbean", in: 45 *Ocean and Coastal Management* (2002), pp. 905–924.

[46] Mark J. Valencia, "Regional maritime regime building: prospects in northeast and southeast Asia", in: 31-3 *Ocean Development and International Law Journal* (2000), p. 237. Also *see* Lewis M. Alexander, *Marine Regionalism in the Southeast Asian Seas* (East-West Center: Honolulu, 1982), pp. 8–12; Ong, *supra* note 61 in Chap. 3.

Indeed, in addition to the general obligations of settling international disputes by peaceful means and refraining from the threat or use of force,[47] States must abstain from practicing any act that may be susceptible of damaging or making any solution or the means selected to achieve such a solution unviable. This obligation is relevant prior and especially from the moment that States are summoned for negotiations concerning the delimitation of maritime boundaries or the implementation of a provisional arrangement.[48]

4.1.5.2 Obligation to Share Information

The principle of cooperation extends across international law and is accordingly reflected in several provisions of UNCLOS.[49] The relation between the principle of cooperation and the use of natural resources was first considered in the context of the principle of peoples' permanent sovereignty over natural resources,[50] which recognized the right to hold the effective control of natural resources and the need for States to cooperate, in particular by sharing knowledge and resources, as well as optimizing development,[51] and consequently promoting the internationalization of resources.[52]

In what concerns joint development agreements, the principle of cooperation of shared resources is also applicable to upstream, midstream and downstream activities, in particular in what the exploration of resources and sharing of information are concerned. However, although the principle of cooperation may have a

---

47 Article 2(3) and (4) of the UN Charter, *supra* note 87 in Chap. 3.

48 Lagoni, *supra* note 70 in Chap. 2 p. 358.

49 Articles 63(3), 64(1), 65, 66(2) and (4), 69(3), 74(3), 83(3), 100, 117, 118, 119, 123, 194(1), 197, 200, 242, 266, 270 and 273 of UNCLOS. Also *see* Resolutions (UNGA) 2996 (XXVII), 2997 (XXVII) and 2295 (XXVII), 15 December 1972, and Resolution (UNGA) 3129 (XXVIII), 13 December 1973; UNEP Doc. GC.6/CRP.2, 19 May 1978.

50 Schrijver, *supra* note 101 in Chap. 3; Becker-Weinberg, supra note 104 in Chap. 3, pp. 373–398.

51 Resolution (UNGA) 2158 (XXI), 25 November 1966.

52 The report of the United Nations prepared by the *Division for Ocean Affairs and the Law of the Sea* does not mention cooperation regarding non-living resources. It also does not exclude it, although it does mention expressly fishing and environmental cooperation *see* United Nations, *The Law of the Sea. Practice of States at the time of entry into force of the United Nations Convention on the Law of the Sea* (UN Publications: New York, 1994), pp. 19–21. Also *see* Resolutions (UNGA) n. 1803 (XVII), 14 December 1962, n. 3201 (S-VI), 1 May 1974, n. 1514 (XV), 14 December1960, n. 1515 (XV), 15 December 1960, and n. 626 (VII), 21 December 1952, n. 2625 (XXV), 24 October 1970, *Declaration on Principles of International Law Concerning Friendly Relations and Cooperation Among States in Accordance with the Charter of the United Nations*; Resolution (UNGA) 3362 (S-VII), 16 September 1975, *Development and International Economic Cooperation*. The latter resolution establishes that cooperation and sharing of technology and equipment, as well as knowledge, expertise, natural resources, amongst other, are the means to achieve the *New International Economic Order*. Also *see Preamble* and Articles 1(3), 13(1) lit. b) and 55 of the UN Charter, *supra* note 87 in Chap. 3.

significant role and eventually also guide States' actions regarding common hydrocarbon deposits, it should not be mistaken with the implementation of a joint development agreement, nor should the different nature of both concepts be confused. Cooperation between States regarding the implementation of a legal regime that includes the protection of collective or diffuse interests of mankind is not the scope of joint development agreements.[53] Joint development agreements seek to fulfil States' needs regarding the development of resources for their own benefit, without prejudice of these States having due regard for the rights of third States, or the fact that these regimes may also contribute towards promoting ventures between developed and developing States regarding common natural resources.[54]

International doctrine and jurisprudence have generally accepted that the principle of cooperation regarding shared or transboundary resources requires neighbouring coastal States sharing information on the respective existence, as well as on the intention of a State to develop common resources or any resource that may have an impact on the interests of another State.[55] For example, as previously mentioned, the Arbitral Tribunal in the second stage of proceedings concerning maritime delimitation between Eritrea and Yemen considered that these countries were bound to inform and consult one another on any hydrocarbon

---

[53] On the need for a holistic approach towards the management of marine natural resources *see* Elisabeth Mann Borgese, "Pacem in Maribus XIX-background paper", in: *Ocean Governance: Sustainable Development of the Seas*, edited by Peter Bautista Payoyo (United Nations University Press: Tokyo, New York, Paris, 1994), p. 355; *Pacem in Maribus* (Dodd, Mead and Company: New York, 1972), p. xi; "The New International Economic Order and the Law of the Sea", in: *The Law of the Sea. Issues in Ocean Resource Management*, edited by Don Walsh (Praeger Publishers: New York, 1977), pp. 82–111; "The role of the Sea-bed Authority in the '80 s and '90 s. The common heritage of mankind", in: *La Gestion des Ressources pour l'Humanité: Le Droit de la Mer/The Management of Humanity's Resources: The Law of the Sea*, edited by René-Jean Dupuy (Martinus Nijhoff Publishers: The Hague, Boston, London, 1982), pp. 35–48; *The Future of the Oceans. A Report of the Club of Rome* (Harvest House: Montreal, 1986), pp. 13–42. Also *see* Peter Bautista Payoyo, *Cries of the Sea. World Inequality, Sustainable Development and the Common Heritage of Humanity*, edited by Shigeru Oda (Martins Nijhoff Publishers: The Hague, London, Boston, 1997), p. 155; Edward Wenk, *The Politics of the Ocean* (University of Washington Press: 1972), p. 428; *Ibid*, *supra* note 14 in Chap. 2, pp. 3–20; Jessup, *supra* note 13 in Chap. 2, pp. 405–508.

[54] Eyal Benvenisti, *Sharing transboundary resources. International Law and Optimal Resource Use* (Cambridge University Press: Cambridge, 2002), pp. 21, 43.

[55] Rodman R. Bundy, "Natural resource development (oil and gas) and boundary disputes", in *The Peaceful Management of Transboundary Resources*, edited by Gerald H. Blake, William J. Hildesley, Martin A. Pratt, Rebecca J. Ridley and Clive H. Schofield (Graham and Trotman/Martinus Nijhoff: London, Dordrecht, Boston, 1995), pp. 36, 39; Masahiro Miyoshi, *Considerations of Equity in the Settlement of Territorial and Boundary Disputes*, in: *International Law in Japanese Perspective*, vol. 2 (Martinus Nijhoff Publishers: Dordrecht, Boston, London, 1993), pp. 133–134; *Ibid*, "Considerations of equity in maritime boundary cases before the International Court of Justice", in: *Liber Amicorum Judge Shigeru Oda*, vol. 2, edited by Nisuke Ando, Edward McWhinney and Rüdiger Wolfrum (Kluwer Law International: The Hague, London, New York, 2002), pp. 1087–1101; Valencia, *supra* note 46, p. 224.

or mineral resources that may straddle the maritime boundary or that lay in its immediate vicinity.[56]

The duty of information is also founded in the consolidated principle of *bona fide*,[57] as every State is bound to inform and has the right to be informed on the existence, location and characteristics of resources that might be susceptible of being found in an area subject to its sovereignty or jurisdiction, or where the claims of two or more States overlap.[58] As previously mentioned, in the absence of such an obligation, it would be clearly disadvantageous for the State unaware of the said information during negotiations and could most probably produce an unbalanced outcome.[59]

---

[56] *The Eritrea—Yemen Arbitration (second phase)*, *supra* note 23, para. 86.

[57] Article 2(2) UN Charter, *supra* note 87 in Chap. 3, and Article 300 of UNCLOS.

[58] Lagoni, *supra* note 63 in Chap. 2, p. 237; Shihata/Onorato, *supra* note 71 in Chap. 2, p. 435.

[59] Benedetto Conforti, "La délimitation et les méthodes pratiques. Equité, délimitation et critères de délimitation", in: *The Continental Shelf and the Exclusive Economic Zone/Le Plateau Continental et la Zone Économique Exclusive*, edited by Donat Pharand and Umberto Leanza (Martinus Nijhoff Publishers: Dordrecht, Boston, London), pp. 55–56.

Chapter 5
Joint Development Agreements in Cases of Transboundary Hydrocarbon Deposits

5.1 Basis for Joint Development After the Delimitation of Maritime Boundaries

When analysing the overall purpose of joint development and the characteristics of hydrocarbon deposits in Chap. 2, reference was made that achieving resource efficiency constitutes a fundamental aspect of cooperation regarding the common exploration and exploitation of hydrocarbon deposits that straddle across a boundary line. It was also mentioned that competitive drilling is contrary to international law and that it leads to the waste of resources, even though States are not legally required to adopt conservation measures, such as unitization, or to engage in cooperative production of common offshore hydrocarbon deposits under the umbrella of a joint development agreement.

International law and resource efficiency notwithstanding, the implementation of joint development agreements after the delimitation of maritime boundaries provides States with the opportunity of having access to resources that would otherwise be off-limits, benefiting from their production and sharing the respective revenues. Cooperation also allows for the reinforcement of capabilities through the exchange of technology and know-how.

Joint development regimes can further contribute to the delimitation of maritime boundaries when included as an integral part of a delimitation treaty, namely by unblocking negotiations that are at a stalemate when States have already agreed in principle on the delimitation line because of the potential presence of straddling hydrocarbon deposits, or when States fail altogether to settle the delimitation of the maritime boundaries of a certain area due to the possible existence of these resources. In both instances, joint development regimes create the legal framework for the exploration and exploitation of hydrocarbon resources, while allowing States to reach, as result, a compromise on the delimitation of maritime boundaries.

In this regard, Robin Churchill and Lowe Vaughn distinguished between creating joint development areas for seabed resources that straddled a boundary line as part of

V. Becker-Weinberg, *Joint Development of Hydrocarbon Deposits in the Law of the Sea*, Hamburg Studies on Maritime Affairs 30, DOI 10.1007/978-3-662-43570-0_5

a delimitation settlement, and those arrangements that simply envisaged the exploitation of transboundary oil and gas fields.[1] As for the first category, these authors considered that the inclusion of such arrangements could ease the drawing of a boundary line by enabling a more equitable sharing of resources. They also referred that it would increase the number of variables involved in boundary negotiations, avoiding focusing exclusively on matters pertaining to common resources.[2]

There is a difference between joint development regimes implemented as an integral part of a delimitation treaty or after the settlement of maritime boundaries, and joint development regimes included in delimitation treaties but that are applicable to maritime areas where States have not reached an agreement on the allocation of the respective joint development area. While the latter regimes constitute provisional arrangements and are subject to the legal framework that will be analysed in Chap. 6, joint development agreements signed after the delimitation of maritime boundaries refer to the exploration and exploitation of hydrocarbon deposits that straddle an already established maritime boundary. States may simultaneously agree on the delimitation of maritime boundaries and the implementation of a provisional arrangement under the legal framework of a general settlement. This is the case, for example, of the agreements signed between Japan and the Republic of Korea,[3] Spain and France,[4] Columbia and Jamaica,[5] and the Netherlands and Germany.[6]

Another, although uncommon, possibility for joint development after the delimitation of maritime boundaries can take place when States agree to jointly develop the resources found in maritime areas that are exclusively subject to the sovereignty and jurisdiction of one State. This could be the case, for example, if

---

[1] Churchill/Lowe identified four types of "*co-operative arrangements*": (a) before, or (b) after the drawing of a boundary line, and in respect of (c) living, or (d) non-living marine resources. Churchill/Lowe, *supra* note 68 in Chap. 2, p. 198.

[2] *Ibid.*

[3] *See* notes 108 in Chap. 6 and 39 in Chap. 7.

[4] *Supra* note 2 in Chap. 4.

[5] Article 3 of the *Maritime Treaty between Jamaica and the Republic of Colombia*, done at Kingston on 12 November 1993 and entered into on 14 March 1994 [online: http://www.un.org/depts/los/LEGISLATIONANDTREATIES/PDFFILES/TREATIES/JAM-COL1993MD.PDF (accessed on August 2013)], reproduced in: *International Maritime Boundaries*, vol. 3, edited by Jonathan I. Charney and Lewis M. Alexander (Martinus Nijhoff Publishers: The Hague, Boston, London, 2004), pp. 2200–2204.

[6] *Supplementary Agreement to the Treaty Concerning Arrangements for Co-operation in the Ems Estuary (Ems-Dollard Treaty)*, signed between the Kingdom of The Netherlands and the Federal Republic of Germany, done at Bennekom on 14 May 1962 and entered into force on 1 August 1963, published at *Bundesgesetzblatt* 1963 II S. 652 and 509 U.N.T.S. 140. Following the signing of this treaty, the two States entered into an agreement regarding the lateral delimitation of the continental shelf, the *Treaty between the Kingdom of the Netherlands and the Federal Republic of Germany concerning the lateral delimitation of the continental shelf in the vicinity of the coast*, done at Bonn on 1 December 1964 and entered into force on 18 September 1965, published at 550 U.N.T.S 123, reproduced in: *International Maritime Boundaries*, vol. 2, *supra note 125 in Chap. 3*, pp. 1841–1842; and the *Treaty concerning the Delimitation of the Continental shelf Under the North Sea*, *supra* note 3 in Chap. 4.

States would agree keeping in force, with the necessary adjustments, joint development agreements signed before the delimitation of maritime boundaries or prior to the succession of any of the relevant States. In both situations, the decision to keep in force these regimes might result from the consideration of the economic relevance of the exploration and exploitation activities for both States, as well as the safeguarding of operators' rights.

The joint development agreement implemented by Saudi Arabia and Kuwait resulted in these countries accepting the terms of the maritime delimitation determined by the Al Uqair Conference and, consequently, maintaining the exploration and exploitation activities within the Neutral Zone that had been created as a result, in order to maintain the economic benefits deriving from the *status quo* prior to their independence.[7] Although Kuwait and Saudi Arabia eventually decided on the equal partition of the Neutral Zone that had been, until then, a common area for both States, they agreed that the exploration and exploitation rights previously granted would be protected and that each State would undertake all obligations deriving from the concession agreements.[8]

States can also agree to enter into joint development agreements after the delimitation of maritime boundaries, envisaging the exploration and exploitation of resources found in a maritime area only subject to the jurisdiction of one State, in the event that the latter does not have the capability or know-how to undertake these activities. Indeed, the first joint development agreement, which was signed by Bahrain and Saudi Arabia, also established the maritime boundaries between both countries. This joint development agreement determined that the oil resources found in one of the joint development areas would be explored by Saudi Arabia in accordance with the latter's discretion, under the condition that half of the benefits resulting from such exploration would be given to Bahrain, without interfering with Saudi Arabia' sovereignty over this area.

---

[7] Husain M. Albaharna, "A note on the Kuwait-Saudi Arabia Neutral Zone Agreement of July 7th, 1965, relating to the partition of the zone", in: 17 *International and Comparative Law Quarterly* (July: 1968), pp. 730–735.

[8] Article 5 of the *Agreement between the State of Kuwait and the Kingdom of Saudi Arabia Regarding the Submerged Zone Contiguous to the Partitioned Zone*, done at Kuwait on 2 July 2000, reproduced in: *International Maritime Boundaries*, vol. 4, edited by Jonathan I. Charney and Robert W. Smith (Martinus Nijhoff Publishers: The Hague, London, New York, 2002), pp. 2837–2840. Also *see* the *Agreement between the State of Kuwait and the Kingdom of Saudi Arabia relating to the partition of the Neutral Zone*, done 7 July 1965 and entered into force on 25 July 1966, reproduced in: Fereidun Fesharaki, "Joint development of offshore petroleum resources: the Persian Gulf experience", in: *The South China Sea. Hydrocarbon Potential & Possibilities of Joint Development*, edited by Mark J. Valencia (Pergamon Press: Oxford, New York Toronto, Sydney, Paris, Frankfurt, 1981), pp. 1330–1334; The study of the BIICL considers that this agreement creates a co-ownership, condominium, co-tenancy or resource-sharing in the Neutral Zone, "*whereby the sovereignty rights of each State are merged into the new legal structure in exchange for defined rights and obligations*" *see* Fox (et al.), *supra* note 111 in Chap. 3, p. 48. Also *see* Isa Huneidi, "The Saudi/Kuwait joint development areas of the neutral zone, onshore and offshore", in: *Joint Development*, vol. 2, *supra* note 70 in Chap. 2, pp. 77–87.

Overall, in cases where joint development regimes are implemented as an integral part of a delimitation treaty or after the settlement of maritime boundaries, the definition of States' jurisdiction over the seabed and the identification of the applicable law to exploration and exploitation activities and to offshore installations and structures derive from the delimitation of maritime boundaries. In these cases, States should agree on the applicable law or attempt to harmonize the relevant legal regimes in order to avoid conflicts of law, namely, regarding jurisdiction and enforcement over installations and structures, as well as States' liability for pollution from seabed activities. These matters will be addressed in Sects. 5.4 and 5.5 of this Chapter.

5.2 States' Obligations Regarding Transboundary Hydrocarbon Deposits

In Chap. 4 reference was made to States' obligations in the absence of mineral resources clauses. It was mentioned that States are subject to the duties to negotiate, of mutual-restraint and to share information on offshore transboundary hydrocarbon deposits. States are also bound to certain obligations regarding seabed activities in maritime areas under their jurisdiction or where they exercise rights of sovereignty, including having due regard for the rights and freedoms of third States, as well as for the protection and preservation of the marine environment, particularly concerning the construction and use of offshore installations and structures.

Without foregoing the analysis in the following Chapter of provisional arrangements, if there is no obligation for States to engage in the joint development of offshore hydrocarbon deposits found in disputed maritime areas, it would not seem coherent to sustain an analogous obligation for transboundary offshore hydrocarbon deposits, seeing that the right to develop the resources of the continental shelf is inherent to the coastal State and is founded on its entitlement.[9] Indeed, as previously referred in Chap. 3 in respect of the duty to cooperate regarding shared natural resources, international law has not considered that such an obligation has developed for hydrocarbon deposits that straddle a boundary line. As then referred, the ILC recognized that the development of these resources would be more efficiently dealt with by the States themselves, due to the unique nature of hydrocarbon deposits and the circumstances in which the respective exploration and exploitation activities take place. Robin Churchill and Lowe Vaughn also noted that:

> The system adopted in such cross-boundary agreements is entirely a matter for the agreement of the parties. Customary international law does not yet seem to yield any precise rules applicable in the absence of such agreement, although some writers have advanced general solutions based on a mixture of basic principles of law, previous treaty practice and robust expediency.[10]

---

[9] Article 77(2) and (4) of UNCLOS.

[10] Churchill/Lowe, *supra* note 68 in Chap. 2, p. 200.

With respect to the obligation of mutual-restraint regarding transboundary offshore hydrocarbon deposits, it should also be recalled that the exploration and exploitation activities of transboundary mineral resources would constitute a breach of the right of the neighbouring State to these resources, causing harm and potentially their exhaustion.

In the event that a State would breach the exclusive sovereign right of another State to explore and exploit the resources of the continental shelf, the unlawful development of transboundary resources would be an internationally wrongful act and the responsible State bound to cease the illegal conduct.[11] In the cases where restitution does not correct the harm done, or when it is not possible as a result of the depletion of transboundary mineral resources, the responsible State would have to make full reparation for the damages caused as a result of the unlawful exploration and exploitation activities.[12]

5.3 Rights and Freedoms of Other States

In addition to the general obligation for States to fulfil their obligations in good faith and to exercise their rights, jurisdiction and freedoms in a manner that does not constitute an abuse of right,[13] UNCLOS provides that, should the interests of coastal Sates—such as regarding the construction and operation of installations and structures—interfere with those of other States in the EEZ, these conflicts shall "*be resolved on the basis of equity and in the light of all the relevant circumstances, taking into account the respective importance of the interests involved to the parties as well as to the international community as a whole.*"[14]

Additionally, the exercise of coastal States' rights in the continental shelf and in particular, those regarding the exploration and exploitation of its resources "*must not infringe or result in any unjustifiable interference with navigation and other rights and freedoms of other States*", namely the right to lay submarine cables and pipelines, the freedom of navigation or the activities regarding international cooperation and promotion of marine scientific research.[15] In fact, the 1958 Convention on the Continental Shelf already provided that:

---

[11] Article 77 of UNCLOS.

[12] *See Draft Articles on Responsibility of States for Internationally Wrongful Acts, with commentaries*, 2001, pp. 35, 87, 9, 98. Also *see* Resolution (UNSC) n. 687/1991, 2981st Meeting, 3 April 1991 (S/RES/687, 1991), para. 16; UN Security Council, *Decision taken by the Governing Council of the United Nations, Compensation Commission during its third session at the 18*th *meeting held on 28 November 1991, as revised at the 24*th *Meeting, 16 March 1992* (S/AC.26/1991/7/Rev.1, 17 March 1992), *Criteria for additional Categories of Claims: United Nations Compensation Commission, Governing Council, Fifth Session, Geneva, 16–20 March 1992*, para. 35.

[13] Article 2(3), 56(2), 78(2), 138 and 147(1), and 300 of UNCLOS.

[14] Article 59 of UNCLOS.

[15] Article 78(2) of UNCLOS.

> "[T]he exploration of the continental shelf and the exploitation of its natural resources must not result in any unjustifiable interference with navigation, fishing or the conservation of the living resources of the sea, nor result in any interference with fundamental oceanographic or other scientific research carried out with the intention of open publication," adding that "[n]either the installations or devices, nor the safety zones around them, may be established where interference may be caused to the use of recognized sea lanes essential to international navigation."[16]

A joint development regime signed after the delimitation of maritime boundaries must also not limit the fishing rights granted to third States as provided by the EEZ regime, since the rights of coastal States in the continental shelf do not alter the legal status of the superjacent waters.[17] Likewise, in the event that an EEZ has not been created by the coastal State, these arrangements must have due regard for the freedom of fishing pursuant to the high seas regime.[18] This is evidently also applicable to agreement signed before the delimitation of maritime boundaries.

5.4 Legal Regime of Offshore Installations and Structures

5.4.1 Installations and Structures Within and Beyond National Jurisdiction

Until the entering into force of UNCLOS, the legal regime applicable to offshore installations and structures was only mentioned in Article 5 of the 1958 Convention on the Continental Shelf. This provision determined that the right of coastal States to construct and maintain or operate installations and other devices for the purposes of undertaking exploration and exploitation activities, including the right to establish safety zones around them, could not unjustifiably interfere with the rights of other States, such as the freedoms of navigation or fishing. Coastal States also had the obligation to give due notice of the construction of offshore installations and to maintain "*permanent means for giving warning of their presence*." Another aspect under this provision was the obligation of coastal States to remove "*abandoned or disused*" installations.

Similarly to the 1958 Convention of the Territorial Sea and the Contiguous Zone, UNCLOS does not expressly mention that coastal States have jurisdiction over offshore installations and structures found in the territorial sea as part of their sovereignty over this maritime area, its bed and subsoil.[19] It does, however, refer

---

16 Article 5(1) and (7) of the 1958 Convention on the Continental Shelf.

17 Articles 62 to 72 of UNCLOS.

18 Article 87(1) lit. (e) of UNCLOS. Rainer Lagoni, in referring to provisional arrangements pursuant to Articles 74(3) and 83(3) of UNCLOS, underlines the safeguarding of other States' "*long-standing fishery rights*" to the disputed maritime area. *See supra* note 70 in Chap. 2, p. 358.

19 Article 2(2) of UNCLOS.

that coastal States have the right to adopt laws and regulations in relation to the exercise of innocent passage in the territorial sea for the purpose of the protection of installations and structures and of cables and pipelines.[20] These laws and regulations may also concern the safety of navigation and of maritime traffic around installations and structures for the purpose of preventing collisions.[21]

In the EEZ, coastal States enjoy sovereign rights for the purpose of exploring and exploiting the resources of the seabed and to undertake other economic activities.[22] They also have jurisdiction with regard to the establishment and use of installations and structures.[23] On the other hand, while the rights of coastal States over offshore installations and structures in the continental shelf are subject to the EEZ regime,[24] the sovereign rights of coastal States over the non-living resources of the seabed and its subsoil are to be exercised pursuant to the continental shelf regime.[25]

As for installations and structures in maritime areas beyond national jurisdiction, UNCLOS stipulates that these shall be used exclusively for "*peaceful purposes*" and no State may claim sovereignty over any part of the high seas or the Area as a result of constructing or operating installations and structures in these maritime areas.[26] The high seas regime also establishes that the legal framework applicable to offshore installations and structures shall be that of the continental shelf, which, in turn, refers to the EEZ regime, consistent with the reference made in Article 80 of UNCLOS.[27] The same is applicable to the freedom to lay cables and pipelines constructed or used in connection with the exploration of the bed of the high seas beyond the continental shelf or the exploitation of its resources or the operations of installations and structures.[28]

Installations and structures in the Area are subject to certain conditions and specifically to the regulations and procedures determined by the Authority. Similarly to installations and structures in other maritime areas, due notice must be given and permanent warning of their presence must be maintained. They may also not interfere with the use of recognized sea lanes essential to international navigation, or in areas of intense fishing activity. Installation and structures in the Area shall have safety zones around with appropriate markings to ensure the safety of both navigation and the installations.[29]

---

[20] Article 21(1) lit. (b) and (c) of UNCLOS.

[21] Article 21(1) lit. (a), (3) and (4), and Article 22(3) lit. (a) of UNCLOS. *See* IMO resolution A.572(14), 20 November 1985, *General Provisions on Ships' Routeing*.

[22] Article 60(1) lit. (b) of UNCLOS.

[23] Article 56(1) lit. (a) and (b) (i) of UNCLOS.

[24] The text of Article 80 of UNCLOS reads: "*Article 60 applies mutatis mutandis to artificial islands, installations and structures on the continental shelf*".

[25] Articles 56(3) and 77(1) of UNCLOS. Also *see* Article 81 of UNCLOS.

[26] Articles 87(2), 88 and 89, and 137(1) of UNCLOS.

[27] Articles 87(1) lit. (c) and (d) of UNCLOS.

[28] Article 87(1) lit. (c) and (d), and 112(1) of UNCLOS. Also *see* Article 79(4) of UNCLOS.

[29] Article 147(2) lit. (c) of UNCLOS.

As it already occurred in the 1958 Convention on the Continental Shelf,[30] UNCLOS provides that offshore installations and structures within and beyond national jurisdiction do not have the status of islands or posses a territorial sea, nor does their presence affect the delimitation of maritime boundaries.[31] UNCLOS also determines that installations used for marine scientific research shall only be placed in the EEZ or the continental shelf subject to the consent of the relevant coastal State.[32]

5.4.2 Classification of Oil Rigs

The location of offshore installations and structures is an essential aspect for determining jurisdiction over them and consequently the applicable law. This is true in respect of fixed offshore installations and structures, for example, concerning the protection of installations and structures. Indeed, as a result of the *Protocol for the Suppression of Unlawful Acts Against the Safety of Fixed Platforms Located on the Continental Shelf*[33] and the *2005 Protocol*[34] that extends the applicability of the *Convention for the Suppression of Unlawful Acts Against the Safety of Maritime Navigation* to oil platforms, States-parties should establish jurisdiction and implement a legal regime concerning criminal offences for acts committed against fixed platforms. Yet, despite the fragmentation of the legal system of the law of the sea on matters such as the protection of installations and structures or the protection and preservation of the marine environment, there is coherence between most international legal instruments regarding the treatment of different offshore installations and structures, particularly, for the purpose of determining jurisdiction and the applicable law.

Just as the 1958 Convention on the Continental Shelf before, UNCLOS does not give a definition of offshore installations or structures—or of ships, for that

[30] Article 5(4) of the 1958 Convention on the Continental Shelf.

[31] Article 60(8) *ex vi* Articles 80 and 87(1) lit. (d), and Article 147(2) lit. (e) of UNCLOS. Also *see* Article 259 of UNCLOS.

[32] The text of Article 246(5) lit. (c) of UNCLOS reads: "*Coastal States may however in their discretion withhold their consent to the conduct of a marine scientific research project of another State or competent international organization in the exclusive economic zone or on the continental shelf of the coastal State if that project: (…) (c) involves the construction, operation or use of artificial islands, installations and structures referred to in articles 60 and 80.*" Also *see* Articles 87(1) lit. (f) and 143(1) of UNCLOS.

[33] *Protocol for the Suppression of Unlawful Acts Against the Safety of Fixed Platforms Located on the Continental Shelf*, done in Rome on 10 March 1988 and entered into force on 1 March 1992, published at 1678 U.N.T.S. 304, reproduced in: 27 *International Legal Materials* 685 (1988).

[34] *Protocol for the Suppression of Unlawful Acts against the Safety of Maritime Navigation Fixed Platforms Located on the Continental Shelf*, done at London on 14 October 2005 and entered into force on 28 July 2008, IMO Doc. LEG/CONF.15/22.

matter, which are also referred to as vessels—and does not extend the concept of flag State jurisdiction to them.[35] UNCLOS also does not distinguish between fixed or movable installations, even though this distinction was discussed during the preparatory work.[36]

The broad reference to installations, structures and devices in UNCLOS,[37] particularly when referred together with the exploration and exploitation of mineral resources, allows for the consideration of different realities and not only of fixed offshore platforms, even though these are perhaps the majority and consequently the first being considered when referring to offshore installations and structures. On the other hand, despite many international legal instruments dealing with marine pollution by ships specifically or tacitly excluding installations or structures,[38] several provisions of UNCLOS and of other specialized international legal instruments are applicable to pollution from offshore installations. This is the case, for example, of the rules on pollution by dumping included in UNCLOS, the *1972 London Convention* and the *1996 Protocol*—which do not limit their application to ships[39]—the *1990 International Convention on Oil Pollution Preparedness, Response and Co-operation* (OPRC 1990),[40] or the *International Convention for the Preservation of Pollution from Ships* (MARPOL 73/78).[41] In the case of mobile installations and structures, such as Mobile Offshore Drilling

[35] Articles 90 and 91 of UNCLOS.

[36] *See* Centre for Oceans Law and Policy, University of Virginia School of Law, *United Nations Convention on the Law of the Sea 1982: A Commentary,* vol. 2, edited by Satya N. Nandan and Shabtai Rosenne (Martinus Nijhoff Publishers: Dordrecht, Boston, London), pp. 570–593.

[37] *Ibid*, p. 584.

[38] This is the case, for example, of the *1954 International Convention for the Prevention of Pollution of the Sea by Oil*; *1969 International Convention on Civil Liability for Oil Pollution Damage* and *1971 International Convention on the Establishment of an International Fund for Compensation for Oil Pollution Damage* (and *1984*, *1992* and *2003 Protocols*); *1972 Convention on the International Regulations for Preventing Collisions at Sea* (and *1987*); *1969 International Convention relating to Intervention on the High Seas in cases of Oil Pollution Casualties Act*; *1996 International Convention on Liability and Compensation for Damage in connection with the Carriage of Hazardous and Noxious Substances by Sea.*

[39] Articles 1(5), 210 and 216 of UNCLOS.

[40] *International Convention on Oil Pollution Preparedness, Response and Co-operation*, done at London on 30 November 1990 and entered into force on 13 May 1995, published at 1891 U.N.T.S. 51 and 30 *International Legal Materials* 733 (1990) [OPRC 1990].

[41] *International Convention for the Preservation of Pollution from Ships*, done at London on 2 November 1973 and entered into force on 2 October 1983, as amended by the *Protocol of 1978 relating to the International Convention for the Prevention of Pollution from Ships, 1973*, published at 1340 U.N.T.S. 61 [MARPOL 73/78].

Units (MODU), which are regulated in the *1989 International Maritime Organisation MODU Code*,[42] the applicable law is that of the respective flag State.[43]

In accordance with the legal framework of OPRC 1990, States shall require operators of oil platforms under their jurisdiction to have oil pollution emergency plans and to comply with oil pollution reporting procedures, including procedures for reporting any observed event at sea involving a discharge of oil or the presence of oil. In addition, upon receiving an oil pollution report, States shall inform all States whose interests are affected or likely to be affected by such oil pollution incident. OPRC 1990 further provides that each State shall establish a national system for responding promptly and effectively to oil pollution incidents and, amongst other relevant procedures, allow the use of its territory for the departure and arrival of transportation engaged in responding to an oil pollution incident. States shall also cooperate in technical cooperation and research and development programmes regarding oil pollution, as well as promote bilateral and multilateral cooperation in preparedness and response.[44]

As for MARPOL 73/78, it includes floating platforms in its definition of ship and consequently extends to oil platforms all those provisions that are susceptible of being applicable to these installations.[45] In this respect, the IMO has approved guidelines that apply Annex I of MARPOL 73/78 to floating platforms that are stationary when operating, such as Floating Production, Storage and Offloading Facilities (FPSOs), used for the offshore production and storage of oil, and Floating Storage Units (FSU), used for the offshore storage of produced oil, even though they do not have the role of transporting oil and do not lie within the definition of oil tanker but are considered as such for the purpose of applying

[42] *Code for the Construction and Equipment of Mobile Offshore Drilling Units* adopted by IMO resolution A.649(16), 19 October 1989 [1989 MODU Code] and concerns MODUs built after 1 May 1991. The 1989 MODU Code superseded the 1979 MODU Code adopted by IMO resolution A.414(11), 15 November 1979, and was updated and revised by the 2009 MODU Code adopted by IMO resolution A.1023(26) *Code for the Construction and Equipment of Mobile Offshore Drilling Units*. The new Code is expected to apply to units whose construction begins on or after 1 January 2012.

[43] Mobile Offshore Unit is a vessel which can be readily relocated and performs an industrial function involving offshore operations other than those traditionally provided by vessels covered by Chapter I of the *International Convention for the Safety of Life at Sea* [1974 SOLAS]. MODU is a unit capable of engaging in drilling operations for the exploration for, or exploitation of, resources beneath the seabed such as liquid or gaseous hydrocarbons, sulphur or salt. *See* IMO resolution A.891(21), 25 November 1999, *Recommendations on Training of Personnel on Mobile Offshore Units (MOUs)*.

[44] Articles 3(2), 4(1) lit. (a) (ii) and lit. (b) (ii), 5(1) lit. (c), 6, 7(3), 8, 9 and 10 of OPREC 1990.

[45] Article 2(4) of MARPOL 73/78.

MARPOL 73/78.[46] If engaged in international voyage under its own propulsion and are flying the flag of a State other than the relevant coastal State, FPSOs and FSUs should also be subject to the legal regime established in the *International Convention for the Safety of Life At Sea*,[47] the *International Convention on Standards of Training, Certification and Watchkeeping for Seafarers*,[48] and the *International Convention on Load Lines*,[49] in addition to MARPOL 73/78.[50] As a result, States shall enforce the obligations undertaken within MARPOL 73/78 to floating platforms operating under their authority, without prejudice of their sovereign rights over the seabed and subsoil adjacent to their coasts for the purpose of developing its natural resources.

However, even though certain obligations were in fact only construed having ships in mind, some important provisions are applicable and their enforcement may significantly contribute towards the protection and preservation of the marine environment from pollution by oil platforms. This is the case, for example, of the obligation of monitoring and reporting incidents and conducting investigations, procedures for the settlement of disputes, communication of information and promotion of technical cooperation.[51] In this respect, Regulation 39 of Annex I of MARPOL 73/78 concerning categories of discharges that may be associated with the operation of oil platforms, refers that only the discharge of machinery space drainage and contaminated ballast of these installations should be subject to MARPOL 73/78, being all others referred therein not included within its scope.

---

[46] IMO resolution MEPC/Circ.406, 10 November 2003, *Guidelines for Application of MARPOL Annex I Requirements to FPSOs and FSUs*; IMO resolution MEPC.139(53), 22 July 2005, *Guidelines for the Application of the Revised MARPOL Annex I Requirements to Floating Production, Storage and Offloading Facilities (FPSOs) and Floating Storage Unites (FSUs)*; IMO resolution MEPC.142(54), 24 March 2006, *Amendments to the Guidelines for the Application of the Revised MARPOL Annex I Requirements to Floating Production, Storage and Offloading Facilities (FPSOs) and Floating Storage Units (FSUs) (Resolution MEPC.139(53)*. Also *see* IMO resolution A.765(18), 4 November 1993, *Guidelines on the Safety of Towed Ships and Other Floating Objects, including Installations, Structures and Platforms at Sea.*

[47] Article 2 and Regulation 2 of the *International Convention for the Safety of Life At Sea*, done at London on 1 November 1974 and entered into force on 25 May 1980, published at 1184 U.N.T.S. 3.

[48] *International Convention on Standards of Training, Certification and Watchkeeping for Seafarers*, done at London on 7 July 1978 and entered into force on 28 April 1984, published at 1361 U.N.T.S. 190.

[49] *International Convention on Load Lines*, done at London on 5 April 1966 and entered into force on 21 July 1968, published at 640 U.N.T.S. 133.

[50] IMO MSC-MEPC.2/Circ.9, 25 May 2010, *Guidance for the Application of Safety, Security and Environmental Protection provisions to FPSOs and FSUs*. Also *see* IMO MSC/Circ. 1111, 7 June 2004, *Guidance Relating to the Implementation of SOLAS Chapter XI-2 and the ISPS Code*.

[51] Articles 2(4) and 3(1) lit. (b) and (2), 8(1), 10, 11, 12 and 17 of MARPOL 73/78.

5.4.3 Construction and Operation

The provisions of UNCLOS applicable to offshore installations and structures determine that coastal States have the exclusive right to construct and regulate the construction, operation and use of offshore installations and structures in the continental shelf, as well as of offshore installations and structures for economic purposes and those that may interfere with the exercise of the rights of the coastal State in the EEZ.[52] Moreover, coastal States have jurisdiction regarding offshore installations and structures used for marine scientific research in the continental shelf and the EEZ.[53] Coastal States are also entitled to adopt and enforce laws and regulations on matters concerning customs, fiscal, health, safety and immigration regarding offshore installations and structures in the EEZ and continental shelf, as well as in the territorial sea.[54]

In exercising these rights, coastal States must give due notice of the construction of offshore installations or structures and provide and maintain permanent means for giving warning of their presence. Coastal States may establish safety zones around offshore installations or structures, not exceeding a distance of 500 meters, except as authorized by generally accepted international standards, or as recommended by the competent international organization.[55] This limits the possibility of States having excessive discretion regarding the establishment of safety zones, however, without preventing the creation of larger safety zones when required, as long as in conformity with international standards and guidelines.

The international standards, guidelines and recommendations on offshore installations and structures regarding construction, operation, removal and the determination of the breadth of their safety zones, and also for the prevention, reduction and control of pollution by dumping, are those provided by the IMO being the competent international organization, bearing in mind its global mandate.[56] Although UNCLOS does not mention the IMO expressly in Article 60, during the

---

[52] Article 60(1) lit. (b) and (c) of UNCLOS. *A contrario*, coastal States do not have jurisdiction over installations and structures that are not being used for marine scientific research, pursuant to Part XIII of UNCLOS, or for economic purposes, such as military installations, so long as they do not interfere with the exercise of the rights of the coastal State in this maritime area.

[53] Articles 56(1) lit. (b) (ii), and 246(1) and 258 of UNCLOS.

[54] Articles 60(2) and 80, and 21(1) lit. (f) and (h) of UNCLOS. The European Union has adopted two Directives on minimum requirements for improving the safety and health protection of workers in the mineral-extracting industries, which are also applicable to offshore installations and structures. *See* Council Directives 92/91/EEC, of 3 November 1992, and 92/104/EEC, of 3 December 1992, both amended by Directive 2007/30/EC, of the European Parliament and of the Council, of 20 June 2007.

[55] Article 60(5) of UNCLOS. Also *see* Article 260 of UNCLOS regarding the safety zones around scientific research installations.

[56] Article 60(4) to (7) of UNCLOS. *See* IMO LEG/MISC.6, 10 September 2008, *Implications of the United Nations Convention on the Law of the Sea for the International Maritime Organization*, pp. 7–9.

preparatory work certain proposals did in fact refer to the "*applicable standards in existence or to be established by the Inter-Governmental Maritime Consultative Organization regarding the establishment and breath of safety zones.*"[57]

Prior to the entry into force of UNCLOS, the 1958 Convention on the Continental Shelf merely referred that coastal States should establish safety zones around installations and structures up to a distance of 500 meters, without predicting any exception to this rule.[58]

Initially, the ILC found that it was not essential to specify the size of safety zones, even though it did consider 500 meters a reasonable distance.[59] Eventually, the ILC did recognize the inadequacy of the alternative use of a vague criterion, such as that of 'reasonable distance', which would allow States a considerable degree of discretion in determining the breath of the safety zones around installations and structures, potentially leading to conflicts and unreasonably interfering with the rights and freedoms of third States, namely, navigation and fishing.[60] This would also be contrary, to a certain extent, to the principle that installations and structures do not generate adjacent maritime areas, since, if no limit had been defined, States could through the right to establish safety zones, allow that these or a combination of several result in having de facto control over the respective maritime areas by not allowing or severely limiting other States from exercising their rights and freedoms in these areas.

The obligation of giving due notice of safety zones and of their extent is a fundamental duty of coastal States in order to ensure the protection of offshore installations and structures under their jurisdiction and the safety of other uses and activities, namely navigation and fishing. Accordingly, offshore installations and structures and their safety zones must comply with generally accepted international standards and may not interfere with the use of recognized sea lanes essential to international navigation.[61] This is an indispensable aspect of the prevailing freedom of navigation that is taken into consideration also regarding the removal of offshore installations.[62] On the other hand, flag States must take all the necessary steps to ensure that ships flying their flag will comply with the relevant international standards applicable to navigation, namely by having regard for the safety zones

---

[57] The only reference to the IMO is in Article 2(2) of Annex VIII of UNCLOS. *See UNCLOS: A Commentary*, *supra* note 36, p. 577, and pp. 585 and 586 on the same point.

[58] Article 5(3) of the 1958 Convention on the Continental Shelf.

[59] *Yearbook of the International Law Commission, supra note 56 in Chap. 3*, p. 48; *Yearbook of the International Law Commission, 1956, vol. 2, Documents of the eighth session, including the report of the Commission to the General Assembly*, p. 299.

[60] *Ibid.*

[61] IMO resolution A.671(16), 19 October 1989, *Safety Zones and Safety of Navigation around Offshore Installations and Structures.*

[62] The text of Article 60(7) of UNCLOS reads: "*Artificial islands, installations and structures and the safety zones around them may not be established where interference may be caused to the use of recognized sea lanes essential to international navigation.*".

established around offshore installations and structures.[63] Flag States must also take action in the case of infringement of safety zones around offshore installations or structures by ships flying their flag.[64] This concern for the safety of navigation is also present in the exercise of the right of coastal States to adopt laws and regulations on the exercise of the right of innocent passage through the territorial sea.[65]

Although the obligation of coastal States to give due notice or publicity is not specified in UNCLOS, the fundamental principle underlying the protection of offshore installations and structures and the safety of navigation is that all those concerned, namely, coastal and flag States and those operating ships and offshore installations, are aware of the existing sea lanes and traffic separation schemes, the location of installations and structures, and the safety zones around them, thus preventing the infringement of the costal States' laws and regulations.[66]

5.4.4 Removal and Decommissioning

Offshore installations and structures that are no longer in use or that are abandoned constitute a great concern for the safety of navigation and represent a threat for the marine environment, as well as for other important activities, such as fishing. Indeed, after installations and structures have completed their functional purpose, coastal States must, nonetheless, maintain the necessary safety zones and warning signals, in addition to giving publicity to their depth, position and dimensions.[67]

Contrary to the 1958 Convention on the Continental Shelf, which included a legally binding obligation for the removal of abandoned installations or that were no longer in use,[68] UNCLOS has left to the discretion of coastal States the decision on creating a removal obligation.[69] In fact, the preparatory work of Article 60 of

---

[63] Articles 56(1) lit. (b) (ii), 60(4) and (6), 260 to 262 of UNCLOS.

[64] IMO resolution A.671(16), *supra* note 61; IMO resolution A.572(14), 20 November 1985, *General Provisions on Ships' Routeing.*

[65] Article 21(3) of UNCLOS. Also *see* Articles 41, 42 and 45 of UNCLOS regarding straits used for international navigation and Articles 52 to 54 of UNCLOS in respect of passage though archipelagic waters. .

[66] Articles 22(1) and (4), and 24(2) of UNCLOS. Also *see* IMO resolution A.671(16), *supra* note 61.

[67] Article 60(3) of UNCLOS.

[68] The text of Article 5(5) of the 1958 Convention on the Continental Shelf reads: "*Any installations which are abandoned or disused must be entirely removed.*" This provision resulted from a proposal of the United Kingdom see *United Nations Conference on the Law of the Sea, Official Records, Volume VI: Fourth Committee (Continental Shelf), Summary records of meetings and Annexes, 24 February-27 April 1958,* Doc: A/CONF.13/42, p. 4, para. 14.

[69] Articles 60(3), 147(2) lit. (a) and 249(1) lit. (g) of UNCLOS. *See* Edward Duncan Brown, "The significance of a possible EC EEZ for the law relating to artificial islands, installations, and structures, and to cables and pipelines, in the exclusive economic zone", in: 23 *Ocean Development and International Law* (1992), pp. 126–132.

UNCLOS mentions the "*enormous costs*" that would be implicit in a generalized removal obligation were of great concern and taken into consideration.[70]

Despite this disparity between the two Conventions and since the removal obligation included in the 1958 Convention on the Continental Shelf did not become part of customary law and develop into a binding obligation for all States,[71] UNCLOS prevails over the 1958 Convention on the Continental Shelf in those cases where States are parties to both international legal regimes.[72]

The removal obligation notwithstanding, the 1958 Convention on the Continental Shelf and later also UNCLOS, were not precise in the reference to "*abandoned or disused installations*", namely, in what ought to be considered being included in these two concepts and when does the obligation to remove an offshore installation or structure begin.[73]

Since offshore installations and structures require the use of submarine cables and pipelines, the question of the extension of this regime to the latter is also not clear in UNCLOS. The Convention, similar to the 1958 Convention on the Continental Shelf, makes no reference to the removal of submarine cables and pipelines. Yet, on the other hand, Article 208 of UNCLOS does not exclude pollution from submarine cables and pipelines in respect of pollution from seabed activities, even though Article 79(2) of UNCLOS envisages that States may adopt laws and regulations for the prevention, reduction and control of pollution from pipelines. Nevertheless, it is generally considered that Article 60 of UNCLOS is not applicable to submarine cables and pipelines.[74] Indeed, as Rainer Lagoni expertly concluded:

---

[70] *UNCLOS: A Commentary*, *supra* note 36, p. 582.

[71] Paul Peters, Alfred H. A. Soons and Lucie A. Zima, "Removal of installations in the Exclusive Economic Zone", in: 15 *Netherlands Yearbook of International Law* (1984), pp. 167–207. These authors concluded that in the light of State practice since 1958, the rule of removal had not become part of customary international law and therefore would not be binding upon a State not party to the 1958 Convention on the Continental Shelf *see ibid*, pp. 192–193.

[72] Article 311(1) of UNCLOS.

[73] "*"Abandonment", in the legal sense, is more than merely walking away from a thing. It means the voluntary relinquishment of all right, title, claim and possession of a thin, with the intention of not reclaiming it; or giving up of a thing absolutely, without reference to any particular person or purpose, as vacating property with the intention of not returning to it. The term "disuse" is less easy to define in precise terms; however, in this context it is clear that a degree of permanency of non-use is to inferred: an installation disused merely temporarily, e.g. while a well is shut in, cannot have been intended. It will be difficult, therefore, to establish at what precise point in time an installation or device is "abandoned or disused", i.e. when the obligation to remove is born. Even when a field is "exhausted" and production stops, it may be the intention of the licensee to keep the installation in place, in case oil production can be resumed at a later date on the basis of secondary or tertiary recovery techniques. Even when the license expires, it may not necessarily be assumed that the installation is permanently abandoned.*" Peters/Soons/Zima, *supra* note 71, p. 173.

[74] "*Although there is an amazing variety in the terminology used in the Convention, the lack of consistency can be ascribed to the somewhat unorthodox method of treaty-making that was adopted by UNCLOS-III and one must be hesitant to ascribe too much significance to the choice*

> Cables and pipelines are not installations or structures in the sense of the Convention; their laying and maintenance are not regulated in Article 60 of the Law of the Sea Convention.[75]

UNCLOS uses a wide variety of terminology and often interchanges between related concepts, failing to define many of them. In what concerns the removal of offshore installations and structures or devices, the reference to decommissioning of installations and structures would have been more appropriate, for example, by allowing States to reuse them for other purposes.[76] This way, instead of requiring their removal, either completely or partially, UNCLOS would have included the possibility of rehabilitating or transforming offshore installations and structures, such as using them as artificial reefs for coastal management purposes.[77] The main

(Footnote 74 continued)
of expressions used in different contexts. It is clear from the above that, both from a historical and from a systematic interpretation, one must come to the conclusion that pipelines are not included in "installations and structures" in the context of Article 60.3. (…) *Thus we maintain the conclusion that no pipelines—whatever their length or purpose—are subject to the provisions of Article 60; neither the removal obligation, nor the provision regarding safety zones accordingly applies to them.*" *Ibid*, p. 191. Also *see* Brown, *supra* note 69, p. 138: "*There appears to be general agreement that submarine pipelines were not intended to be included among the continental shelf "installations and devices" referred to in Article 5 of the Geneva Convention on the Continental Shelf or the installations of the continental shelf and the EEZ referred to in Articles 60 and 80 of the UN Convention. Neither the IMO Guidelines and Standards nor the Oslo Commission Guidelines refer to pipelines. There seems, therefore, to be no specific requirement in international conventional law for their removal after abandonment or disuse*".

[75] Rainer Lagoni, "Cables and pipeline survey at sea", in: *Coexistence, Cooperation and Solidarity, Liber Amicorum Rüdiger Wolfrum*, vol. 1, edited by Holger P. Hestermeyer, Doris König, Nele Matz-Lück, Volker Röben, Anja Seibert-Fohr, Peter-Tobias Stoll and Silja Vöneky (Martinus Nijhoff Publishers: Leiden, Boston, 2012), p. 935. Also *see* Lagoni, *supra* note 46 in Chap. 2, p 203.

[76] *See* OSPAR [*Convention for the Protection of the Marine Environment of the North-East Atlantic*] *Decision 98/3 on the Disposal of Disused Offshore Installations, 22–23 July 1998* [online: http://www.ospar.org (accessed on August 2013)].

[77] *See* J. M. MacDonald, "Artificial reef debate: habitat enhancement or waste disposal?", in: 25–1 *Ocean Development and International Law* (1994), pp. 87–118. Also *see* OSPAR *Guidelines on Artificial Reefs in relation to Living Marine Resources*, 2012; OSPAR *Assessment of Construction or Placement of Artificial Reefs*, 2009; *2009 London Convention and Protocol/UNEP Guidelines for the Placement of Artificial Reefs, 30*th *Consultative Meeting of Contracting Parties to the 1972 Convention on the Prevention of Marine Pollution by Dumping of Wastes and Other Matter, 27–31 October 2008*, and *3*rd *Meeting of Contracting Parties to the 1996 Protocol to the 1972 Convention on the Prevention of Marine Pollution by Dumping of Wastes and Other Matter on 31 October 2008*. These Guidelines define an artificial reef as "*a submerged structure deliberately constructed or placed on the seabed to emulate some functions of a natural reef such as protecting, regenerating, concentrating, and/or enhancing populations of living marine resources. Objectives of an artificial reef may also include the protection, restoration and regeneration of aquatic habitats, and the promotion of research, recreational opportunities, and educational use of the area. The term does not include submerged structures deliberately placed to perform functions not related to those of a natural reef—such as breakwaters, mooring, cables, pipelines, marine research devices or platforms—even if they incidentally imitate some functions of a natural reef.*" (pp. 4 and 5).

concern in UNCLOS is not that offshore installations and structures are removed, but rather that they do not constitute a danger. Consequently, as long as there is no risk of such interference, there should not be an obligation of removal.[78]

One of the aspects of the discretionary nature of the removal obligation under UNCLOS is that it also takes into account the situations in which the removal of the offshore installations and structures is neither required nor feasible. The decision facing States regarding installations and structures that are no longer in use vary in each case and must take into consideration practicable and technical aspects, as well as the protection of the environment and of human health and safety, and also the very high costs of removal operations. Only after this assessment can States determine the reuse, recycle or disposal of an offshore installation or structure. In fact, the removal could threaten the safety of navigation, fishing activities, or the protection and preservation of the marine environment. The reference to decommissioning would have served the overall purpose of eliminating the danger posed by abandoned offshore installations and structure, without requiring their removal. Notably, UNCLOS already envisages in Article 60(3) *in fine* the possibility of only partially removing an offshore installation or structure, as it may be required.

The wording of this provision does further enclose a relevant distinction for the purpose of consideration of the said discretionary nature of the removal obligation of abandoned or disused offshore installations and structures, or what Edward Duncan Brown referred to as "*qualified-removal rule*"[79]:

> Any installations or structures which are abandoned or disused shall be removed to ensure safety of navigation, taking into account any generally accepted international standards established in this regard by the competent international organization. Such removal shall also have due regard to fishing, the protection of the marine environment and the rights and duties of other States.

Accordingly, it establishes a preference with regard to the safety of navigation concerning the removal of abandoned or disused offshore installations or structures, which is not identical in respect of fishing, for example.[80] It has also been noted that the absence of a reference to 'international standards' and 'competent international organization' related to fishing or the protection and preservation of the marine environment is "*unfortunate*", placing under the umbrella of the IMO the consideration of all other interests, which may not be very effective considering the limitations deriving from the powers of this Organization.[81]

The existing guidelines and standards for the removal of offshore installations and structures provided by the IMO are not binding and merely determine that coastal States having jurisdiction over such installations and structures should ensure their total or partial removal in the case of abandonment or permanent

---

[78] Peters/Soons/Zima, *supra* note 71, p. 173.

[79] Brown, *supra* note 26 in Chap. 3, p. 263.

[80] *Ibid*, pp. 266–267.

[81] *Ibid*, 267.

disuse, or notify the IMO of their decision of non-removal or partial removal.[82] According to these guidelines and standards, the decision of the coastal State should be based on a case-by-case evaluation, taking into account the safety of navigation or of other uses of the sea, the rate of deterioration of the material and the impact on the marine environment, the effect on living resources, the costs and technical feasibility, and the determination of a new use or other reasonable justification for allowing the installations or structure to remain in the seabed.[83]

There is, therefore, no international binding regime dealing with the removal or decommissioning of offshore installations and structures in the EEZ and continental shelf. There is too no removal obligation of installations and structures in the territorial sea, archipelagic waters or internal waters, these being maritime areas under the sovereignty of coastal States. In fact, it may occur that the safe disposal in situ of an offshore installation or structure constitutes the best alternative to its removal, without representing, however, an act of pollution by dumping pursuant to Article 210 of UNCLOS, or causing the negative impacts inherent to dumping of an offshore installation or structure.[84] This would be the case of an installation and structure being reused for the purpose of in situ conservation. Accordingly, Article 8 lit. (i) of the *Convention on Biological Diversity*[85] provides that States shall, as far as possible and as appropriate, endeavor to create the conditions needed for the compatibility between uses and the conservation of biological diversity. This Convention further determines that its provisions will supersede those of any international legal regime in case of serious damage or threat to biological diversity, including the removal obligation as established in UNCLOS.[86]

The definition of dumping provided in Article 1(5) lit. (b) (ii) of UNCLOS does not include the "*placement of matter for a purpose other than the mere disposal thereof, provided that such placement is not contrary to the aims of this Convention.*"[87] In this case, one of the aims to be taken into account is foremost the general obligation for all States to protect and preserve the marine environment

---

[82] IMO resolution A.672(16), 19 October 1989, *Guidelines and Standards for the Removal of Offshore Installations and Structures on the Continental Shelf and in the Exclusive Economic Zone.*

[83] The IMO Guidelines and Standards refer that: "*All abandoned or disused installations and structures emplaced on the seabed on or after 1 January 1998, standing less that 100 m of water and weighing less than 4.000 tonnes in air, excluding the deck and superstructure, should be entirely removed*".

[84] For example *see* UNEP (DEC)/MED IG.15/Inf. 13, 1 October 2003, *Guidelines Dumping of Platforms and Other Man-Made Structures at Sea.*

[85] *Convention on Biological Diversity*, done at Rio de Janeiro on 5 June 1992 and entered into force 29 December 1993, published at 1760 U.N.T.S. 79 [Convention on Biological Diversity].

[86] Article 22(1) of the Convention on Biological Diversity.

[87] Also *see* Article 3(4) lit. (b) of the *Prevention of Pollution of the Mediterranean Sea by Dumping from Ships and Aircraft*, done at Barcelona on 16 February 1976 and entered into force on 12 February 1978, published at 1102 U.N.T.S. 1, renamed *Protocol for the Prevention and Elimination of Pollution of the Mediterranean Sea by Dumping from Ships and Aircraft or*

referred to in Article 192 of UNCLOS. This is also the fundamental nature of the obligation of prevention of pollution by dumping[88] and of the specific obligation to adopt the measures necessary to minimize pollution from offshore installations and devices used in seabed activities.[89] However, should the decommissioning of an offshore installation or structure be considered dumping, the conditions established in Articles 210 and 216 of UNCLOS would apply, although the final decision would also rest with the respective coastal State.

The referred Article 210(4) of UNCLOS further provides that States shall endeavour to establish global and regional rules, standards and recommended practices and procedures to prevent, reduce and control pollution of the marine environment by dumping, such as those provided under the 1972 London Convention.[90] Accordingly, in the event that dumping of an offshore installations or structure is allowed in accordance with the 1972 London Convention, the relevant State-party must obtain the necessary permit from the IMO before dumping any wastes or matter as prescribed by this legal instrument. It must also inform this organization without delay, if dumping of an offshore installation or structure appears to be the only way of averting the threat to the safety of navigation, amongst other concerns that too must be taken into account. States shall be responsible for damages to the environment of other States or to any other area of the environment caused by dumping.[91]

States must provide, therefore, under their legal regime the conditions for the removal or decommissioning of installations and structures within their national jurisdiction, either by means of the applicable law or of the relevant license or concession agreement, or other form of granting operators the right to explore and exploit offshore hydrocarbon deposits. As for joint development agreements after the delimitation of maritime boundaries, the respective legal framework must establish similar obligation or determine the applicable law and jurisdiction, even though, as with joint development agreements signed before the delimitation of maritime boundaries, the agreements known to date do not include similar provisions.

(Footnote 87 continued)
Incineration at Sea, after the amendments adopted on 10 June 1995, Doc. UNEP(OCA)/MED IG.6/7, not yet in force.

[88] Article 210(1) of UNCLOS.

[89] Article 194 (3) lit (c) of UNCLOS.

[90] IMO LEG/MISC.6, *supra* note 56, p. 65.

[91] *Convention on the Prevention of Marine Pollution by Dumping of Waste and Other Matter*, done at London on 29 December 1972 and entered into force on 30 August 1975, published at 1046 U.N.T.S. 120 [1972 London Convention]; *Protocol to the 1972 London Convention*, done at London on 7 November 1996 and entered into force on 24 March 2006, published at 36 *International Legal Materials* 1 (1997). Also *see Specific Guidelines for Assessment of Platforms or Other Man-made Structures at Sea*, adopted at the *Twenty-second Consultative Meeting of Contracting Parties to the 1972 London Convention, from 18–22 September 2000.*

5.5 States' Responsibility for Pollution from Seabed Activities Subject to National Jurisdiction

The protection and preservation of the marine environment in UNCLOS is an obligation that all States are bound to, and one that is applicable to all maritime areas, including both within and beyond national jurisdiction.[92] Coastal States are in particular responsible for pollution caused by the exploration and exploitation activities that take place in maritime areas within national jurisdiction.[93] UNCLOS establishes a direct connection between, on the one hand, the right of coastal States to construct and operate offshore installations and structures for the purpose of exploring and exploiting the non-living resources of the continental shelf, and, on the other hand, the obligation to adopt laws and regulations to prevent, reduce and control pollution from such installations and structures.[94] In this respect Article 193 of UNCLOS provides that:

> States have the sovereign right to exploit their natural resources pursuant to their environmental policies and in accordance with their duty to protect and preserve the marine environment.

Coastal States may adopt laws and regulations to prevent, reduce and control pollution of the marine environment from installations and structures under their jurisdiction and from seabed activities.[95] States may also proceed accordingly regarding installations under their control and used in the activities in the Area,[96] in addition to undertaking the necessary measures to prevent, mitigate or eliminate any danger to their coastline or related interests from pollution from said activities.[97] Still within the Area regime, the Authority has the right to adopt the appropriate measures for the protection of the marine environment,[98] while the Seabed Disputes Chamber of the International Tribunal for the Law of the Sea (Seabed Disputes Chamber) has jurisdiction on disputes concerning the liability of the Authority for wrongful acts in the exercise of its powers and functions.[99]

The Convention governs the relation between the rights of coastal States regarding the protection and preservation of the marine environment and the

---

[92] Article 192 of UNCLOS.

[93] ITLOS considered that the term 'responsibility' referred "*to the primary obligation whereas the term 'liability' refers to the secondary obligation, namely, the consequences of a breach of the primary obligation*". *See Request for Advisory Opinion submitted to the Seabed Disputes Chamber, supra note 77 in Chap.* 2, para. 66. On States' liability for environmental damage *see* Philippe Sands, *Principles of International Environmental Law* (Cambridge University Press: Cambridge, 2003), pp. 871–939.

[94] *See* IMO LEG/MISC.6, *supra* note 56.

[95] Articles 56(1) lit. (b) (iii), 194(3) lit. (c), 208(1) and 214 of UNCLOS.

[96] Article 209(2) of UNCLOS.

[97] Articles 142(3), 209, 215 of UNCLOS.

[98] Article 145 of UNCLOS.

[99] Article 187 lit. (e) of UNCLOS and Article 22 of Annex III of UNCLOS.

exercise of their rights of sovereignty and jurisdiction, namely, in connection with the internal waters, where States have full sovereignty,[100] or with the exercise of the rights of innocent passage through the territorial sea[101] and of transit passage in straits used for international navigation,[102] as well as regarding the exercise of the powers of enforcement in respect of pollution from foreign ships in the EEZ.[103] It also determines that, in certain circumstances, States are subject to a reinforced obligation of cooperation regarding the protection and preservation of the marine environment, such as in respect of enclosed or semi-enclosed seas.[104]

UNCLOS further provides that the provisions regarding the protection and preservation of the marine environment pursuant to Part XII are without prejudice to specific obligations assumed by States in other international legal instruments concluded previously to the Convention in respect to the protection and preservation of the marine environment, as well as in instruments which may be concluded in furtherance of the general principles set forth in UNCLOS.[105] These include, for example, the legal instruments applicable to offshore installations and structures that were previously referred. However, certain obligations envisaged by UNCLOS regarding seabed activities often rely excessively on the discretion of coastal Sates to enforce compliance by operators.[106] This is the case, for example, of the Convention determining that States must provide recourse for "*prompt and adequate compensation or other relief in respect of damage caused by pollution of the marine environment by natural or juridical persons under their jurisdiction,*" as well as that they shall cooperate in the implementation and development of existing international law and regarding responsibility and liability.[107]

---

[100] Article 8 of UNCLOS. *See Military and Paramilitary Activities in und against Nicaragua (Nicaragua v. United States of America), Merits, Judgment, I.C.J. Reports 1986*, paras. 212, 213, 251.

[101] Articles 21(1) lit. (f) and 220(3) of UNCLOS.

[102] Articles 42(1) lit. (b) and 43 lit. (b) of UNCLOS.

[103] Article 220(3) of UNCLOS.

[104] Article 123 lit. (b) of UNCLOS.

[105] Articles 237 and 311 of UNCLOS. *See* Allan Boyle, "Relationship between international environmental law and other branches of international law", in: *The Oxford Handbook of International Environmental Law*, edited by Daniel Bodansky, Jutta Brunnée and Ellen Hey (Oxford University Press: Oxford, 2007), p. 135.

[106] On the national implementation of international environmental law *see* Catherine Redgwell, "National implementation", in: *The Oxford Handbook of International Environmental Law*, edited by Daniel Bodansky, Jutta Brunnée and Ellen Hey (Oxford University Press: Oxford, 2007), pp. 929–946. Also *see Ibid*, "From permission to prohibition: the 1982 Convention on the Law of the Sea and Protection of the Marine Environment", in: *The Law of the Sea: Progress and Prospects*, edited by David Freestone, Richard Barnes, David Ong (Oxford University Press: Oxford, 2006), pp. 188–189. In referring to pollution by dumping, Catherine Redgwell concludes that UNCLOS "*does not set absolute standards but rather adopts a form of international minimum harmonization based on obligation of result (ie national standards shall be 'no less effective' than global rules and standards, contained within and without the LOSC, in preventing, reducing and controlling pollution of the marine environment by dumping)*".

[107] Articles 235(2) and (3) of UNCLOS.

The Seabed Disputes Chamber in the Advisory Opinion concerning the *Responsibilities and obligations of States sponsoring persons and entities with respect to activities in the Area* referred that the responsibility of sponsoring States to ensure that activities in the Area are carried out pursuant to Part XI of UNCLOS, is an obligation existing under international law. It also clarified that the 'mechanism' through which the rules of UNCLOS become effective for sponsored contractors under the domestic law of the sponsoring State "*consists in the creation of obligations which States Parties must fulfil by exercising their power over entities of their nationality and under their control*".[108] The Chamber further concluded that the violation, by an act or an omission, of the said obligation of sponsoring States would entail liability, underlining, however, that not every violation of an obligation by a sponsored contractor automatically gives rise to the liability of the respective sponsoring State. The Chamber also noted that the 'obligation to ensure' is an obligation of conduct, or of means, and not of result, and that liability would be limited to the States' failure to meet their 'obligation to ensure' compliance by sponsored contractors.[109]

Based on this Advisory Opinion, it is possible to infer valuable guiding principles that are also applicable to States' obligations regarding pollution from seabed activities in maritime areas under national jurisdiction. Indeed, the Advisory Opinion observed that:

> The expression "to ensure" is often used in international legal instruments to refer to obligations in respect of which, while it is not considered reasonable to make a State liable for each and every violation committed by persons under its jurisdiction, it is equally not considered satisfactory to rely on mere application of the principle that the conduct of private persons or entities is not attributable to the State under international law (...).[110]

Another well-known example which was also referred by the Chamber can be found in Article 194(2) of UNCLOS. Accordingly, States undertake to adopt all measures necessary to ensure that activities under their jurisdiction are carried out in a manner as not to cause damage by pollution, including those designed to minimize pollution from offshore installations used in seabed activities.[111] In the case of joint development agreements signed after the delimitation of maritime boundaries, the relevant States should harmonize and enforce environmental standards and rules in their own jurisdiction to achieve a comprehensive and an effective protection and preservation of the marine environment. Amongst these, States should monitor the risks and effects of pollution of the marine environment from seabed activities taking place also under the framework of a joint development agreements signed after the delimitation of maritime boundaries.[112]

---

108 *Request for Advisory Opinion submitted to the Seabed Disputes Chamber, supra note 77 in Chap. 2*, para. 108.

109 *Ibid*, paras. 109–110, 177.

110 *Ibid*, para. 112.

111 Article 194(3) lit. (c) of UNCLOS.

112 Article 204 of UNCLOS.

Another example is the obligation for States to assess the potential effects on the marine environment of "*planned activities*" under their jurisdiction or control.[113] Although not expressly referred in the Convention, this is the case of the obligation to conduct an environmental impact assessment that has been considered to constitute a general obligation under customary international law, whenever there is a risk that the proposed activity may have a significant adverse impact in a transboundary context, in particular, on a shared resource.[114]

Coastal States are, therefore, responsible for pollution from seabed activities subject to national jurisdiction and may be considered liable in the event that they do not adopt the necessary measures pursuant to international law to ensure that operators comply, in their jurisdiction, with the relevant laws and regulations to prevent, reduce and control pollution of the marine environment from installations and structures and from seabed activities.

Lastly, in the event of a dispute regarding pollution from seabed activities, UNCLOS determines that States may follow one of the compulsory mechanisms for the settlement of disputes set out in the Convention for the violation of "*specified international rules and standards for the protection and preservation of the marine environment*," provided that these are applicable to the relevant coastal State and have been established by UNCLOS, or through a competent international organization or diplomatic conference in accordance with the Convention.[115] UNCLOS also provides that the International Tribunal for the Law of the Sea (ITLOS) may prescribe appropriate provisional measures to prevent serious harm to the marine environment.[116]

---

[113] Article 206 of UNCLOS.

[114] *Request for Advisory Opinion submitted to the Seabed Disputes Chamber, supra note 77 in Chap. 2*, paras. 145–149.

[115] Article 297(1) lit (c) of UNCLOS.

[116] Article 290(1) of UNCLOS.

Chapter 6
Joint Development Agreements in Areas of Overlapping Claims: Legal Regime

6.1 Economic Activities in Disputed Maritime Areas

The delimitation of maritime boundaries is a complex and on many occasions, a lengthy procedure that may very well prove to be, depending on the circumstances, an unattainable task when States make overlapping maritime claims. Likewise, territorial disputes regarding islands or their importance for the purpose of delimitation may also contribute towards the difficulty in settling maritime boundaries.

In the event that States should fail to settle their maritime boundaries, either by means of a delimitation treaty or through a compulsory settlement mechanism, they may implement provisional arrangements pursuant to Articles 74(3) and 83(3) of UNCLOS.[1] The text of both Articles reads:

> Pending agreement as provided for in paragraph 1, the States concerned, in a spirit of understanding and cooperation, shall make every effort to enter into provisional arrangements of a practical nature and, during this transitional period, not to jeopardize or hamper the reaching of the final agreement. Such arrangements shall be without prejudice to the final delimitation.

The possibility of States implementing interim measures pending the delimitation of the EEZ and the continental shelf between States with opposite or adjacent coasts was a novel aspect introduced by UNCLOS and one which was included in the compromise reached during negotiations in three outstanding issues[2]:

---

[1] On the drafting of Articles 74 and 83 of UNCLOS *see* Tullio Scovazzi, "The evolution of international law of the sea: new issues, new challenges", in: 286 *Recueil des Cours de l'Académie de Droit International* (2001), pp. 194–200; Tanja, *supra* note 60 in Chap. 3, pp. 92–116; Oude Elferink, *supra* note 60 in Chap. 3, pp. 27–33; Brown, *supra* note 60 in Chap. 3, pp. 331–350; Lagoni, *supra* note 70 in Chap. 2 pp. 349–354.

[2] *Report of the Chairman of negotiating group 7*, Document A/CONF.62/L.47, UNCLOS III Off. Rec., V. XVIII, pp. 76–77, para. 5 [online: http://untreaty.un.org/cod/diplomaticconferences/lawofthesea-1982/lawofthesea-1982.html (accessed on August 2013)].

V. Becker-Weinberg, *Joint Development of Hydrocarbon Deposits in the Law of the Sea*, Hamburg Studies on Maritime Affairs 30, DOI 10.1007/978-3-662-43570-0_6

1. The question of the criteria to be applied for the delimitation of economic zones or continental shelves adjacent or opposite to each other.
2. The question of interim measures to be applied pending final delimitation.
3. The question of the settlement of delimitation disputes.[3]

These three issues were settled together as part of a package solution, which essentially resulted in the consensus that delimitation should be effected by agreement and based on international law, without adopting any of the criteria put forward for the delimitation of the EEZ and the continental shelf, namely the equidistance rule and the delimitation in accordance with equitable principles.[4] The package solution further recognized that the adoption of interim measures could have a significant influence on the negotiation of maritime delimitation. As a result, it too should be dependent on the final delimitation and therefore should adopt a more neutral formulation.[5] This is also consistent with the guiding principle included in the *Charter of the United Nations* (UN Charter)[6] regarding provisional measures, which determines that they shall be without prejudice to the rights, claims, or position of the relevant States.[7]

The drafting group concluded that, although States agreed that the adoption of interim measures should be encouraged, particularly to avoid any aggravation of a delimitation conflict, States did not approve that the adoption of interim measures should be mandatory. The drafting group additionally recognized the importance given by several States that provisional arrangements should be based upon criteria different from those to be prescribed for the delimitation of maritime boundaries.[8] There was also a general agreement to the effect that any provisional arrangements should be without prejudice to the final delimitation and that mutual-restraint ought to be exercised. Several States also considered that the lack of agreement on the delimitation of maritime areas should not hinder the possibility of undertaking economic activities in the disputed maritime area, such as regarding its use or of the marine natural resources

---

[3] *Report by the Chairman of Negotiating Group 7 on the work of the Group at its 17th–27th meetings*, NG7/24, 14 September 1978, Document A/CONF.62/RCNG/2, UNCLOS III Off. Rec. Volume X, pp. 170–172 [online: *ibid*].

[4] *Ibid.*

[5] *Ibid*; *Report of the Chairman of the Negotiating Group 7*, Document NG7/45, UNCLOS Off. Rec. Volume XII [online: *ibid*].

[6] *Supra* note 87 in Chap. 3.

[7] Article 40 of the UN Charter.

[8] The early versions of Articles 74(3) and 83(3) of UNCLOS read: "*Pending agreement or settlement, the States concerned shall make provisional arrangement, taking into account the provisions of paragraph 1.*", which in turn stated that "[t]*he delimitation of the* [EEZ/continental shelf] *between adjacent or opposite States shall be effected by agreement in accordance with equitable principles, employing, where appropriate, the median or equidistance line, and taking into account of all relevant circumstances.*" ICNT, Document A/CONF.62/WP.10, UNCLOS III Off. Rec. Volume VIII [online: http://untreaty.un.org/cod/diplomaticconferences/lawofthesea-1982/lawofthesea-1982.html (accessed on August 2013)]. Also *see* Articles 62 and 71 of the RSNT (part II), Document A/CONF.62/WP.8/Rev.1/PartII, UNCLOS III Off. Rec. V [online: *ibid*].

found therein.[9] As a matter of fact, in 2007, the Arbitral Tribunal in the case between Guyana and Suriname underlined that the obligation contained in Article 83(3) of UNCLOS "*constitutes an implicit acknowledgment of the importance of avoiding the suspension of economic development in a disputed maritime area, as long as such activities do not affect the reaching of a final agreement.*"[10]

It has been noted, however, that despite encouraging restraint and practical co-operation pending final delimitation, in light of the negotiating history of Article 83(3) of UNCLOS it may "*be difficult to see that any great purpose is served by this provision.*"[11] In fact, Rainer Lagoni considered this provision and Article 74(3) of UNCLOS as not being "*very enlightening.*"[12]

Articles 74 and 83 of UNCLOS are undeniably not of much use for the delimitation of the EEZ or the continental shelf between States with opposite or adjacent coasts, nor for the assessment of the scope and content of provisional arrangements. Nonetheless, as it will be examined in Chaps. 7 and 8, State practice and international jurisprudence have considered the implementation of provisional measures a valuable legal option pending the delimitation of maritime boundaries, particularly joint development agreements of offshore hydrocarbon deposits for their functional characteristics. Also see Bay of Bengal Maritime Boundary Arbitration between Bangladesh and India (Bangladesh v. India), P.C.A. Award 7 July 2014, para. 508.

6.2 Provisional Arrangements of a Practical Nature (Article 83(3) of UNCLOS)

6.2.1 Relevant Aspects

6.2.1.1 Conflicting Titles

The ability for two or more States to enter into a joint development agreement of offshore hydrocarbon deposits and to implement a cooperative regime in a specific

[9] *Report by the Chairman of Negotiating Group 7 on the work of the Group at its 17th–27th meetings*, NG7/24, 14 September 1978, Document A/CONF.62/RCNG/2, UNCLOS III Off. Rec. Volume X, p. 171 [online: *ibid*]. Also see *Report of the Chairman of the negotiating group 7*, Document NG7/45, UNCLOS Off. Rec. V. XII [online: *ibid*]; *Reports of the Committees and Negotiating Groups on negotiations at the resumed seventh session contained in a single document both for the purposes of record and for the convenience of delegations*, Document A/CONF.62/RCNG/1, UNCLOS III Off. Rec. Volume X., pp. 123–124 [online: *ibid*].

[10] *Guyana/Suriname, supra note 25 in Chap. 2*, para. 460.

[11] Brown, *supra* note 26 in Chap. 3, p. 159. Edward Duncan Brown underlines that "*it can certainly do no harm to encourage restraint and practical co-operation pending the establishment of boundaries.*" On the drafting history of Article 83(3) of UNCLOS *see UNCLOS: A Commentary, supra note 36 in Chap. 5*, pp. 948–985. It underlines the concern manifested by certain States regarding the possible interpretation of this provision as introducing a moratorium prohibiting economic activities in the disputed maritime area (p. 975).

[12] Lagoni, *supra* note 70 in Chap. 2 p. 358.

maritime area pending the delimitation of maritime boundaries is strictly related with the legal title upon which these States substantiate their respective claims over the relevant continental shelf. States may only undertake activities in disputed maritime areas that they are legally entitled to. Any regime of internationalization of marine natural resources depends first and foremost on States being entitled to develop such resources.

In order to verify whether a maritime claim of one State is objected by other States, thus revealing the existence of a conflict as a result of opposing claims, it is not sufficient that a State declares that a dispute exists or that the interests of two or more States are in conflict. Likewise, the mere denial of a dispute does not demonstrate its non-existence, nor is the unilateral maritime delimitation of one State opposable to other States.[13] It is necessary that the claim of one State is positively opposed by that of another State, without implying the assertion of an express contradiction. An objection may result from the conduct of a State or through the failure to react on an occasion that called for a reaction in order to affirm or preserve title in the face of an obvious rival claim.[14] States asserting or putting forward facts and contentions to support their claims have the respective burden of proof.[15] In other words, "[a] *claim concerning a continental shelf area is an assertion to have a legal title to this particular area*".[16]

As for acts and activities developed for the purpose of improving or creating the appearance of a legal title or evidence of sovereignty or jurisdiction over maritime features, the ICJ has reiterated that these can only be considered if the other State can not make out a superior claim.[17] The Court considered that such acts and activities included, inter alia, legislative acts or acts of administrative control, acts relating to the application and enforcement of criminal or civil law, acts regulating immigration, acts regulating fishing and other economic activities, naval patrols, as

---

[13] *Delimitation of the Maritime Boundary in the Gulf of Maine Area, supra note 60 in Chap. 3*, paras. 87, 89, 112–113.

[14] *Applicability of the Obligation to Arbitrate under Section 21 of the United Nations Headquarters Agreement of 26 June 1947, Advisory Opinion, I.C.J. Reports 1988*, paras. 37–38; *Interpretation of Peace Treaties, Advisory Opinion: I.C.J. Reports 1950*, p. 74; *Case concerning Northern Cameroons (Cameroon v. United Kingdom), Preliminary Objections, Judgment of 2 December 1963: I.C.J. Reports 1963*, p. 27; *Case of the Mavrommatis Palestine Concessions, supra note 39 in Chap. 4*, pp. 11–12; *Case concerning the Temple of Preah Vihear, supra note 103 in Chap. 3*, pp. 30–33; *Aegean Sea Continental Shelf case, supra note 53 in Chap. 3*, paras. 84–85.

[15] *Case concerning the Temple of Preah Vihear, ibid*, p. 16.

[16] Rainer Lagoni, "Overlapping claims to continental shelf areas (preconditions and possible solutions)", in: *The Aegean issues: Problems and Prospects* (Foreign Policy Institute: 1989), p. 149.

[17] *Legal status of Eastern Greenland, Judgment, 1933 P.C.I.J. Series A/B, No. 53*, p. 46.

well as search and rescue operations. It also concluded that sovereignty over minor maritime features may be established on the basis of a "*modest display of State powers in terms of quality and quantity*".[18]

In this respect, the Court in the territorial and maritime dispute between Nicaragua and Colombia ruled that most acts and activities performed by the latter, in particular, legislation relating to territorial organization, regulation of fishing activities and related measures of enforcement, maintenance of lighthouses and buoys, and naval visits, were exercised in the maritime area that included all the disputed features and the respective surrounding maritime area. Furthermore, the Court noted that over the years Colombia had continuously, consistently and publicly displayed an unopposed exercise of sovereign authority in respect of the maritime features in dispute and that these facts provided very strong support for Colombia's claim of sovereignty over the same.[19] The Court also considered that although third State practice and map evidence could not be regarded as recognition of Colombia's sovereignty over the maritime features in dispute, it afforded some measure of support for its claim.[20]

A dispute of overlapping entitlements occurs when the valid legal titles upon which States sustain their respective maritime claims are in conflict, given that the recognition of a claim of one State would necessarily imply the exclusion of part or the totality of the claim put forward by another State. For example, in the case of joint development agreements, the regime initially considered by Japan, South Korea and Taiwan (that does not have statehood) could not be implemented due to China's sovereignty claim over part of the continental shelf included in the proposed joint development area. China believed to be the only holder of the right to enter into negotiations and agreements regarding the development of resources found in the Taiwanese continental shelf and that Taiwan's participation would represent the recognition of its sovereign rights.[21] Consequently, Taiwan withdrew from this tri-lateral attempt of a joint development agreement, allowing Japan and

[18] *Territorial and Maritime Dispute between Nicaragua and Honduras in the Caribbean Sea (Nicaragua v. Honduras), Judgment, I.C.J. Reports 2007 (II),* paras. 174, 176–208; *Territorial and Maritime Dispute (Nicaragua v. Colombia) 2012, supra note 53 in Chap. 3,* para. 80.

[19] *Territorial and Maritime Dispute (Nicaragua v. Colombia) 2012, ibid,* paras. 80–84. Also *see Sovereignty over Pulau Ligitan and Pulau Sipadan (Indonesia/Malaysia), Judgment, I.C.J. Reports 2002,* para. 135.

[20] *Territorial and Maritime Dispute (Nicaragua v. Colombia) 2012, ibid,* paras. 95, 102, 103. The Court eventually acknowledged that Colombia had sovereignty over the islands at Alburquerque, Bajo Nuevo, East-Southeast Cays, Quitasueño, Roncador, Serrana and Serranilla.

[21] On China's claims supporting the right to offshore development of oil and gas *see* Paul C. Yuan, "China's jurisdiction over its offshore petroleum resources"; in: 12-2 *Ocean Development and International Law* (1982), pp. 191–208. Also *see* Choon-ho Park, "Joint development of mineral resources in disputed waters: the case of Japan and South Korea in the East China Sea", in: *The South China Sea. Hydrocarbon Potential & Possibilities of Joint Development,* edited by Mark J. Valencia (Pergamon Press: Oxford, New York, Toronto, Sydney, Paris, Frankfurt, 1981), p. 1337.

South Korea to initiate negotiations towards a bilateral joint development agreement, without including any part of the conflicting continental shelf.[22]

International jurisprudence has established that "*the coast, in order to be considered as relevant for the purpose of the delimitation, must generate projections which overlap with projections from the coast of the other* [State]" and that, therefore, all parts of the coast of one State that do not overlap with the extension of the coast of the other State, due to its geographic situation, should not be considered for the purpose of delimitation.[23] Geographical and geological characteristics are also relevant factors that may further have an impact on the designation of a joint development area.[24]

Consequently, the legal entitlement of States to maritime areas and the rights resulting thereof, particularly the exclusive right to develop the resources found in the continental shelf, are attached and directly derive from the sovereignty of the State over the territory to which the continental shelf is adjacent to.[25] Indeed, the continental shelf is a legal concept connected with the principle that the land dominates the sea and that the rights of exploration and exploitation of the continental shelf derive *ispo jure* from coastal States' sovereignty over the land mass.[26] Adjacency is the paramount criterion for determining the legal status of the continental shelf, independently of the fact that the notion of natural prolongation is not considered as a legal title for the purpose of maritime delimitation, but rather as one of several circumstances that should be taken into account in order to achieve an equitable solution.[27]

---

[22] *Supra* note 3 in Chap. 5.

[23] *Case Concerning Maritime Delimitation in the Black Sea, supra note 53 in Chap. 3*, para. 99. Also *see Continental Shelf (Tunisia/Libya), supra note 10 in Chap. 3*, para. 75.

[24] Francisco Orrego Vicuña, "Regional cooperation in non-living resources: joint management zones", in: *Ocean Governance: Sustainable Development of the Seas*, edited by Peter Bautista Payoyo (United Nations University Press: Tokyo, New York, Paris, 1994), pp. 172–174.

[25] *North Sea Continental Shelf cases, supra note 46 in Chap. 3*, paras. 39–43; *Aegean Sea Continental Shelf case, supra note 53 in Chap. 3*, para. 86; *Continental Shelf (Tunisia/Libya), supra note 10 in Chap. 3*, para. 48; *Continental Shelf (Libya/Malta), supra note 60 in Chap. 3*, para. 49.

[26] *North Sea Continental Shelf cases*, *ibid*, para. 96; *Aegean Sea Continental Shelf* case, *ibid*, para. 86; *Maritime Delimitation and Territorial Questions between Qatar and Bahrain (Qatar v. Bahrain), Merits, Judgment, I.C.J. Reports 2001*, para. 185. On this last case *see* Barbara Kwiatkowska, "The Qatar v. Bahrain maritime delimitation and territorial questions case", in: 33–34 *Ocean Development and International Law Journal* (2002), pp. 227–262; *Ibid*, "The Qatar v. Bahrain maritime delimitation and territorial questions case", in: 3–6 *Maritime Boundaries* (2003), pp. 1–44; Yoshifumi Tanaka, "Reflections on maritime delimitation in the Qatar/Bahrain case", in: 52 *International and Comparative Law Quarterly* (January 2003), pp. 53–80; Maurice Mendelson, "The curious case of Qatar v. Bahrain in the International Court of Justice", in: 72 *British Yearbook of International Law* (2001), pp. 183–211.

[27] *North Sea Continental Shelf cases, supra note 46 in Chap. 3*, paras. 43, 96; *Continental Shelf (Tunisia/Libya), supra note 10 in Chap. 3*, paras. 68, 70, 73, 75; *Continental Shelf (Libya/Malta), supra note 60 in Chap. 3*, paras. 61–62; *Case Concerning Maritime Delimitation in the Black Sea, supra note 53 in Chap. 3*, paras. 77, 99.

6.2.1.2 Obligation to Make Every Effort and of Mutual-Restraint

When referring to the drafting history of Articles 74(3) and 83(3) of UNCLOS reference was made that States' duty to make every effort to enter into provisional arrangements should not be construed as an obligation to implement these interim measures.[28] Instead, States are merely required to engage in negotiations towards reaching a provisional solution in accordance with the principles of cooperation and good faith. However, despite the fact that Article 83(3) of UNCLOS merely includes a reinforced obligation to negotiate, does not necessarily imply that this obligation is void of any real legal content.[29]

The Arbitral Tribunal in the case between Guyana and Suriname considered that the duty to negotiate in good faith required States to have "*a conciliatory approach to negotiations, pursuant to which they would be prepared to make concessions in the pursuit of a provisional arrangement*", particularly in view of the fact that such provisional arrangements are by definition temporary and without prejudice of final delimitation.[30]

In this particular case, the Arbitral Tribunal concluded that both countries had violated their obligation to make every effort to enter into provisional arrangements and consequently for having jeopardised a final delimitation agreement. Specifically, Guyana for not having informed Suriname of the intention to explore the resources of the disputed maritime area and failing to provide official detailed notice of the planned activities, as well as for not having engaged Suriname earlier on in discussions concerning drilling activities. In what concerns Suriname, the Arbitral Tribunal considered that it had not engaged in dialogue with Guyana in the spirit of understanding and cooperation as required by UNCLOS, resorting instead to a threatening conduct towards the latter.[31]

The Arbitral Tribunal also concluded that within such efforts to enter into provisional arrangements, Guyana should have further attempted to seek the cooperation of Suriname regarding the undertaking of activities. Accordingly, Guyana should have offered to share the results of the exploration and given Suriname the opportunity to observe these activities, as well as to share the financial benefits gained.[32] The Arbitral Tribunal also considered that only acts

---

[28] Ian Townsend-Gault and William G. Stormont, "Offshore petroleum joint development arrangements: functional instrument? Compromise? Obligation?", in: *The Peaceful Management of Transboundary Resources*, edited by Gerald H. Blake, William J. Hildesley, Martin A. Pratt, Rebecca J. Ridley and Clive H. Schofield (Graham and Trotman/Martinus Nijhoff: London, Dordrecht, Boston, 1995), p. 58.

[29] Cameron, *supra* note 36 in Chap. 4, pp. 565–567. Also *see* Ian Townsend-Gault, "Petroleum development offshore: legal and contractual issues", in: *Petroleum Investment Policies in Developing Countries*, edited by Nicky Beredjick and Thomas Wälde (Graham and Trotman: London, 1988), p. 145; Lagoni, *supra* note 63 in Chap. 2, p. 243.

[30] *Guyana/Suriname, supra note 25 in Chap. 2*, para. 461.

[31] *Idem*, paras. 471–486, 488.

[32] *Idem*, para. 477.

that do not cause a physical change to the marine environment, such as seismic surveys, would generally be regarded as acts that, although unilateral, would not have the effect of jeopardizing or hampering a final agreement on maritime delimitation.[33]

It would seem, however, that this attempt to contextualize the obligation of States to make every effort to enter into provisional arrangements should foremost be consistent with States' discretionary power to negotiate and enter into agreements, the scope of which was analysed in Chap. 3. In fact, the discretionary nature of States' powers to enter into and establish the content and scope of provisional arrangements in conformity with international law is an essential characteristic of these regimes, in addition to their temporary character and not hindering the final delimitation of maritime boundaries.

In consideration of these observations and without foregoing States' obligation of mutual-restraint pending the final delimitation of maritime boundaries, the obligation breached in the case between Guyana and Suriname regards precisely the duty to enter into meaningful negotiations and provide information, rather than an implied recognition of an obligation to implement provisional arrangements.

6.2.1.3 Final Delimitation of Maritime Boundaries

Joint development agreements are an alternative to maritime boundary delimitation, allowing States to take part in a joint endeavour for the purpose of exploration and exploitation of hydrocarbon deposits, sharing the resources found in a specific maritime area where States' claims overlap.[34] They are essentially functional and pragmatic arrangements that do not hinder such claims.[35] Indeed, even before the entering into force of UNCLOS, the ICJ had already recognized that the

---

[33] *Idem*, paras. 466–467, 480–481.

[34] In this respect Zou Keyuan, concluding that "*joint development is a most feasible mechanism to shelve disputes and pave the way for cooperation pending settlement of territorial and/or maritime disputes over certain sea area die to overlapping claims*", identified the following characteristics included in its concept of joint development: "*1. It is an arrangement between two countries; 2. It is usually concerned with an overlapping maritime boundary area; 3. It can be used as a provisional arrangement pending the settlement of the boundary delimitation disputes between the countries concerned; 4. It is designed to jointly develop the mineral resources in the disputed area or in a defined area shared by two countries.*" Zou Keyuan, "A new model of joint development for the South China Sea", in: *Recent Developments in the Law of the Sean and China*, edited by Myron H. Nordquist, John Norton Moore and Kuen-chen Fu (Martinus Nijhoff Publishers: Leiden, Boston, 2006), p. 157; *Ibid*, *supra* note 66 in Chap. 2, p. 155.

[35] "*Nevertheless, if the interested countries have a will to set aside the formidable issue of delimitation for a while in favour of prospective economic profits to accrue from a provisional compromise settlement, they have a chance to devise a joint development scheme. It is also possible that they may defuse their tense relations by such a provisional measure for at least a certain period of time*" Miyoshi, *supra* note 2 in Chap. 4, pp. 3, 41. Also *see* Masahiro Miyoshi, "The basic concept of joint development of hydrocarbon resources on the continental shelf", in: 3-1 *International Journal of Estuarine and Costal Law* (February 1988), p. 5.

conclusion of interim agreements did not prevent the Court from rendering judgment or compelled it to dismiss a claim.[36]

UNCLOS makes a clear distinction between the implementation of provisional measures and the delimitation of maritime boundaries. The development of common offshore hydrocarbon deposits and the delimitation of maritime boundaries should not be confused, nor ought the correlation between these two considered as choosing one over the other. The assumption that delimitation is a precedent or a precondition for the development of common marine resources is not accurate. Indeed, joint development agreements signed before the delimitation of maritime boundaries do not represent a renunciation to or the attainment of any right or claim over part or the totality of a disputed maritime area, or the suspension of any right to the same, namely the sovereign and exclusive right of the coastal State to develop the resources of the continental shelf.[37] In fact, similarly to the joint development regimes included in delimitation treaties examined in Chap. 5, the successful implementation of provisional measures regarding the development of offshore hydrocarbon deposits may eventually help States achieve an agreement on the delimitation of maritime boundaries, since they no longer have to address the complex issue of exploration and exploitation of common resources. In some cases, maritime boundaries disputes eventually cease to exist or become dormant as States look towards provisional measures that do not impede their respective claims.

As previously mentioned, the lack of agreement on the delimitation of maritime boundaries does not hinder the possibility of States engaging in economic activities in disputed maritime areas. In this respect and considering that UNCLOS does not determine the form or content of provisional arrangements—particularly concerning exploration and exploitation activities, the management of the relevant maritime area, or the rights and obligations of States and of third States, before and after the delimitation of maritime boundaries—States are free to choose and adopt the provisional arrangement that best serves their purposes, as long as it does not jeopardize or hamper the final delimitation of maritime boundaries.[38]

The interim and practical nature notwithstanding, provisional arrangements are subject to international law depending on States consenting to be bound by international law. As mentioned in Chap. 3, the relevant principles of the law of treaties do not determine that the use of a more informal or, for that matter,

---

[36] *Fisheries Jurisdiction case, supra note 40 in Chap. 3*, paras. 37, 41.

[37] Alexandros Yannis, "The concept of suspended sovereignty in International Law and its implications in international politics", in: 13-5 *European Journal of International Law* (2002), pp. 1038–1052; Jonathan I. Charney, "International maritime boundaries for the continental shelf: the relevance of natural prolongation", in: *Liber Amicorum Judge Shigeru Oda*, vol. 2, edited by Nisuke Ando, Edward McWhinney and Rüdiger Wolfrum (Kluwer Law International: The Hague, London, New York, 2002), pp. 1028–1029.

[38] On provisional measures under UNCLOS *see* Natalie Klein, "Provisional measures and provisional arrangements in maritime boundary disputes", in: 21 *International Journal of Marine and Coastal Law* (2006), pp. 444–449, 460.

practical composition, has the effect of provisional arrangements pursuant to Article 83(3) of UNCLOS not being considered a treaty, i.e., "*an international agreement concluded between States in written form and governed by international law.*"[39] The practical nature of provisional arrangements does imply that these are non-binding or that claiming States and other States are not bound to certain obligations or entitled to exercise their rights and freedoms in disputed maritime areas.

6.2.2 Rights and Duties of Claiming States

The legal entitlement of States to maritime areas and the exclusive right to develop the resources found in the continental shelf are attached and directly derive from the sovereignty of States over the territory to which the continental shelf is adjacent to, regardless of its prior proclamation or occupation.[40] Only these States are entitled to explore and exploit the offshore hydrocarbon deposits found in the respective adjacent maritime areas.[41] Should a State decide not to explore and exploit these resources, no other State may do so without the latter's consent.[42] Therefore, States must make their claims known, as well as the respective legal title upon which they support their claims.

In addition to the rights of claiming States that derive from their legal entitlement to the continental shelf, too the respective duties are entailed. These include, inter alia, the obligation to protect and preserve the marine environment that will be subsequently analysed.

6.2.3 Rights and Duties of Other States

States may not implement a joint development regime that undermines the rights of other States or that constitutes a de facto appropriation of the disputed maritime areas, for example, by exercising rights that are intrinsically related with the delimitation of maritime boundaries and that do not result from States' entitlement to the disputed maritime area. This would be the case of several of the rights that costal States enjoy under the EEZ regime and that depend of the delimitation of maritime boundaries, or of claims that are not based on a valid legal title under international law.

---

[39] *Supra* note 79 in Chap. 3.

[40] Article 77 (2) *in fine*, (3) of UNCLOS.

[41] Articles 77(2), (4) and 81 of UNCLOS.

[42] *Ibid.*

Provisional arrangements pursuant to Article 83(3) of UNCLOS are subject to the limits of maritime jurisdiction and primarily those deriving from the rights of other States in disputed maritime areas, more so when there is no allocation of the disputed maritime areas to any of the claiming States. They include the legal status of the superjacent waters and of the air space above those waters, as well as the freedoms of navigation, overflight and of laying submarine cables and pipelines, or any other lawful use of the seas related to these freedoms.[43] Joint development regimes of offshore hydrocarbon deposits signed before the delimitation of maritime boundaries do not grant the respective claiming States the possibility of implementing in the respective disputed maritime areas a more stringent legal regime than that which they could in maritime areas after the delimitation of maritime boundaries, particularly regarding the aforementioned rights of other States.

Therefore, in disputed maritime areas, claiming States exercise their respective rights subject to the limits established under international law and in particular the law of the sea and which should be given full weight when implementing a joint development regime.

6.2.4 Settlement of Disputes (Part XV of UNCLOS)

One of the three basic elements previously referred of the final compromise package of Articles 74 and 83 of UNCLOS concerns the settlement of maritime delimitation disputes.[44] The outcome of the negotiations of the Convention resulted in the possibility of States excluding the submission of any dispute regarding the interpretation and application of Articles 74 and 83 of UNCLOS to the compulsory dispute settlement mechanism identified in Part XV of UNCLOS.[45] This exclusion also include the disputes concerning the non-compliance with the obligation to make every effort to enter into provisional arrangements of a practical nature and the obligation not to take unilateral action in areas of overlapping claims that could jeopardize or hamper the reaching of the final agreement on maritime delimitation. A State can jeopardize the final delimitation of maritime boundaries by engaging in drilling activities in disputed maritime areas. In this case, it would be possible to argue the latters' responsibility for breach of the duty to fulfil, in good faith, the obligations assumed under UNCLOS and to act in a manner that does not constitute an abuse of right.[46]

---

[43] Articles 58(1), 78 and 79 of UNCLOS.

[44] *Report of the Chairman of the Negotiating Group 7*, Document NG7/45, UNCLOS Off. Rec. Volume XII [online: http://untreaty.un.org/cod/diplomaticconferences/lawofthesea-1982/lawofthesea-1982.html (accessed on August 2013)].

[45] Article 298(1) lit. (a) of UNCLOS.

[46] Article 300 of UNCLOS.

Regarding the possibility of replacing the role of provisional arrangements under Articles 74(3) and 83(3) of UNCLOS with those envisaged in Article 290 of UNCLOS,[47] Natalie Klein considered that it would be doubtful that International Courts or Tribunals "*would have sufficient information and adequate expertise to create a detailed joint exploration or exploitation regime that could operate on a transitional basis.*" Klein questioned whether joint development imposed under Article 290 of UNCLOS would safeguard the rights of the relevant States, concluding that such a possibility would go beyond and contradict the substantive obligation included in Articles 74(3) and 83(3) of UNCLOS. Accordingly, International Courts and Tribunals may make recommendations to enter into joint development, reinforcing States' obligation to make every effort to enter into provisional arrangements, but not that an agreement be reached.[48] As a result, States' consent remains an indispensable element for the implementation of any provisional arrangement.

6.3 Identifying the Joint Development Area

The possibility for States to enter into a joint development agreement is closely connected with the identification of the relevant maritime area where States' claims overlap.[49] However, the overlap of entitlements may not always result in the straightforward identification of the joint development area. One of the most complex situations, for example, concerns the overlap of claims in maritime areas adjacent to islands that are disputed by two or more States, particularly as UNCLOS does not provide rules applicable to the settlement of land disputes. The legal definition and regime provided by Article 121(1) of UNCLOS[50] merely presents an unclear formula applicable to the characterization of islands and the

[47] The text of Article 290(1) of UNCLOS reads: "*If a dispute has been duly submitted to a court or tribunal which considers that prima facie it has jurisdiction under this Part or Part XI, Sect. 5, the court or tribunal may prescribe any provisional measures which it considers appropriate under the circumstances to preserve the respective rights of the parties to the dispute or to prevent serious harm to the marine environment, pending the final decision.*"

[48] Klein, *supra* note 38, pp. 452–453. Also *see* Igor V. Karaman, *Dispute Resolution in the Law of the Sea* (Martinus Nijhoff Publishers: Leiden, 2012), pp. 198–199.

[49] Mark J. Valencia, "Taming troubled waters: joint development of oil and mineral resources in overlapping claim areas", in: 23-3 *San Diego Law Review Association* (May–June 1986), p. 670; M.C.W. Pinto, "Maritime boundary issues and their resolution an overview", in: *Liber Amicorum Judge Shigeru Oda*, vol. 2, edited by Nisuke Ando, Edward McWhinney and Rüdiger Wolfrum (Kluwer Law International: The Hague, London, New York, 2002), pp. 1115–1142; Fox (et al.), *supra* note 111 in Chap. 3, p. 315; Lagoni, *supra* note 46 in Chap. 2, p. 281; Johnston/Valencia, *supra* note 38 in Chap. 4, p. 37, note 41.

[50] The text of Article 121(1) of UNCLOS reads: "*An island is a naturally formed area of land, surrounded by water, which is above water at high tide.*"

indication of the maritime areas they are allowed as a result.[51] The ambiguity of this formula is consistent with the wording used in Article 10(1) of the 1958 Convention on the Territorial Sea and Contiguous Zone, which has been considered being part of customary international law.[52]

On many occasions, territorial disputes regarding islands are motivated by the ambition of States to secure access to the marine natural resources found in the respective adjacent maritime areas. It may also be the case that islands are instrumental and, indeed, decisive to achieve a more advantageous outcome in the delimitation of maritime boundaries. In this respect, international jurisprudence has considered in several occasions that small islands should not be taken into account or given their full potential if they produce a disproportionate effect on the delimitation line under consideration and consequently have an inequitable outcome.[53] Under certain circumstances, islands may be given a partial effect in order to meet these requirements.[54]

---

[51] *See* Myron H. Nordquist, "Textual interpretation of article 121 in the UN Convention on the Law of the Sea", in: *Coexistence, Cooperation and Solidarity. Liber Amicorum Rüdiger Wolfrum*, vol. 1, edited by Holger P. Hestermeyer, Doris König, Nele Matz-Lück, Volker Röben, Anja Seibert-Fohr, Peter-Tobias Stoll and Silja Vöneky (Martinus Nijhoff Publishers: Leiden, Boston, 2012), pp. 991–1036; Clive Schofield, "The trouble with islands: the definition and role of islands and rocks in maritime delimitation", in: *Maritime Boundary Disputes, Settlement Processes, and the Law of the Sea*, edited by Seoung-Yong Hong and Jon Van Dyke (Martinus Nijhoff Publishers: 2009), pp. 21–37; Victor Prescott and Clive Schofield, *The Maritime Political Boundaries of the World* (Martinus Nijhoff Publishers: Leiden, Boston, 2005), pp. 57–89; Barbara Kwiatkowska and Alfred H.A. Soons, "Entitlement to maritime areas of rocks which cannot sustain human habitation or economic life of their own", in: 21 *Netherlands Yearbook of International Law* (1990), pp. 139–181.

[52] In the case between Qatar and Bahrain the ICJ did not specifically refer to Article 121(3) of UNCLOS but did observed that the entitlement to maritime rights accorded to an island in paragraph 2 is expressly limited by reference to paragraph 3. In the case between Nicaragua and Colombia, the Court considered that the legal regime of islands was an indivisible regime and had the status of customary international law. *See Maritime Delimitation and Territorial Questions between Qatar and Bahrain, supra note 26 in Chap. 6*, paras. 167, 185 and 195; *Territorial and Maritime Dispute (Nicaragua v. Colombia) 2012, supra note 53 in Chap. 3*, para. 139.

[53] *North Sea Continental Shelf cases, supra note 46 in Chap. 3*, para. 57; *Continental Shelf (Libya/Malta), supra note 60 in Chap. 3*, paras. 53–54, 64–73; *Maritime Delimitation and Territorial Questions between Qatar and Bahrain, supra note 26 in Chap. 6*, paras, 185, 219; *Case Concerning Maritime Delimitation in the Black Sea, supra note 53 in Chap. 3*, paras. 102, 166–168, 179–188. Also *see Maritime Delimitation in the Area between Greenland and Jan Mayen, supra note 33 in Chap. 4*, paras. 68, 69, and *Separate Opinion of Judge Schwebel*, para. 128; *Delimitation of the Maritime Boundary in the Gulf of Maine Area, supra note 60 in Chap. 3*, para. 157; *Land and Maritime Boundary between Cameroon and Nigeria, supra note 83 in Chap. 3*, para. 301.

[54] *Continental Shelf (Tunisia/Libya), supra note 10 in Chap. 3*, paras. 129–131. Also *see Arbitration between the United Kingdom of Great Britain and Northern Ireland and the French Republic on the Delimitation of the Continental Shelf, Decision of the Court of Arbitration dated 30 June 1977*, reproduced in: 54 *International Law Reports* 1979, p. 6, paras. 194–203. On the relevance given to islands for the purpose of maritime delimitation *see* Derek W. Bowett, "The arbitration between the United Kingdom and France concerning the continental shelf boundary in the English Channel and South-Western approaches", in: 49 *British Yearbook of International Law*

Additionally, not every feature constitutes an island and consequently is permitted full maritime areas.[55]

As a result, International Courts and Tribunals are entrusted with the task of determining what constitutes an island or a rock, and what should their relevance be for the purpose of delimitation. For example, the Arbitral Tribunal in the arbitration between the United Kingdom and France concluded that:

> The case of the Channel Islands must, in view of the Court, be differentiated from that of the rocks or small islands which figure in some of the precedents canvassed by the Parties in their pleadings. Possessing a considerable population and a substantial agricultural and commercial economy, they are clearly territorial and political units which have their own separate existence, and which are of a certain importance in their own right separately from the United Kingdom.[56]

This Arbitral Award notwithstanding, it is not completely unlikely that International Courts and Tribunals would avoid the difficult issue of determining the legal status of a feature, as it happened in the *Case Concerning Maritime Delimitation in the Black Sea (Romania v. Ukraine)* in which the ICJ considered that the confirmation of whether Serpents' Island was in fact an island was not relevant for the outcome of the case.[57]

In what concerns disputes between entitlements of different natures, such as the overlap between the territorial sea-entitlement derived from an island and the entitlement to a continental shelf and EEZ, the ICJ has considered that, as coastal States have sovereignty over the seabed and water column in its territorial sea and

(Footnote 54 continued)
(1978), pp. 1–29. Also *see* Robert Beckman and Clive Schofield, "Moving beyond disputes over island sovereignty: ICJ decision sets stage for maritime boundary delimitation in the Singapore Strait", in: 40 *Ocean Development and International Law* (2009), pp. 1–35; Clive Schofield and Dustin Kuan-Hsiung Wang, "The regime of islands under UNCLOS: implications for the South China Sea", in: *Maritime Energy Resources in Asia: Legal Regimes and Cooperation*, edited by Clive Schofield (The National Bureau of Asian Research: Seattle, 2012), pp. 61–77.

[55] The text of Article 121(2) and (3) of UNCLOS reads: "*2. Except as provided for in paragraph 3, the territorial sea, the contiguous zone, the exclusive economic zone and the continental shelf of an island are determined in accordance with the provisions of this Convention applicable to other land territory. 3. Rocks which cannot sustain human habitation or economic life of their own shall have no exclusive economic zone or continental shelf.*" On the interpretation of this provision, see Yann-Huei Song, "Article 121(3) of the Law of the Sea Convention and the disputed offshore islands in East Asia: a tribute to Judge Choon-Ho Park", in: Governing Ocean Resources. New Challenges and Emerging Regimes. A Tribute to Judge Choon-Ho Park, edited by Jon M. Van Dyke, Sherry P. Broder, Seokwoo Lee and Jun-Hyun Paik (Martinus Nijhoff Publishers: Leiden, Boston, 2013), pp. 61–97.

[56] *Arbitration between the United Kingdom of Great Britain and Northern Ireland and the French Republic*, *supra* note 54, para. 184.

[57] *Case Concerning Maritime Delimitation in the Black Sea, supra note 53 in Chap. 3*, paras. 180, 184, 187. Also *see Territorial and Maritime Dispute (Nicaragua v. Colombia) 2012, supra note 53 in Chap. 3*, para. 180; *Aegean Sea Continental Shelf case, supra note 53 in Chap. 3*, para. 83; *Maritime Delimitation and Territorial Questions between Qatar and Bahrain, supra note 26 in Chap. 6*, paras. 191–195, 201, 219.

merely enjoy specific rights, rather than sovereignty, with respect to the continental shelf and EEZ; in situations of overlap between different titles, more weight should be given to the sovereignty over the territorial sea, rather than to the sovereign rights and jurisdiction in its EEZ and continental shelf.[58]

Also in these cases, joint development agreements can exert particular relevance. For example, the dispute between Iceland and Norway regarding the delimitation of maritime boundaries in the area between Iceland and Jan Mayen had in its core the divergence regarding the characterization of Jan Mayen. While Iceland enacted legislation in 1979 that established a 200 nm continental shelf and EEZ close to Jan Mayen; Norway, the following year, approved a law that established a fishing zone with a maximum extension of 200 nm from Jan Mayen. Both States eventually managed to agree on a fishing regime, which resulted in Iceland having a fishing area with 200 nm where Norway could not fish, even though without prejudice of its fishing rights in Jan Mayen. These States also agreed on establishing a Conciliation Commission for the purpose of making recommendations on the delimitation of the continental shelf between Iceland and Jan Mayen.[59] As a result and subsequent to the recommendations of the Commission, Iceland and Norway signed an agreement which established a 200 nm continental shelf in favour of Iceland and roughly 90 nm for Jan Mayen.[60]

---

[58] Bangladesh/Myanmar, *supra* note 71 in Chap. 3, para. 169; *Territorial and Maritime Dispute (Nicaragua v. Colombia) 2012, supra note 53 in Chap. 3*, para. 178.

[59] *Agreement between Iceland and Norway concerning Fishery and Continental Shelf questions between Iceland and Jan Mayen*, done on 28 May 1980 and entered into force on 13 June 1980; *Report and Recommendations to the Governments of Iceland and Norway of the Conciliation Commission on the Continental Shelf Area between Iceland and Jan Mayen*, done on June 1981 [online: http://untreaty.un.org/cod/riaa/cases/vol_XXVII/1-34.pdf (accessed on August 2013)], reproduced in: 20 *International Legal Materials* (July 1981), pp. 797–842. *See International Maritime Boundaries*, vol. 2, *supra note 125 in Chap. 3*, pp. 1755–1760. Edward Duncan Brown underlines the fact that the Commission recommended a joint development area instead of a line of delimitation, which was in fact the Commissioń's mandate, concluding that "[i]*t would be a mistake to treat this case as being akin to a delimitation case decided judicially in accordance with international law*" and that "*it would be more accurate to regard the whole settlement process a seamless web which started in direct diplomatic negotiations, continued through conciliation process in which nominees of the parties were joined by an independent chairman, and ended in a bilateral treaty based on the fruits of the conciliation process.*" Adding, that "*this case offers a good example of the flexibility of the conciliation process, where a commission is free to draw upon approaches and methods sanctioned by law but equally free to consider extra-legal methods where a recommendation ex aequo et bono might lead to what the parties will consider to be an acceptable equitable solution.*" Brown, *supra* note 26 in Chap. 3, pp. 177–179.

[60] *Agreement on the Continental Shelf between Iceland and Jan Mayen*, done at Oslo on 22 October 1981 and entered into force on 2 June 1982, reproduced in: *International Maritime Boundaries*, vol. 2, *supra note 125 in Chap. 3*, pp. 1762–1765. This Agreement favoured Iceland due to the economic differences between the latter and Norway at the time of the execution of the agreement and also considering Norway's advantage regarding know-how of offshore production acquired with the production of oil and gas in the North Sea. The Agreement established that the costs of surveys would be solely borne by Norway and that the implementation of the gathered data towards the creation of the joint development area would be carried out by Norwegian

The agreement further implemented the recommendations of the Commission regarding geological, seismic and magnetic survey and joint development of hydrocarbon deposits found in an area located almost in its entirety beyond the median line between Iceland and Jan Mayen, and significantly beyond the maritime boundary delimitated in the 1981 agreement.[61]

Another example of attempts made to implement joint development regimes in maritime areas adjacent to disputed islands is the case of the Memorandum of Understanding (MoU) signed by Iran and Sharjah. This agreement laid the conditions for the joint occupation of the island of Abu Musa and the delimitation of a territorial sea of 12 nm.[62] This MoU also recognized the rights granted by Sharjah for the exploration of the oil resources of the seabed and subsoil of the territorial sea of Abu Musa and established the equal sharing of the respective revenues between the two countries, as well as that nationals of Iran and Sharjah would have equal fishing rights in the territorial sea of Abu Musa.[63]

The prospects of joint development of offshore hydrocarbon deposits also provided the United Kingdom and Argentina a way forward and an opportunity to resume relations following the 1982 armed conflict over the sovereignty of the Falkland Islands (Malvinas) and the islands of South Georgia and South Sandwich. On 27 September 1995, the two countries signed a joint declaration that established cooperation on offshore activities in the Southwest Atlantic Ocean, while excluding the maritime areas of South Georgia and South Sandwich islands and adopting the sovereignty formula approved during the diplomatic meetings held in Madrid, in 1989. The scope of this declaration was to encourage offshore coordinated activities regarding the exploration and exploitation of hydrocarbon deposits, according to the best practices and commercial principles of the oil

(Footnote 60 continued)
authorities. In the event that any of the information collected during the aforementioned surveys would have economic value, the benefits deriving from such interest would be shared by the states in the agreed proportion. *See* Willy Østreng, "Reaching agreement on international exploitation of ocean mineral resources (with special reference to the joint development area between Jan Mayen and Iceland)", in: *Geology and Hydrocarbon Potential of the South China Sea and Possibilities of Joint Development*, edited by Mark J. Valencia (Pergamon Press: New York, Oxford, Toronto, Sydney, Paris, Frankfurt, 1985), pp. 555–571; Elliot L. Richardson, "Jan Mayen in perspective", in: 82 *American Journal of International Law* (1988), pp. 443–458.

[61] On the agreement *see* Miyoshi, *supra* note 2 in Chap. 4, pp. 34–35.

[62] *Memorandum of Understanding between Iran and Sharjah concerning the island of Abu Musa*, done on 24 November 1971, reproduced in: Ali A. El-Hakim, *The Middle Eastern States and the Law of the Sea* (Manchester University Press: Manchester, 1979), pp. 208–211. On the memorandum *see* Miyoshi, *ibid*, pp. 10–12.

[63] Iran occupied a group of islands in the Persian Gulf in 1971, including the island of Abu Musa which was also partially occupied by the emirate Sharjah. This situation resulted in both countries claiming sovereignty on the adjacent maritime areas of Abu Musa. The Iranian law on exploration and exploitation of the continental shelf, dated 19 May 1949, declared Iranian jurisdiction over all natural resources of the seabed and subsoil of the continental shelf of the Persian Gulf and the Gulf of Oman. *See* Kamyar Mehdiyoun, "Ownership of oil and gas resources in the Caspian Sea" *in* 94-1 *American Journal of International Law* (January 2000), p. 181.

industry. For this purpose, Argentina and the United Kingdom created a joint commission composed of delegations of both countries that would meet at least twice yearly to decide, by consensus, on matters regarding the implementation of such cooperative activities. These included making recommendations on the protection of the marine environment, analyse requests by operators for the exploration and exploitation of hydrocarbon deposits and agree on the tax regime applicable to these activities, as well as on the promotion of joint ventures between operators of both countries and proposing the unitization of oil fields. However, after a considerable reflection period that started in the year 2000, Argentina withdrew from the joint statement in March 2007 due to differences regarding the maritime area where such activities should take place.[64]

An important development in State practice regarding potential areas for joint development is the agreement implemented by Mauritius and Seychelles in respect of mineral resources found in the continental shelf beyond 200 nm. After a decade of negotiations on maritime boundaries and the signing of the agreement on the delimitation of the EEZ, these States agreed on entering into a provisional arrangement pursuant to Article 83(3) of UNCLOS.[65] Simultaneously, these States signed a treaty that established the outer limit boundaries of the extended continental shelf.[66] The latter agreement followed the successful joint submission of Mauritius and Seychelles to the Commission on the Limits of the Continental Shelf (CLCS),[67] which resulted in establishing joint jurisdiction over an additional area

---

[64] On the conflict between Argentina and the United Kingdom regarding the Falkland Islands (Malvinas) and the precedents and prospects for joint development *see* Vasco Becker-Weinberg, "Recalling the Falkland Islands (Malvinas) Sovereignty Formula", in: 27 *Ocean Yearbook* (2013), pp. 411–433.

[65] *Treaty Concerning the Joint Management of the Continental Shelf in the Mascarene Plateau Region between the Government of the Republic of Seychelles and the Government of the Republic of Mauritius*, done at Vacoas on 13 March 2012 and entered into force on 18 June 2012, reg. n. 49783 U.N.T.S., 11 July 2012 [online: http://www.mfa.gov.sc/uploads/files/filepath_45.pdf (accessed on August 2013)].

[66] *Treaty Concerning the Joint Exercise of Sovereign Rights Over the Continental Shelf in the Mascarene Plateau Region*, done at Vacoas on 13 March 2012 and entered into force on 18 June 2012, reg. n. 49782 U.N.TS., 11 July 2012 [online: http://www.mfa.gov.sc/uploads/files/filepath_44.pdf (accessed on August 2013)].

[67] On 1 December 2008, Mauritius and the Seychelles jointly submitted to the CLCS information on the limits of their continental shelf beyond 200 nm *see Receipt of the Joint Submission made by the Republic of Mauritius and the Republic of Seychelles to the Commission on the Limits of the Continental Shelf, 12 December 2008*, UN Doc. CLCS.14.2008.LOS (Continental Shelf Notification) [online: http://www.un.org/depts/los/clcs_new/submissions_files/musc08/clcs14_2008e.pdf (accessed on August 2013)]. The original joint submission was revised on 12 March 2010 concerning the Mascarene Plateau region [online: http://www.un.org/depts/los/clcs_new/submissions_files/musc08/SMS-ES-DOC_Rev.pdf (accessed on July 2013)]. The revised submission only represented a partial submission in respect of a portion of the extended continental shelf claimed by each of the two States. The submissions for the other portions will be made by each of the two States individually. The CLCS on 30 March 2011 adopted the recommendations on the limits of the continental shelf and recognized the legal entitlement of Mauritius and the Seychelles to delineate the outer limits of the continental shelf

of 396,000 km^2 in the Indian Ocean.[68] These agreements did not settle the maritime boundary dispute between the two States, seeing that Mauritius and Seychelles still have to reach an agreement on the final delimitation of the maritime boundaries in the continental shelf beyond 200 nm.

Another significant factor to consider regarding the possibility of joint development in respect of mineral resources found in the continental shelf beyond 200 nm, is the recognition by ITLOS in its first maritime delimitation case, between Myanmar and Bangladesh, that it would not be "*unusual*" for States to enter into agreements involving transboundary resources in areas of overlapping claims beyond 200 nm in the outer continental shelf.[69] In fact, the Tribunal's decision included the delimitation of maritime boundaries in the continental shelf beyond 200 nm, even before these States had made their submissions to the CLCS. In this respect, ITLOS considered that the institutional framework established in UNCLOS entailed that the activities of the different bodies were complementary to each other, to ensure coherent and efficient implementation of the Convention.[70] The Tribunal concluded that:

> There is a clear distinction between the delimitation of the continental shelf under article 83 and the delineation of its outer limits under article 76. Under the latter article, the Commission is assigned the function of making recommendations to coastal States on matters relating to the establishment of the outer limits of the continental shelf, but it does so without prejudice to delimitation of maritime boundaries. The function of settling disputes with respect delimitation of maritime boundaries is entrusted to dispute settlement procedures under article 83 and Part XV of the Convention, which include international courts and tribunals.[71]
>
> Just as the functions of the Commission are without prejudice to the question of delimitation of the continental shelf between States with opposite or adjacent coasts, so the exercise by international courts and tribunals of their jurisdiction regarding the delimitation of maritime boundaries, including that of the continental shelf, is without prejudice to the exercise by the Commission of its functions on matters related to the delineation of the outer limits of the continental shelf.[72]
>
> In the view of the Tribunal, it would be contrary to the object and purpose of the Convention not to resolve the existing impasse. Inaction in the present case, by the Commission and the Tribunal, two organs created by the Convention to ensure the

(Footnote 67 continued)
beyond their 200 nm limits in the Mascarene Plateau region, thus recognizing that these two countries had an additional 396,000 km^2 of the Indian Ocean *see Summary of the Recommendations of the Commission on the Limits of the Continental Shelf in regard to the Joint Submission made by Mauritius and Seychelles concerning the Mascarene Plateau region on 1 December 2008*, p. 5, para. 19 and p. 21, paras. 59 to 62 [online: http://www.un.org/depts/los/clcs_new/submissions_files/musc08/sms08_summary_recommendations.pdf (accessed on July 2013)].

68 *See Statement by the Ministry of Foreign Affairs of the Seychelles* [online: http://www.mfa.gov.sc/static.php?content_id=36&news_id=278 (accessed on July 2013)].

69 Bangladesh/Myanmar, *supra* note 71 in Chap. 3, para. 472.

70 *Ibid*, para. 373.

71 *Ibid*, para. 376.

72 *Ibid*, para. 379.

> effective implementation of its provisions, would leave the Parties in a position where they may be unable to benefit fully from their rights over the continental shelf.[73]
>
> The Tribunal observes that the exercise of its jurisdiction in the present case cannot be seen as an encroachment on the functions of the Commission, inasmuch as the settlement, through negotiation, of disputes between States regarding delimitation of the continental shelf beyond 200 nm is not seen as precluding examination by the Commission of the submissions made to it or hindering it from issuing appropriate recommendations.[74]

Therefore, the joint development agreement between Mauritius and Seychelles, following the successful submission and the decision of ITLOS in the Myanmar and Bangladesh case, constitutes an important development in international law and in particular in State practice regarding the consideration of possible areas for joint development.

6.4 Protection and Preservation of the Marine Environment in Disputed Maritime Areas

6.4.1 Pollution from Seabed Activities

Until the entering into force of UNCLOS, the protection and preservation of the marine environment had not received the legal substance that is currently recognizable in the transversal and multidisciplinary character of its normative framework.[75] The Convention altered the perception of the relation between States and

---

[73] *Ibid*, para. 392.

[74] *Ibid*, para. 393.

[75] *See* Catherine Redgwell, "International Environmental Law", in: *International Law*, edited by Maclom D. Evans (Oxford University Press: Oxford, 2010), pp. 687–721; Elisa Morgera, *Corporate Accountability in International Environmental Law* (Oxford University Press: Oxford, 2009), pp. 11–18; Peter H. Sand, "The evolution of international environmental law", in: *The Oxford Handbook of International Environmental Law*, edited by Daniel Bodansky, Jutta Brunnée and Ellen Hey (Oxford University Press: Oxford, 2007), pp. 30–43; Malgosia Fitzmaurice, "International Protection of the Environment", in: 293 *Recueil de Cours de l'Académie de Droit Internationale de La Haye* (2001), pp. 22–47; Hugo Caminos, "Enforcement jurisdiction under the United Nations Convention on the Law of the Sea. An overview", in: *Coexistence, Cooperation and Solidarity. Liber Amicorum Rüdiger Wolfrum*, vol. 1, edited by Holger P. Hestermeyer, Doris König, Nele Matz-Lück, Volker Röben, Anja Seibert-Fohr, Peter-Tobias Stoll and Silja Vöneky (Martinus Nijhoff Publishers: Leiden, Boston, 2012), pp. 737–776; David Ong, "The progressive integration of environmental protection within offshore joint development agreements", in: *Exploitation of Natural Resources in the 21st Century*, edited by Malgosia Fitzmaurice and Milena Szuniewicz (Kluwer Law International: The Hague, London, New York, 2003), pp. 113–141; Tullio Treves, "Codification du droit international et pratique des États dans le droit de la mer", in: 223 *Recueil de Cours de l'Académie de Droit Internationale de La Haye* (1990), pp. 171–176; Jon Van Dyke, "Giving teeth to the environmental obligations in the LOS Convention", in: *Oceans Management in the 21st Century: Institutional Frameworks and Responses*, edited by Alex G. Oude Elferink and Donald R. Rothwell (Martinus Nijhoff Publishers: Leiden/Boston: 2004), pp. 167–186; Rüdiger Wolfrum, "Means of ensuring

the seas and oceans, giving way to the awareness of the interconnectivity of different marine ecosystems and the understanding of the duty to cooperate as a fundamental principle in the prevention of pollution of the marine environment. In its *Preamble*, UNCLOS refers to a legal order for the seas and oceans that will simultaneously promote the efficient utilization of marine natural resources and the protection and preservation of the marine environment, entrusting all States to cooperate to this end on a global and regional basis.[76] Article 192 of UNCLOS specifically provides that all "*States have the obligation to protect and preserve the marine environment*", without making any distinction between States or maritime areas, namely those within and beyond national jurisdiction, or those that are disputed by two or more States. Moreover, UNCLOS introduced the concern for the adoption of preventive measures and the control of the sources of pollution, rather than only dealing with its consequences and the responsibility for damaging or causing harm to the marine environment.

In contradistinction, the 1958 Convention on the Continental Shelf merely provided that the exploration of this maritime area and the exploitation of its resources should not unjustifiably interfere with the conservation of living resources and that the coastal State was obliged to undertake in the safety zones created around offshore installations all appropriate measures for the protection of the living resources of the sea from harmful agents.[77] With the 1958 Convention on the High Seas there was a clear reference to the prevention of pollution by discharge of oil from ships or pipelines or resulting from the exploration and exploitation of the seabed and subsoil. This Convention also referred to the adoption of measures and international cooperation to prevent pollution from dumping of radioactive waste.[78]

Over the years, several non-binding and programmatic instruments have contributed to increase awareness for the need to protect and preserve the marine environment, essentially by making recommendations and providing guidelines

(Footnote 75 continued)
compliance with and enforcement of international environmental law", in: 272 *Recueil de Cours de L'Académie de Droit Internationale de La Haye* (1998), pp. 9–154; Alexander Yankov, "The concept of protection and sustainable development in the marine environment", in: 18 *Ocean Yearbook* (2004), pp. 267–283; Sandrine Maljean-Dubois and Lavanya Rajamani, *La mise en oeuvre du droit international de l'environnement/Implementation of International Environmental Law* (Martinus Nijhoff Publishers: Leiden, Boston, 2011), pp. 484–488; Sands, *supra* note 93 in Chap. 5, pp. 3–122, 391–458; Rothwell/Stephens, *supra* note 26 in Chap. 3, pp. 338–382; Churchill/Lowe, *supra* note 68 in Chap. 2, pp. 328–399; Tanaka, *supra* note 23 in Chap. 3, at 253–311; Brown, *supra* note 26 in Chap. 3, at 336–397.

[76] *See* Articles 197–199, 204, 208(1), (2), (4) and (5), 211(1), 214, and 217(1) of UNCLOS. Also *see The MOX Plant Case (Ireland v. United Kingdom), Request for provisional measures, I.T.L.O.S. No.10, 2001,* para. 82; *Case Concerning Land Reclamation by Singapore in and around the Straits of Johor (Malaysia v. Singapore), Provisional Measures, I.T.L.O.S. No.12, 2003*, para. 92; *Southern Bluefin Tuna Cases (New Zealand v. Japan; Australia v. Japan), Requests for Provisional Measures, Order, I.T.L.O.S. Nos. 3 and 4, 1999*, para. 78.

[77] Article 5(1) and (7) of the 1958 Convention on the Continental Shelf.

[78] Articles 24 and 25 of the 1958 Convention on the High Seas.

and also promoting the adoption of regional and global measures for the safeguard and management of the marine environment. Most of these instruments had a ground-breaking effect and received a wide acceptance due to their non-binding nature, thus allowing for swift adoption and also the avoidance of a complex ratification procedure. Some of the guidelines included in these instruments eventually contributed towards the formation of international customary law regarding the protection and preservation of the marine environment and eventually were integrated in binding international legal instruments, such as in Part XII of UNCLOS, which received a significant degree of acceptance.[79] This is the case, for example, of the *Declaration of the 1972 United Nations Conference on the Human Environment* which affirmed in Principle 21 the sovereign right of every State to exploit their resources pursuant to their own environmental policies. It further recognized the responsibility of every State to ensure that activities within their jurisdiction or control would not cause damage to the environment of other States or the areas beyond the limits of national jurisdiction. This Declaration also stated in Principles 22 and 24 that States should cooperate to further develop international law regarding liability and compensation for the victims of pollution and other environmental damage caused by activities within the jurisdiction or control of such States to areas beyond their jurisdiction, as well as to effectively control, prevent, reduce and eliminate adverse environmental effects, having due regard for the sovereignty and interests of all States. The said Declaration additionally emphasized the importance of sharing "*in the best spirit of co-operation and good-neighbourliness*"[80] technical data relating to the work to be carried out by States within their national jurisdiction with a view of avoiding significant harm that may occur in the environment of the adjacent areas.[81]

---

[79] Patricia Birnie, Alan Boyle and Catherine Redgwell, *International Law and the Environment* (Oxford University Press: Oxford, New York, 2009), p. 387; Pierre-Marie Dupuy, "Formation of customary international law and general principles", in: *The Oxford Handbook of International Environmental Law*, edited by Daniel Bodansky, Jutta Brunnée and Ellen Hey (Oxford University Press: Oxford, 2007), pp. 455–456. Other examples after the entry into force of UNCLOS include: the 1995 Fish Stocks Agreement and the reference made in its *Preamble* to Chap. 17 of *Agenda 21*, concerning the sustainable use and conservation of marine living resources of the high seas (*see* Sects. 17.44–17.69); the *Protocol to the Convention on the Prevention of Marine Pollution by Dumping of Wastes and Other Matter of 29 December 1972*, done at London on 7 November 1996; and the Convention on Biological Diversity.

[80] Resolution (UNGA) n. 2995 (XXVII), 15 December 1972. Also *see* Article 74 of the UN Charter.

[81] The Declaration was approved during the first global environment conference, the 1972 *United Nations Conference on the Human Environment* held in Stockholm *see* A/CONF.151/26 (Vol. I), *Report of the United Nations Conference on the Human Environment, Stockholm, 5–16 June 1972* (United Nations publication, Sales No. E.73.II.A.14. and corrigendum), chap. I. Also *see* Resolution (UNESC) n. 1346 (XLV), 30 July 1968 and Resolution (UNGA) n. 2398 (XXIII), 3 December 1968, recommending and calling the United Nations to convene a *Conference on the Human Environment*; and Article 3 of the *Charter of Economic Rights and Duties of States*, adopted by the Resolution (UNGA) n. 3281 (XXIX), 12 December 1974.

The said Principle 21 was reaffirmed by Resolution (UNGA) n. 44/228, adding that States should contribute towards the preservation and protection of the global and regional environment in accordance with their capacities and specific responsibilities, and that they are responsible for the damage to the environment and natural resources caused by activities within their jurisdiction or control through transboundary interference. To this end, the Resolution supported the notion of action plans for the common management of eco-systems and critical environmental problems at the national, regional and global levels.[82]

In 1992, the United Nations held the *Conference on the Environment and Development* in Rio de Janeiro approving the well-known *Agenda 21* and which essentially provided an outline for sustainable development. Chapter 17 of *Agenda 21* included non-binding recommendations and guidelines relating to the pollution of the marine environment from various sources, as well as on how to develop strategies and plans to protect and preserve the marine environment at the global, regional and national levels.[83] The major contribution of *Agenda 21* was that it emphasized the need for an integrated approach to marine and coastal areas, having coastal States committed themselves to an integrated management and sustainable development of coastal areas and the marine environment under their national jurisdiction, particularly through international and regional cooperation and coordination.[84] However, in addition to not addressing the issue of the sustainable development of disputed maritime areas, like most non-binding instruments, *Agenda 21* did not include any reference to dispute settlement mechanism in the event of environmental damage. It merely referred in Sect. 17.30(C) that

[82] Section I(7), (8), (12) of the Resolution (UNGA) n. 44/228, 22 December 1989. This Resolution decides to convene the 1992 *United Nations Conference on Environment and Development*. Also *see* Resolutions (UNGA) n. 1803 (XVII), 14 December 1962; n. 3201 (S-VI), 1 May 1974; n. 3362 (S-VII), 16 September 1975; n. 1514 (VX), 14 December 1960; n. 1515 (XV), 15 December 1960; and n. 626 (VII), 21 December 1952.

[83] Chapter 17 of *Agenda 21* (*Protection of the Oceans, all kinds of Seas, Including Enclosed & Semi-enclosed Seas, & Coastal Areas & the Protection, Rational Use & Development of their Living Resources*). *Agenda 21* was adopted during the 1992 *United Nations Conference on Environment and Development* that took place in Rio de Janeiro, of which also resulted the *Rio Declaration on Environment and Development* (A/CONF.151/26, Vol. I), the *Statement of Forest Principles*, the *United Nations Framework Convention on Climate Change* (A/CONF.151/26, Vol. III) and the Convention on Biological Diversity. In 1997 the United Nations adopted the *Programme for the Further Implementation of Agenda 21*, (UNGA Doc. A/RES/S-19/2, 19 September 1997).

[84] *Agenda 21* was followed by the *Johannesburg Declaration on Sustainable Development* adopted at the *2002 World Summit on Sustainable Development* and the *Plan of Implementation of the World Summit on Sustainable Development*, which recognized that ensuring the sustainable development of the oceans requires an effective global and regional coordination and cooperation, acknowledging in this respect the importance of States ratifying or acceding and implementing UNCLOS and also the relevance of Chap. 17 of *Agenda 21* for achieving sustainable development. *See Report of the World Summit on Sustainable Development, Johannesburg, South Africa, 26 August–4 September 2002* (United Nations publication, Sales No. E.03.II.A1 and corrigendum), chap. I, resolution 1, annex. Also *see* Resolution (UNGA) n. 55/1999, 20 December 2000.

States should assess existing regulatory measures to address discharges, emissions and safety and the need for additional measures regarding the degradation of the marine environment from offshore oil and gas platforms.

UNCLOS refers in Articles 208 and 209 to the regime regarding pollution from seabed activities subject to national jurisdiction and in the Area. According to Article 208(1) of UNCLOS, "[c]*oastal States shall adopt laws and regulations to prevent, reduce and control pollution of the marine environment arising from or in connection with sea-bed activities subject to their jurisdiction and from artificial islands, installations and structures under their jurisdiction, pursuant to articles 60 and 80.*" However, similarly to the high seas, there is no body of law that establishes jurisdiction over offshore oil rigs that are located in disputed maritime areas. In fact, there are no exceptions to the rule of exclusive jurisdiction of States over installations and structures located in the high seas, as it occurs, for example, regarding ships and vessels.[85] As a result, the applicable law should be that of the authorizing State exercising its freedom to construct installations and structures.[86] Indeed, since States may not appropriate themselves of disputed maritime areas pending the delimitation of boundaries, jurisdiction regarding oil rigs located in these areas ought to be based on the entitlement of the relevant States to grant the right to develop the resources of the seabed, including the right to consent, authorize and regulate the construction, operation, use and decommission of oil rigs and other installations necessary for the exploration and exploitation activities.

In addition, in disputed maritime areas, the exclusive effect of the said underlying rule of entitlement should also be given, mutatis mutandis, some consideration,[87] particularly concerning the non-admissibility of neighbouring States intervening in the area lawfully claimed by one or more States. Correspondingly, there is a correlative obligation of coastal States to protect and

---

[85] Article 87(1) lit. (d) of UNCLOS.

[86] *S.S. Lotus (France v. Turkey), 1927 P.C.I.J., Series A, No. 10,* pp. 18–19: "*Now the first and foremost restriction imposed by international law upon a State is that—failing the existence of a permissive rule to the contrary—it may not exercise its power in any form in the territory of another State. In this sense jurisdiction is certainly territorial; it cannot be exercised by a State outside its territory except by virtue of a permissive rule derived from international custom or from a convention.*" *See Award between the United States and the United Kingdom relating to the rights of jurisdiction of United States in the Bering's sea and the preservation of fur seals, decision of 15 August 1893*, reproduced in: *United Nations Reports of International Arbitral Awards, vol. XXVIII*, pp. 263–276. Also *see* Hossein Esmaeili, *The Legal Regime of Offshore Oil Rigs in International Law* (Dartmouth/Ashgate: Aldershot, Burlington USA, Singapore, Sydney, 2001), p. 88.

[87] *Case concerning Right of Passage over Indian Territory (Merits), Judgement of 12 April 1960, I.C.J. Reports 1960*, p. 11: "*Whereas the international legal system is essentially based upon mutual respect of sovereignties; Whereas, in order to justify the discretionary power which the Union of India claims to possess in respect of Portuguese transit, it would be necessary to accept that, while recognizing the sovereignty of Portugal over the enclaves, it tacitly reserved a right at its will to render impossible the exercise of that sovereignty; Whereas such a reservation cannot logically be admitted and would be contrary to the elementary requirements of good faith.*"

preserve the marine environment in these areas in the terms previously described.[88] It is the case where, in disputed maritime areas, neighbouring States are subject to the maxim "*cui non licet quod est minus, utique non licet quod est plus*" (who cannot do less, cannot do more).

Therefore, States that authorize and regulate the construction, operation, use and decommission of oil rigs and other installations necessary for the exploration and exploitation of offshore hydrocarbon deposits in disputed maritime areas, should also enforce the relevant obligations regarding the protection and preservation of the marine environment from pollution caused by seabed activities, including those resulting from international legal instruments that the relevant States are party to, such as those previously examined in Chap. 5. It would be contrary to the general principles of international law and constitute an abuse of rights if States were not bound to the obligation to protect and preserve the marine environment in disputed maritime areas but were, nonetheless, entitled to develop its resources. This would result in coastal States being subject to a more stringent legal regime in areas included in their national jurisdiction, than in those areas that are not subject to a designated coastal State.

6.4.2 Obligation to not Cause Transboundary Harm or Damage

Under UNCLOS, the protection and preservation of the marine environment is a general but not a vague obligation. This is also the case of transboundary effects of pollution and the obligation of States to prevent transboundary harm or damage.[89]

It has been the standing of international law that no State has the right to use or permit the use of its territory in such a manner as to cause injury to the territory of another State,[90] including, to a certain extent, when States claim that they were unaware that their territory had been used in such a manner.[91] Thus the adage: *sic*

---

[88] *The Island of Palmas Case (or Miangas) (US v. Netherlands), Award of the Tribunal, P.C.A., 4 April 1928*, para. 839: "*Territorial sovereignty, as has already been said, involves the exclusive right to display the activities of a State. This right has as corollary a duty: the obligation to protect within the territory the rights of other States, in particular their right to integrity and inviolability in peace and in war, together with the rights which each State may claim for its nationals in foreign territory. Without manifesting its territorial sovereignty in a manner corresponding to circumstances, the State cannot fulfil this duty. Territorial sovereignty cannot limit itself to its negative side, i.e. to excluding the activities of other States; for it serves to divide between nations the space upon which human activities are employed, in order to assure them at all points the minimum of protection of which international law is the guardian.*"

[89] Article 194(2) of UNCLOS.

[90] *Trail Smelter Case (United States, Canada) 16 April 1938 and 11 March 1941*, reproduced in: *United Nations Reports of International Arbitral Awards, vol. III, 1965*. Also *see Corfu Channel case, Judgment of April 9th, 1949, I.C.J. Reports 1949*, p. 22; *Case concerning Right of Passage, supra* note 87, pp. 19, 43.

[91] *Corfu Channel case, ibid*, p. 18.

utere tuo alienum non laedas (one must use his own so as not to damage that of another).

UNCLOS did not specifically regulate the situation of transboundary harm or damage caused by pollution from seabed activities in disputed maritime areas, nor is there any rule regarding the responsibility for the prevention of environmental damages that may occur therein.[92] There is, however, as previously mentioned, a connection between the obligation of States to protect and to preserve the marine environment and the exercise of their sovereign rights and jurisdiction over a certain maritime area. Should there be a disassociation between, on the one hand, the development of marine natural resources or, for that matter, the undertaking of other activities such as marine scientific research in disputed maritime areas,[93] and on the other hand, the obligation to protect and to preserve the marine environment as a result of the pending delimitation of maritime boundaries, there would inevitably be a situation where neighbouring coastal States could only do very little to avoid that activities taking place in disputed maritime areas do not cause environmental damage to its territory and coastal communities. The obligation to protect and preserve the marine environment requires that States undertake all measures to ensure that activities under their jurisdiction or control are carried out and operate in a manner not to cause damage or pollution, and without unjustifiably interfering with the activities of other States.[94] States must monitor the risks and effects of the activities occurring under their control and inform other States of these risks and effects and of possible consequences that such activities might have, as well as determine the appropriate means to deal with them.[95]

In the event that damage or pollution occurs, the relevant States must ensure that it does not spread beyond the maritime areas where States exercise sovereign rights, in order to mitigate or limit the damage sustained.[96] States are responsible for failing to adopt the necessary measures to prevent or mitigate damages, or to inform a neighbouring State of the existence of a risk of transboundary harm or damage, without prejudice of the liability of States for damages resulting from internationally wrongful acts or which might result from establishing a link between the non-fulfilment of the previously mentioned obligations and the extent of damages, thus confirming the objective and subjective elements of State

---

[92] Article 221 of UNCLOS extends the jurisdiction of coastal States regarding the adoption of enforcement measures beyond the territorial sea to protect their coastline from pollution or the threat of pollution resulting from maritime casualties.

[93] Articles 240 lit. (d) and 263(1), (3) of UNCLOS.

[94] Article 194(2) and (4) of UNCLOS. Also *see Legality of the Threat or Use of Nuclear Weapons, Advisory Opinion, I.C.J. Reports 1996,* para. 29.

[95] *The MOX Plant Case, supra* 76, paras, 82, 84, 89(1). Also *see Lac Lac Lanoux Arbitration, supra note 39 in Chap. 4*, p. 133; *Corfu Channel case, supra* note 90, p. 18.

[96] Article 194(2), (3)c) and (4) of UNCLOS. Also *see Gabčíkovo-Nagymaros Project, supra note 40 in Chap. 4*, para. 80.

responsibility.[97] Yet, customary international law does not make damage a requirement for the liability of States. Indeed, a "*State may be held liable under customary international law even if no material damage results from its failure to meets its international obligations.*"[98] As for reparation, UNCLOS does not specify "*what constitutes compensable damage, or which subjects may be entitled to claim compensation*".[99] International law only refers that reparation must, as far as possible, wipe out all the consequences of the illegal act and re-establish the situation which would, in all probability, have existed if that act had not been committed.[100]

Nonetheless, the obligation to protect and preserve the marine environment might prove to be a difficult task if, for example, operators that cause environmental damage as a result of seabed activities undertaken in disputed maritime areas are not subject to and do not have assets in the jurisdiction of the claiming State or the harmed neighbouring coastal State, or even when there is not an effective control of the operator's environmental conduct or the enforcement of international regulations and standards by the home-State.[101] In this case, the extent of the obligations to prevent and to protect the marine environment in disputed maritime areas is essential to determine the possibility of reparation or its success regarding the affected population of the neighbouring coastal State.

The entry into force of UNCLOS recognized and accepted the protection and preservation of the marine environment as a community or collective interest that benefits every State. However, despite the growing awareness of the existence of

---

[97] *Gabčíkovo-Nagymaros Project, supra note 40 in Chap. 4*, para. 152: "*It is a well-established rule of international law that an injured State is entitled to obtain compensation from the State which has committed an internationally wrongful act for the damage caused by it.*" *Phosphates in Morocco (Italy v. Fr.), 1938 P.C.I.J., Series A/B, No. 74*, p. 28: "*That being so, it is in this decision that we should look for the violation of international law—a definitive act which would, by itself, directly involve international responsibility. This act being attributable to the State and described as contrary to the treaty right of another State, international responsibility would be established immediately as between the two States.*" *Dickson Car Wheel Co. Case (1931) (US v. Mexico)*, reproduced in: *United Nations Reports of International Arbitral Awards, vol. IV*, p. 678: "*Under international law, apart from any convention, in order that a State may incur responsibility it is necessary that an unlawful international act be imputed to it, that is, that there exist a violation of a duty imposed by an international juridical standard. The above cited Convention requires further the existence of damage suffered by a national of the claimant Government. It is indispensable therefore, in order that a claim may prosper before this Commission, that two elements coexist: an unlawful international act and a loss or injury suffered by a national of the claimant Government.*" Also *see Yearbook of the International Law Commission, Second Report on State Responsibility, by Mr. Roberto Ago, Special Rapporteur, The Origin of International Responsibility*, UN Doc. A/CN.4/233 (1970), p. 187.

[98] *Request for Advisory Opinion submitted to the Seabed Disputes Chamber, supra note 77 in Chap. 2*, paras. 178, 179, 210.

[99] *Ibid*. paras. 179, 193–197.

[100] *The Factory at Chorzów (Claim for Indemnity), The Merits, 1928 P.C.I.J., Series A, No. 17*, para. 48.

[101] Morgera, *supra* note 75, pp. 25–24, 44–46.

community or collective interest, international law and in particular the law of the sea has not made the right of each State to intercede on behalf of that same interest in areas subject to the jurisdiction or sovereignty of other States, or to which two or more States are entitled to; even if, ultimately, every State has an interest in the protection and preservation of the marine environment due to its interconnectivity and the intergenerational responsibility for securing the oceans and seas for future generations.[102] This is most notably in the high seas where all States have a clear interest regarding the protection and preservation of the marine environment.[103]

Consequently, in disputed maritime areas, neighbouring States may only pursue against any State based on the latter's responsibility for damages suffered and not in representation of any community or collective interest, or on the basis of an *actio popularis* as result of such interest being affected,[104] without prejudice of the

[102] Article 30 of the *Charter of Economic Rights and Duties of States*, *supra* note 81. The ICJ recognized "*that the environment is not an abstraction but represent the living space, the quality of life and the very health of human beings, including generations unborn.*" *See Legality of the Threat or Use of Nuclear Weapons*, *supra* note 94, para. 29. Also *see Gabčíkovo-Nagymaros Project, supra note 40 in Chap. 4*, para 140. In this case, the ICJ stated that "[t]*he Court is mindful that, in the field of environmental protection, vigilance and prevention are required on account of the often irreversible character of damage to the environment and of the limitations inherent in the very mechanism of reparation of this type of damage. Throughout the ages, mankind has, for economic and other reasons, constantly interfered with nature. In the past, this was often done without consideration of the effects upon the environment. Owning to new scientific insights and to a growing awareness of the risks for mankind—for present and future generations—of pursuit of such interventions at an unconsidered and unabated pace, new norms and standards have been developed, set fort in a great number of instruments during the last two decades. Such new norms have to be taken into consideration, and such new standards given proper weight, not only when States contemplate new activities but also when continuing with activities begun in the past. This need to reconcile economic development with protection of the environment is aptly expressed in the concept of sustainable development.*"

[103] Article 87(2) of UNCLOS. *See Request for an Examination of the Situation in Accordance with paragraph 63 of the Court's Judgment of 20 December 1974 in the Nuclear Tests (New Zealand v. France) Case, I.C.J. Reports 1995*, para. 64.

[104] *South West Africa, Second Phase, Judgment, I.C.J. Reports 1966*, para. 88: "*For these reasons the Court, bearing in mind that the rights of the Applicants must be determined by reference to the character of the system said to give rise to them, considers that the "necessity" argument falls to the ground for lack of verisimilitude in the context of the economy and philosophy of that system. Looked at in another way moreover, the argument amounts to a plea that the Court should allow the equivalent of an "actio popularis", or right resident in any member of a community to take legal action in vindication of a public interest. But although a right of this kind may be known to certain municipal systems of law, it is not known to international law as it stands at present, nor is the Court able to regard it as imported by the "general principles of law" referred to in Article 38, paragraph 1 (c), of its Statute.*" *Ibid, Dissenting Opinion of Judge Jessup*, pp. 387–388: "*I must repeat, as indicated above, that the municipal law analogy which I have been discussing is far from perfect and the differences in the international law situation must be clearly noted. I agree that there is no generally established actio popularis in international law. But international law has accepted and established situations in which States are given a right of action without any showing of individual prejudice or individual substantive interest as distinguished from the general interest.*" Also *see South West Africa Cases, supra note 39 in Chap. 4, Separate Opinion of Judge Jessup*, p. 425; *Nuclear Tests*

obligation to undertake all measures necessary to effectively control, prevent, reduce and eliminate adverse environmental effects. Dispute settlement mechanisms included in UNCLOS have maintained the bilateralism that has reflected the standing of international jurisprudence on this subject, thus not allowing a State to pursue a case based on collective or community interests.[105]

6.5 Essential Legal and Functional Aspects of Joint Development

6.5.1 State Participation in Exploration and Exploitation Activities

After the delimitation of maritime boundaries, the exploration and exploitation activities of offshore hydrocarbon deposits that are found in the continental shelf of one coastal State and all matters concerning their management, such as the granting of operating rights, are subject to the law of that State. In the case of transboundary resources or resources found in disputed maritime areas, States must either reach an agreement on the common use of these resources or refrain

(Footnote 104 continued)
(Australia v. *France), Judgment, I.C.J. Reports 1974, Joint Dissenting Opinion of Judges Onyeama, Dillard, Jiménez de Arechaga and Sir Humphrey Waldock,* pp. 369–370, 521–522; *Nuclear Tests (Australia* v. *France), Judgment, I.C.J. Reports 1974, Dissenting Opinion of Judge de Castro*, pp. 387–390. On this subject see the comments to Article 48 of the 2001 ILC *Draft Articles on Responsibility of States for International Wrongful Acts* provides that "[a]*ny State other than an injured State is entitled to invoke the responsibility of another State*" if "*the obligation breached is owed to the international community as a whole*" and thus "*may claim from the responsible State: cessation of the internationally wrongful act, and assurance and guarantees of non-repetition* (…) *and performance of the obligation of reparation* (…) *in the interest of the injured State or of the beneficiaries of the obligation breached.*" Several of these Articles have been considered to reflect customary international law *see Request for Advisory Opinion submitted to the Seabed Disputes Chamber, supra note 77 in Chap.* 2, para. 169.

[105] *See* Malgosia Fitzmaurice, "International responsibility and liability", in: *The Oxford Handbook of International Environmental Law*, edited by Daniel Bodansky, Jutta Brunnée and Ellen Hey (Oxford University Press: Oxford, 2007), pp. 1020–1022; Sands, *supra* note 93 in Chap. 5, pp. 184–191. For an interpretation of Article 288(1) of UNCLOS allowing a State to pursue a case based on community interests *see* Rüdiger Wolfrum, "Enforcing community interests through international dispute settlement: reality or utopia", in: *From Bilateralism to Community Interest. Essays in Honour of Judge Bruno Simma*, edited by Ulrich Fastenrath, Rudolf Geiger, Daniel-Erasmus Khan, Andreas Paulus, Sabine Von Schorlemer and Christoph Vedder (Oxford University Press: Oxford, 2011), pp. 1132–1145; Also *see* Bruno Simma, "From bilateralism to community interest in international law", in: 250 *Recueil de cours de l'Académie de Droit Internationale de La Haye* (1994), pp. 217–384. On this topic also *see* Katrien Beekman, "Transboundary damage to the environment *per se*: remedial measures and standing", in: *Revue Belge de Droit International* (1996-2), pp. 480–483.

from undertaking any activities in contravention to their obligation of mutual-restraint in the terms previously described.

Reaching an agreement on the common use of transboundary resources or resources that are found in disputed maritime areas can be an extremely complex issue for States to tackle, as in its core lays the configuration of the regime applicable to the exploration and exploitation activities and the regulation of other uses of the relevant maritime area.[106] It is also a relevant aspect for operators to consider in order to determine the State or entity that will have the power to grant access to exploration and exploitation activities and the respective applicable law.

As mentioned in Sect. 3.3.2 of Chap. 3, existing joint development agreements signed before the delimitation of maritime boundaries have included a wide variety of different alternatives regarding State participation in exploration and exploitation activities. Douglas M. Johnston and Mark J. Valencia recognized that there were many variations of joint development agreements "*both in pattern of claim and in the pattern of cooperation*" and that even a broad definition could not encompass every joint development arrangement, particularly as the execution and content of such regimes depend on the level of reliability between States and the means of action chosen for the resolution of a territorial dispute.[107] Above all, the existing disparity in State practice is the result of States' discretionary powers to determine the scope and content of the joint development agreement, which is intricately connected with the existing level of commitment between States regarding the implementation of such a cooperative effort.

A more straightforward framework may divide the joint development area so that each State becomes responsible for the management of a part of it. In these cases, States often agree that exploration and exploitation activities shall be undertaken by national concessionaires that are often State-owned oil companies.

Under such straightforward joint development arrangements, States shall apply their laws in the respective part of the joint development area and all revenues gained as a result of the exploration and exploitation activities shall be subject to the applicable national taxation rules, without prejudice of States agreeing that the concessionaires of the other State may operate in all parts of the joint development area. In these cases, States often agree on the sharing of the overall revenues and the harmonisation of taxation rules. States must exchange information on exploration and exploitation activities, such as on the levels of production and revenues to ensure, at least, the reliability and effectiveness of the joint development regime. These regimes may further establish that each State shall be responsible for controlling other uses in its part of the joint development area.

Another characteristic of joint development arrangements where there is a predominant State intervention is the absence of an obligation of compulsory unitization. Instead, States merely envisage a duty of consultation in the event that a hydrocarbon deposit extends between different parts of the joint development

[106] Anderson, *supra* note 71 in Chap. 2, pp. 480–481.

[107] Johnston/Valencia, *supra* note 38 in Chap. 4, p. 36.

area, or between the latter and maritime areas subject to national jurisdiction. In so doing, States avoid being bound ab initio to an obligation to enter into an agreement regarding the development of a common deposit, preferring instead to analyse these situations as they occur.

The agreement between Japan and South Korea is a clear example of a regime with a high level of State participation in almost every phase of the joint development regime: from the granting of concessions and the applicability of taxation rules, to activities undertaken in the joint development area. This agreement created a joint commission where both countries are represented and that should meet at least once a year for consultations and to deal with all the questions that may emerge in respect of the implementation of the joint development regime, even though it has been noted that the said joint commission did not convene as often as it had been initially planned.[108]

Also the agreement between Kuwait and Saudi Arabia included the constitution of the joint permanent committee where States were represented and that had an advisory capacity, being States responsible for the management of resources. States should also consult each other before granting or altering any concessions for the exploration and exploitation of hydrocarbon deposits, although without being bound to implement the outcome of such consultations to any of the concessions made known by the other State.

According to the joint commission created by the joint development regime between the United Kingdom and Argentina, both countries agreed on the creation of a joint entity in which delegations of both countries would be represented. This body would meet at least twice yearly and decide by consensus on matters regarding the implementation of the joint development regime, such as making recommendations on the protection of the marine environment, analyse requests by operators for the exploration and exploitation of hydrocarbon deposits and

[108] Article 24 of the *Agreement between Japan and the Republic of Korea Concerning Joint Development of the Southern Part of the Continental Shelf Adjacent to the Two Countries*, done at Seoul on 30 January 1974 and entered into force on 22 June 1978, published at 1225 U.N.T.S. 114 [online: http://www.un.org/Depts/los/LEGISLATIONANDTREATIES/PDFFILES/TREATIES/jap-kor1974south.pdf (accessed on August 2013)], reproduced in: *International Maritime Boundaries*, vol. 1, *supra* note 42 in Chap. 2, pp. 1063–1089. *See* Masahiro Miyoshi, "The Japan-South Korea agreement on joint development of the continental shelf", in: *Geology and Hydrocarbon of the South China Sea and Possibilities of Joint Development, Proceedings of the Second EAPI/CCOP Workshop. East-West Center, Honolulu, Hawaii, 22–26 August 1983*, edited by Mark J. Valencia (Pergamon Press: New York, Oxford, Toronto, Sydney, Paris, Frankfurt, 1985), p. 548; *Ibid*, "Is joint development possible in the South China Sea?", in: *Sustainable Development and Preservation of the Oceans: The Challenges of UNCLOS and Agenda 21*, edited by Mochtar Kusuma-Atmadja, Thomas A. Mensah and Bernard H. Oxman (The Law of the Sea Institute/University of Hawaii: 1983), pp. 615–616; *Ibid*, "The Japan/South Korea joint development agreement 1974", in: *Joint Development*, vol. 2, *supra* note 70 in Chap. 2, pp. 89–97; Choon-ho Park, *East Asia and the Law of the Sea* (Seoul National University Press: Seoul, 1988), pp. 132–133; *Ibid*, "China and maritime boundary issue", in: 22 *German Yearbook of International Law* (1979), pp. 130–135.

agree on the applicable taxation rules, as well as on the promotion of joint ventures between operators of both countries and proposing the unitization of oil fields.[109]

Although also a straightforward joint development regime, the Ems-Dollard Treaty did not create any joint entity for the purpose of managing the common resources or for the supervision of the exploration and exploitation activities. Nonetheless, it did put forward a tight-knit regime of cooperation whereby States delegated such tasks on the respective national concessionaires.[110] Accordingly, States must cooperate closely in the exploration and exploitation of resources, such as exchanging of all plans for operations in the frontier area (internal waters and territorial sea), as well as all reports on the results of such operations. Furthermore, concessionaires should also sign agreements establishing methods of calculating petroleum and natural gas reserves and their output, and defining the details relating to the sharing of products and costs, auditing procedures, as well as mechanisms for the settlement of disputes.

In the Ems-Dollard Treaty, State intervention is reserved as an instance of appeal for the settlement of disputes between the concessionaires regarding certain aspects of the management of resources and only in the absence of agreement will States submit to an Arbitral Tribunal in accordance with the terms set out in this regime. Moreover, similarly to the agreement between Bahrain and Saudi Arabia, States may also agree that in the joint development area only one State undertakes exploration and exploitation activities under the condition that part of the revenues are to be shared with the other States.

6.5.2 Creation of Joint Entities

In addition to the already referred joint entities where States are represented and that have been constituted mostly for the purpose of overseeing the implementation of the joint development agreement and the exchange of information, there are also agreements that create entities that, to a certain extent, undertake States' role in the management of the joint development area. These entities should have legal personality and the necessary powers and autonomy to grant exploration and exploitation rights and exercise all other competences as agreed by States, such as the collection of revenues and taxes, or the settlement of disputes between operators. In most cases, their level of autonomy may not be very significant, since decisions must be reached by consensus and are only binding when States

---

[109] On the joint commission created by the *Joint Declaration on Cooperation over Offshore Activities in the Southwest Atlantic*, done on 27 September 1995 [online: http://www.falklands.info/history/95agree.html (accessed on August 2013)] *see* Becker-Weinberg, *supra* note 64, pp. 416–417.

[110] On this agreement *see* Hans-Dietrich Treviranus, "Der deutsch-niederländische Ems-Dollart-Vertrag", in: 23 *Zeitschrift für Ausländisches Offentliches Recht und Völkerrecht* (1963), pp. 536–553.

expressly agree to them. Such situations may result in an impasse with immediate repercussion on the exploration and exploitation rights granted to operators in the event that States fail to reach a consensus or an agreement on the relevant matters. In this respect, the creation of a joint entity with the necessary powers to manage the exploration and exploitation activities in the joint development area, would be preferable than, for example, the placing exclusively in the hands of national entities or of one of the States the responsibility for the management of these activities.[111] Evidently, the mere creation of joint entities does not, by itself, ensure the efficient development of common offshore hydrocarbon deposits. A complex organic and legal structure may constitute an obstacle towards the capable development of resources. Consequently, the efficiency of the actions of these entities will essentially depend on the respective level of autonomy and the powers conferred by States regarding the management of the joint development area, even though, as already mentioned, States are in most cases reluctant in granting excessive powers to this end.[112] This was the case of the lengthy negotiations between Thailand and Malaysia regarding the powers of the joint authority for the purpose of developing the resources of the continental shelf of the Gulf of Thailand where sovereignty claims overlapped and that eventually resulted in this entity having fewer powers that it had been initially expected.

Likewise, the joint commission created under Article 4 of the agreement between Colombia and Jamaica has the necessary powers to elaborate the modalities for the implementation of exploration and exploitation activities of living or non-living resources and of any other activities for the economic growth of the joint development area, but only these States may authorize arrangements for leases, licences, and joint ventures regarding such activities.

More complex joint development regimes have created two or more entities with comprehensive autonomy. These agreements instituted, to some extent, a system of 'checks-and-balance' whereby one entity is responsible for the effective management of resources and the other constitutes a representative body whose members are appointed by States and that controls the activity of the latter. However, even in these cases, the disparity between agreements does not allow for the identification of a 'best practice' or standard that could suit a wide range of different situations. The most advanced organizational structures are included in the *Timor Gap Treaty* signed between Indonesia and Australia,[113] the *Timor Sea*

---

[111] Mochtar Kusuma-Atmadja, "Joint development of oil and as by neighboring countries", in: *Sustainable Development and Preservation of the Oceans: The Challenges of UNCLOS and Agenda 21*, edited by Mochtar Kusuma-Atmadja, Thomas A. Mensah and Bernard Oxman (The Law of the Sea Institute, University of Hawaii: 1983), p. 592.

[112] Kim, *supra* note 38 in Chap. 4, p. 149.

[113] Articles 5 to 11 of the *Treaty between Australia and the Republic of Indonesia on the Zone of Cooperation in an Area between the Indonesian Province of East Timor and Northern Australia*, done on 11 December 1989 and entered into force on 9 February 1991, reproduced in: *ibid*, 1256–1328 [Timor Gap Treaty].

Treaty implemented between the latter and East Timor,[114] the Treaty between Nigeria and Sao Tome and Principe,[115] the agreement signed between Senegal and Guinea-Bissau,[116] and also the agreement regarding overlapping areas on the continental shelf beyond 200 nm signed by Mauritius and Seychelles.[117] An interesting aspect of the latter agreement is that no member of any of its "*three-tiered joint administrative structure*", i.e. the Ministerial Council, the Joint Commission and the Designated Authority, may have any financial or personal interest in any of the natural resources project in the joint development area.[118]

Independently of the number of entities created or the scope of their autonomy, their accountability is essential for the efficient implementation of a joint development regime, particularly regarding the possibility of operators being able to react against unforeseen alterations of applicable rules, such as those regarding taxation or payment of revenues, or even the cancelation of exploration and exploitation rights. The lack of harmonisation or clarification of rules regarding dispute settlement mechanisms or the definition of the applicable law and jurisdiction may lead to a deadlock situation. Operators also require the definition of rules and procedures applicable to conflicts that may occur between operators. However, despite the fact that these aspects are closely related with the granting of rights to operators and the definition of the applicable law and jurisdiction and the choice of dispute settlement mechanisms, States fail to include them in most joint development agreements that have been implemented thus far.

---

[114] Article 6 of the *Exchange of Notes Constituting an Agreement between the Government of Australia and the United Nations Transitional Administration in East Timor (UNTAET) concerning the Continued Operation of the Treaty between Australia and the Republic of Indonesia on the Zone of Cooperation in an Area between the Indonesian Province of East Timor and Northern Australia of 11 December 1989*, done on 10 February 2000, [online: http://www.un.org/Depts/los/LEGISLATIONANDTREATIES/PDFFILES/TREATIES/AUS-TLS2002EX.PDF (accessed on August 2013)] [Timor Sea Treaty].

[115] Articles 6 to 14 of the Agreement Nigeria/Sao Tome and Principe, *supra* note 21 in Chap. 4.

[116] Article 4 of the *Management and Cooperation Agreement between the Government of the Republic of Senegal and the Government of the Republic of Guinea-Bissau*, done at Dakar on 14 October 1993 and entered into force 21 December 1995, published at 1903 U.N.T.S. (1996) 34–63 [online: http://www.un.org/depts/los/LEGISLATIONANDTREATIES/PDFFILES/TREATIES/SEN-GNB1993MC.PDF (accessed on August 2013)]; *Protocol of Agreement Relating to the Organization and Operation of the Agency for Management and Cooperation between the Republic of Guinea-Bissau and the Republic of Senegal instituted by the agreement of 14 October 1993*, done at Bissau on 12 June 1995 and entered into force 21 December 1995. The Agreement and Protocol are reproduced in: *International Maritime Boundaries, vol. 3, supra note 5 in Chap. 5*, pp. 2258–2278.

[117] Article 4 of the Agreement Mauritius/Seychelles, *supra* note 66.

[118] Article 4 lit. (e), *idem*.

6.5.3 Access to Operations

The conditions for granting rights of exploration and exploitation is an essential aspect of any joint development agreement in disputed maritime areas, even though reaching an agreement on this issue may not always be very straightforward, particularly if States do not share the same knowledge and experience on offshore activities.

Exploration and exploitation activities are extremely time-consuming and highly costly ventures, commanding vast resources for a long period of time without certainty on their return. In most cases, when determining the essential criteria for the granting of exploration and exploitation rights, States are confronted with the decision on their level of participation in these activities and their share on production and revenues obtained. Often, due to differences in capabilities that allow for the undertaking of exploration and exploitation activities, States rather relinquish a larger share and control of the production and revenues in favour of not partaking on any risk or investment required for such activities. Alternatively, States may prefer to maintain a considerable control over production and resources and have a larger take on production and revenues. In these cases, States should adopt a form of arrangement with operators that will necessarily reflect States' degree of involvement in such activities.[119]

The different understanding between two or more States regarding the procedure and form for the granting of the rights of exploration and exploitation may lead to difficult and lengthy negotiations. This was the case of the previously mentioned circumstances surrounding the implementation of the joint development regime between Thailand and Malaysia, whereby the latter, more used to offshore development, preferred the use of product sharing contracts over Thailand's proposal for the granting of concessions.

In other cases, such as in the Persian Gulf, as a result of the considerable experience in using concessions regarding onshore development, States adopted the same legal model regarding offshore production.

Some States choose to issue licenses as it is evidenced in the *Protocol* signed between Senegal and Guinea-Bissau or in the *Jan Mayen Agreement*, as well as the regimes applicable to the *Murchison Field Reservoir*, the *Markham Field Reservoirs* and the *Frigg Field Reservoir*. While other States favour the use of product sharing contracts, as it is the case of the *Timor Sea Treaty* and of the *Timor Gap Treaty*, which includes a model of product sharing contracts in Annex C, or also of the petroleum development contracts pursuant to Article 23 of the agreement signed between Nigeria and Sao Tome and Principe. Each one of these possibilities represents a unique set of problems and requires multiple approaches in order to safeguard the different interests at hand. The main distinction takes into account

[119] Kamal Hossain, "Choice of petroleum development regime on joint development of offshore oil and gas", in: *Joint Development*, vol. 2, *supra* note 70 in Chap. 2, pp. 72–76.

the level of participation and risk that States are prepared to undertake regarding exploration and exploitation of offshore hydrocarbon deposits.

An important requirement of joint development regimes is that the relevant States are informed and give their consent to the rights granted in the joint development area and agree, if applicable, on the competence of the entity created to that effect under the respective agreement. Moreover, when granting rights to operators, States should agree on the content of operators' obligations and on all other matters that might have implications in the management of the joint development area in order to ensure the efficient development of resources. These include, inter alia, compulsory unitization, protection and preservation of the environment and attention for other uses of the relevant maritime area. The granting of rights to operators could also be subject to the adoption of further requirements, such as the construction and operation of pipelines for the purpose of transportation of resources or the assessment and commercialization of production, as well as the fulfilment of regulations regarding employment and training, health and safety, corporate governance, transparency, sound management and good practices, solvency and availability of adequate funds, revenue schemes, financial and customs provisions, dispute settlement mechanism, applicable law, reporting and publication of relevant information, amongst others.

Most joint development agreements do not include provisions on granting access to operations, they rather establish general provisions. This is the case, for example, of the MoU signed between Angola and Congo whereby both countries agreed to implement a joint development regime in a "*zone of common interest*" in the Lower Congo Basin, pending the delimitation of maritime boundaries between both countries.[120] Accordingly, this MoU merely provides that both States shall agree on the granting of rights to operators and that each country shall appoint experts to prepare the agreements that define the modalities for the joint development of hydrocarbon deposits.[121]

In the absence of provisions on granting access to operations or the division of the joint development area so that each State becomes responsible for the management of a specific part of the same, the assessment of such rules and procedures will depend on the applicable law. However, it is not always possible to determine which law is applicable in the joint development area in the absence of an agreement between the relevant States in that respect.

---

[120] *Memorandum of Understanding for the Common Petroleum Development of the Lower Congo Basin between the Democratic Republic of Congo and the Republic of Angola*, done on 18 June 2003, published in the *Diário da República de Angola*, 1ª S., n. 61, 30 July 2004 (in Portuguese), [online: http://www.minpet.gov.ao/VerLegislacao.aspx?id=276 (accessed on August 2013)]. *See* Resolutions of the Angolan Government n. 3/08 of 21 January 2008, published in the *Diário da República de Angola*, 1ª S., n. 7, 21 January 2008 (in Portuguese); n. 8/08 of 21 January 2008, published in the *Diário da República de Angola*, 1ª S., n. 11, 21 January 2008 (in Portuguese); and n. 33/08, of 23 April 2008 published in the *Diário da República de Angola*, 1ª S., n. 67, 23 April 2008 (in Portuguese).

[121] *Ibid.*

6.5.4 Safeguard of Pre-existing Rights

Time and again, States have unlawfully granted rights to operators and allowed drilling in maritime areas that were disputed by two or more States pending the delimitation of maritime boundaries.[122] Under such similar circumstances, States have often considered the option of entering into joint development agreements a means to achieve the continuation of the respective exploration and exploitation activities in disputed maritime areas, consequently safeguarding operators pre-existing rights.

Additionally, before the delimitation of maritime boundaries and when licenses or concession rights have been granted to operators in disputed maritime areas, States adopted measures to safeguard operators' rights when entering into maritime delimitation agreements.[123] In this respect, Rainer Lagoni refers to the insertion of a "*grandfather clause*" whereby the rights of licensees or concessionaires would be preserved after the delimitation of maritime boundaries.[124] States could also envisage that operators' rights would be subject to termination or reassessment as a result of the delimitation of maritime boundaries.[125]

In the arbitration between Saudi Arabia and Arabian American Oil Company the Tribunal considered that the rights and obligations of the company constituted acquired rights and that consequently could not be modified by Saudi Arabia without its consent.[126] In fact, in the case between Sapphire International Petroleums Ltd. and National Iranian Oil Company, the Arbitral Tribunal recognized that the respect for acquired rights is one of the general principles of law recognized by international Tribunals.[127] In this case, the Tribunal further

---

[122] *See* Fox (et al.), *supra* note 111 in Chap. 3, pp. 214–217; Ian Townsend-Gault, "The impact of a joint development zone on previously granted interests", in: *Joint Development*, vol. 2, *supra* note 70 in Chap. 2, pp. 174–184.

[123] Articles 17 and 29(2) of the *Universal Declaration of Human Rights*, done at Paris on 10 December 1948, Resolution (UNGA) n. 217 A (III), 10 December 1948, United Nations Doc A/810 at 71 (1948). [online: http://www.un.org/en/documents/udhr/ (accessed on August 2013)]. Resolutions (UNGA) 42/114 and 42/115, 7 December 1987; 43/123, 8 December 1988; and 43/124 and 45/98, 14 December 1990. Also *see* Resolution (UNGA) 523 (VI) 12 January, and 629 (VII), 21 December, both of 1952, and Resolution (UNGA) 1803 (XVII), 14 December 1962. On the right to compensation *see* Andrea Giardina, "Nationalisation et indemnisation en Droit International", in: *Le Droit Pétrolier et la Souveraineté des Pays Producteurs* (Librairie Générale de Droit et de Jurisprudence: Paris, 1973), pp. 39–49.

[124] International Law Association, *supra* note 15 in Chap. 2, p. 17.

[125] Lagoni, *supra* note 70 in Chap. 2 pp. 357–358.

[126] *Saudi Arabia v. Arabian American Oil Company (Armaco), Arbitral Award*, 27 *International Law Reports* (1963), pp. 166–167, 227.

[127] *Sapphire International Petroleums Ltd. v National Iranian Oil Company, Arbitral Award,* 35 *International Law Reports* (1967), p. 184. Also *see Barcelona Traction, Light and Power Company, Limited, Judgment, I.C.J. Reports 1970*, para. 33: "*When a State admits into its territory foreign investments or foreign nationals, whether natural or juristic persons, it is bound to extend to them the protection of the law and assumes obligations concerning the treatment to*

acknowledged that the loss of profit as a result of the said interference with acquired rights regarding concession areas not yet developed, entitled the company to compensation, in addition to the expenses incurred.[128]

6.5.5 Taxation, Sharing of Costs and Revenues

The legal terms and conditions of a joint development agreement pending the delimitation of maritime boundaries mirror the atypical and unique circumstances of the context in which these regimes are negotiated.[129]

There is no rule establishing revenue or cost sharing *ratios* as States are free to define the content of the joint development regime that they wish to implement and that best serves their interests. Nonetheless, in the case that States agree on the taxation of exploration and exploitation activities, States should coordinate and establish the necessary taxation rules, as well as the respective cost and revenue sharing-schemes and the appropriate measures to ensure that operators' financial and taxation obligations are enforceable in these States' legal orders.

In the absence of an agreement between the States on the applicable taxation rules, the assessment of such rules will depend on determining the applicable law. In this case, contrary to joint development agreements signed after the delimitation of maritime boundaries, it is not possible to refer to Article 60 of UNCLOS (*ex vi* Article 80 of UNCLOS) seeing that there is more than one State that is entitled to the joint development area. In these cases, the State that has granted the right to the operators would be entitled to apply its own taxation rules.

Most agreements establish the equal share of benefits and costs deriving from the exploration and exploitation activities in the joint development area, even though there are agreements that do not include any reference to revenue sharing, either because of their programmatic nature or simply for the reason that States chose not to disclose this information. In this respect, for example, Sect. 1 of the MoU signed between Angola and Congo merely provides that the resources of the "*zones of common interest*" to be included in the joint development area are considered as being part of the indivisible property of both States and the respective interests shall be equally shared.

(Footnote 127 continued)
be afforded them. These obligations, however, are neither absolute nor unqualified. In particular, an essential distinction should be drawn between the obligations of a State towards the international community as a whole, and those arising vis-à-vis another State in the field of diplomatic protection. By their very nature the former are the concern of all States. In view of the importance of the rights involved, all States can be held to have a legal interest in their protection; they are obligations erga omnes."

[128] *Ibid* pp. 187–188.

[129] *See* Ian Townsend-Gault, "Joint development of offshore mineral resources-progress and prospects for the future", in: 12-3 *Natural Resources Forum* (1988), p. 282.

Other agreements provide for extremely comprehensive taxation rules and complex cost and revenue sharing-schemes, such as the *Timor Sea Treaty* or the agreements signed between Nigeria and Sao Tome and Principe, and Senegal and Guinea-Bissau. According to Article 15 of the latter and apart of the payment of taxes and fees and fines regarding fishing activities, operators shall pay taxes on oil production and an area tax, as well as a tax ad valorem over mining production, royalties on petrol activities, taxes on the benefit due by operators, additional petrol quotas, and a share in the revenues from commercialization carried out by the Agency for Management and Cooperation. In what concerns the distribution of benefits and costs, this agreement establishes in Article 2 that fishing resources shall be equally shared, while the resources of the continental shelf shall be divided in the proportion of 85:15 between Senegal and Guinea-Bissau, notwithstanding the revision of these amounts considering the size of resources eventually found. Distinctively, the *Jan Mayen Agreement* takes into account the economic differences between Iceland and Norway and favours the former regarding the allocation of costs. For example, according to Article 3 of this agreement, Norway shall undertake all the costs related with seismic or magnetic surveys, unless the States agree otherwise.

In what concerns taxation, some agreements adopt taxation regulations that envisage measures for the avoidance of double taxation, prevention of fiscal evasion or exemptions in order to attract investment. This is the case, for example, of the very elaborate customs and taxation regime included in Article 17 lit. (a) of the agreement between Thailand and Malaysia, which provides that the relevant tax authorities shall apply a tax exemption on operators' income during the first 8 years of production. The agreement between Japan and South Korea also includes a tax exemption in Article 18 for certain activities, such as for the importation and exportation of equipment and necessary materials for the development of the resources found in the joint development area. The agreement signed between Nigeria and Sao Tome and Principe exempts operators in Article 22 from the payment of any customs or taxes on all petroleum equipment, which, under the agreement, also includes oil rigs.

There is also the specific case of compatibility between joint development regimes regarding overlapping claims in the continental shelf beyond 200 nm and the rule of revenue-sharing of mineral resources exploited in these areas pursuant to Article 82 of UNCLOS. As mentioned before in this Chapter, the first joint development agreement in the continental shelf beyond 200 nm, before the delimitation of maritime boundaries, was signed by Mauritius and the Seychelles. Although this arrangement only refers that these countries shall equally share the revenues received, they are bound, as States-parties to UNCLOS, to the regime pursuant to Article 82 of this Convention, which requires that coastal States make "*payments or contributions in kind in respect of the exploitation of the non-living resources of the continental shelf beyond 200 nautical miles.*" This provision is the outcome of the compromise reached between, on the one hand, the coastal States entitled to a large continental shelf and, on the other hand, those States that intended to draw the limit at 200 nm, thus avoiding a reduction of the Area. The solution found during the negotiation of the Convention consisted of coastal States being able to claim a continental shelf beyond 200 nm in accordance with Article 76 of UNCLOS subject, however, to making payments to the 'international community' through the Authority. A comprehensive analysis of Article 82 of UNCLOS has identified many of the difficulties regarding the fulfilment of this provision's requirements, particularly, as a result of the ample degree of discretion given to States. Such difficulties include the assessment of the amounts to be paid, the value of contributions in kind, the calculation and the execution of the payments or contributions, the precise role of the Authority and how it is to be performed, particularly, in light of the 'equitable sharing criteria' and its development by the Assembly pursuant to Articles 160(2) lit. (f)(i) and 162(2) lit. (o)(i) of UNCLOS, and also the applicable mechanism for the settlement of disputes regarding Article 82 of UNCLOS.[130]

Although Article 82 of UNCLOS does not make any reference to disputed maritime areas, there is no legal impediment for joint development of mineral resources found in overlapping areas in the continental shelf beyond 200 nm. There is also no reason to consider joint development in these areas as excluded from the scope of Article 83(3) of UNCLOS. However, in the absence of a clear understanding between the relevant States, it may be difficult to ensure the compatibility of the ambiguous regime included in Article 82 of UNCLOS and the rules on sharing of revenues defined in a joint development agreement.

The obligation to make payments or contributions in kind is of the coastal State, even in those cases where joint development agreements create joint entities for the purpose of managing the joint development area and the seabed activities. Since in disputed maritime areas there are, at least, two coastal States that are entitled to the continental shelf and the exploitation of its resources, in the case of joint development agreements, this obligation would have to be considered as being extended to both. Yet, this would necessarily lead to the issue of the responsibility of each State *vis-à-vis* the Authority for the fulfilment of Article 82 of UNCLOS and the amount that each State should pay or contribute in kind. For example, in the case of the joint development agreement between Mauritius and the Seychelles, are the States solely responsible before the Authority for only the respective share of the resources pursuant to the terms of the agreement, i.e. 50:50? What would happen if one State does not fulfil its responsibility or if it falls within the category of "*developing country which is a net importer of a mineral resource produced from its continental shelf*" and, consequently, is exempt from making

---

[130] On Article 82 of UNCLOS *see UNCLOS: A Commentary, supra note 36 in Chap. 5*, pp. 930–947. Also *see* International Law Association, *Third Report of the International Committee on Legal Issues of the Outer Continental Shelf to the Biennial International Law Association Conference in Rio de Janeiro, Brazil, 17–21 August 2008*, by David M. Ong (Seventy-Third Conference: Rio de Janeiro, 2008).

such payments or contributions?[131] Moreover, considering that a joint development agreement may establish that only one of the States undertakes or manages the seabed activities sharing, nonetheless, the respective revenues with the other States-parties to the agreement, what would be the role of the latter under Article 82 of UNCLOS?

Coastal States must, therefore, comply with their obligations on revenue-sharing pursuant to Article 82 of UNCLOS also when implementing a joint development agreement in areas of overlapping claims in the continental shelf beyond 200 nm.

6.5.6 Employment, Health and Safety

UNCLOS provides that States may adopt and implement rules regarding the construction of oil rigs and other installations in the EEZ,[132] as well as in respect of their operation and use, including on health and safety,[133] carrying out of inspections,[134] and decommissioning.[135] However, most joint development agreements signed before the delimitation of maritime boundaries do not include rules and regulations on employment, health and safety. In some cases, the agreements merely create the obligation for States to consult each other on these matters, or that the entities created under these regimes submit to the relevant States recommendations for the adoption of such standards.

More comprehensive joint development agreements, for example, as provided in Articles 24 and 25 of the *Timor Gap Treaty* or Articles 36 and 37 of the agreement between Nigeria and Sao Tome and Principe, contain provisions applicable to employment and training of workers and regarding activities directly or indirectly connected with the management of common resources or the exploration and exploitation activities in the joint development area. These include, for example, the coordination of safety and security measures, rules regarding surveillance, search and rescue missions and air traffic control.

These agreements also often grant the entities established under the same the necessary powers to coordinate or replace States regarding the establishment of rules and regulations on employment, health and safety, as well as on the exercise of jurisdiction, including the ability to carry out inspections and enforcement actions in the joint development area, either applying the regime established for this purpose in the joint development agreement or that of each State, should these agreements create areas of jurisdiction in the joint development area.

---

[131] Article 82(3) of UNCLOS.

[132] Article 56(1) lit. (b), (i) of UNCLOS.

[133] Article 60(1) lit. (b) and (c), (2) of UNCLOS.

[134] Article 73(1) of UNCLOS.

[135] Article 60(3) of UNCLOS.

However, provisions on employment, health and safety often only define standards and guidelines, such as granting preference to the hiring of nationals or allowing for the exchange of workers between the States involved, failing to establish rules on the compatibility between the applicable national legal regimes or harmonising different rules and regulations, or even determining the applicable law and jurisdiction in the joint development area in that respect. Additionally, most joint development agreements fail to include rules regarding employment conditions to be enforced by concessionaires and licensees operating in the joint development area, or for that matter by the States or State-owned entities.

Unitization agreements also establish health and safety standards and the harmonisation of applicable rules and regulations, as well as rules on the exercise of jurisdiction regarding these matters. In some cases, these agreements include the right of inspectors from one State to have access to all relevant information and to any structure or device, as well as to carry out inspections in order to achieve compliance by operators of health, safety and environmental standards and issue orders to operators in the event of imminent danger to the life of a person or of an incident involving serious pollution. This is the case of the agreements concerning the *Frigg Field Reservoir*, *Murchison Field Reservoir* and *Markham Field Reservoir*, as well as the agreement concerning cross-boundary petroleum cooperation between the United Kingdom and Norway and the *Greater Sunrise* unitization agreement signed between Australia and East Timor. This last agreement, for example, established that the law of Australia should be applicable to health and safety matters in the unitization area.[136]

6.5.7 Protection and Preservation of the Marine Environment

A small number of joint development agreements signed before the delimitation of maritime boundaries include provisions on the protection and preservation of the

---

[136] *Agreement between the Government of Australia and the Government of the Democratic Republic of Timor-Leste relating to the Unitization of the Sunrise and Troubadour fields*, done at Dili on 6 March 2003 [online: http://www.un.org/depts/los/LEGISLATIONANDTREATIES/PDFFILES/TREATIES/AUS-TLS2003UNI.PDF (accessed on August 2013). Also *see Memorandum of Understanding between the Government of the Democratic Republic of East Timor and the Government of Australia Concerning an International Unitization Agreement for the Greater Sunrise field*, done at Dili on 20 May 2002 [online: http://www.un.org/depts/los/LEGISLATIONANDTREATIES/PDFFILES/TREATIES/AUS-TLS2002SUN.PDF (accessed on August 2013)]; *Memorandum of Understanding between the Government of the Democratic Republic of Timor-Leste and the Government of Australia relating to the Exploitation of the Sunrise and Troubadour Petroleum Fields in the Timor Sea*, done on 6 March 2003 [online: http://www.anp-tl.org/webs/anptlweb.nsf/vwAll/Resource-SunriseTroubadour%20-%20Memorandum%20of%20Understanding%20/$File/12.Sunrise%20MOU%20-MEMORANDUM%20OF%20UNDERSTANDING.pdf?openelement (accessed on August 2013)].

marine environment and often they merely make a general reference to the subject. This is the case of the joint development agreement between Japan and South Korea:

> Article XX.
> The Parties shall agree on measures to be taken to prevent collisions at sea and to prevent and remove pollution of the sea resulting from activities relating to exploration or exploitation of natural resources in the Joint Development Zone.[137]

Other joint development regimes establish that the relevant States should cooperate to prevent, minimize and eliminate pollution of the marine environment arising from the exploration and exploitation activities in the joint development area, providing assistance in that respect to the relevant entities created under these agreements. This is the case of the agreement signed between Senegal and Guinea-Bissau,[138] as well as of the *Timor Sea Treaty*[139] and *Timor Gap Treaty*,[140] and the agreement between Mauritius and Seychelles.[141]

These agreements also establish that operators shall be liable for damage caused by pollution of the marine environment as a result of the activities undertaken in the joint development area and that the said entities may issue regulations to protect the marine environment in the joint development area and establish a contingency plan for combating pollution from operations.

> (a) Australia and East Timor shall co-operate to protect the marine environment of the JPDA so as to prevent and minimise pollution and other environmental harm from petroleum activities. Special efforts shall be made to protect marine animals including marine mammals, seabirds, fish and coral. Australia and East Timor shall consult as to the best means to protect the marine environment of the JPDA from the harmful consequences of petroleum activities.
> (b) Where pollution of the marine environment occurring in the JPDA spreads beyond the JPDA, Australia and East Timor shall co-operate in taking action to prevent, mitigate and eliminate such pollution.
> (c) The Designated Authority shall issue regulations to protect the marine environment in the JPDA. It shall establish a contingency plan for combating pollution from petroleum activities in the JPDA.
> (d) Limited liability corporations or limited liability entities shall be liable for damage or expenses incurred as a result of pollution of the marine environment arising out of petroleum activities within the JPDA in accordance with:
>
> i. their contract, licence or permit or other form of authority issued pursuant to this Treaty; and
> ii. the law of the jurisdiction (Australia or East Timor) in which the claim is brought.[142]

---

[137] *Supra* note 108.

[138] Article 23 of the Protocol, *supra* note 116.

[139] Article 10 of the Timor Sea Treaty, *supra* note 114.

[140] Article 18 of the Timor Gap Treaty, *supra* note note 113.

[141] Article 12 of the Agreement Mauritius/Seychelles, *supra* note 66.

[142] Article 10 of the Timor Sea Treaty, *supra* note 114.

Powers and Functions of the Designated Authority

1.(h) issuing regulations and giving directions under this Treaty on all matters related to the supervision and control of petroleum activities including on health, safety, environmental protection and assessments and work practices, pursuant to the Petroleum Mining Code;

2. The Joint Commission shall exercise its powers and functions for the benefit of the peoples of Australia and East Timor having regard to good oilfield, processing, transport and environmental practice.[143]

The joint development regime implemented by Nigeria and Sao Tome and Principe further develops these States' obligations regarding the protection and preservation of the marine environment, particularly by giving a definition of pollution[144] and entrusting the joint entity created under this agreement with the management of activities relating to exploration for and exploitation of the resources in the joint development area and the preservation of the marine environment, having regard to the relevant rules of international law.[145] This agreement provides that inspectors may order the immediate cessation of any or all petroleum operations in the joint development area if necessary or expedient to protect the coastline or other maritime interests of either State, including fishing interests, against actual or potential pollution.[146]

38.1 The Authority shall take all reasonable steps to ensure that development activities in the Zone do not cause or create any appreciable risk of causing pollution or other harm to the marine environment.

38.2 In accordance with paragraph 1, the States Parties on the recommendation of the Authority shall agree necessary measures and procedures to prevent and remedy pollution of the marine environment resulting from development activities in the Zone.

38.3 In order to facilitate the effective monitoring of the environmental impact of petroleum activities in the Zone both States Parties shall regularly provide the Authority with such relevant information as they obtain from contractors or inspectors concerning levels of petroleum discharge and contamination. In particular the States Parties shall immediately inform the Authority of the occurrence of the following events:

(a) any petroleum spillage or event likely to cause pollution and requiring remedial measures beyond the capacity of the operator;
(b) discharge into the sea of large quantities of petroleum from an installation or pipeline;
(c) collisions at sea involving damage to an installation or pipeline;

143 Annex C under Article 6 lit. (b), (v) of the Timor Sea Treaty, *ibid.*

144 The text of Article 1(21) of the Agreement Nigeria/Sao Tome and Principe, *supra* note 21 in Chap. 4, reads "*"Pollution" means the introduction of substances or energy into the marine environment, including estuaries, which results or is likely to result in deleterious effects such as harm to living resources and marine life, hazards to human health, impairment of quality for use of sea water or reduction of amenity.*"

145 Article 9.5 lit. (o), *ibid.*

146 Article 30.5 lit. (c), *ibid.*

(d) evacuation of personnel from an installation due to force majeure, distress or other emergency.
The notification shall include any measures taken or proposed with respect to such events.

38.4 Nothing in this Treaty shall prejudice the taking or enforcement by each State Party or by the States Parties jointly of measures in the Zone proportionate to the actual or threatened damage to protect their coastline or exclusive maritime areas from pollution or threat of pollution which may reasonable be expected to result in major harmful consequences.[147]

Likewise, the regime agreed by the United States of America and Mexico provides that States shall adopt common safety and environmental standards and ensure that their respective standards and requirements are compatible. Moreover, these States also recognize the importance of their existing international obligations with respect to oil pollution preparedness, response, and cooperation, and are to review their implementation of such obligations in light of the activity contemplated under this Agreement in order to ensure an appropriate framework for ongoing cooperation.[148]

The agreement also includes the possibility of offshore operations being terminated for the purpose of averting significant damage to the environment:

Article 18
Inspections

1. Subject to applicable national law, each Party shall, under procedures to be developed and agreed under this Agreement, have the right to inspect Facilities in a Unit Area approved pursuant to this Agreement.
2. To enable Inspectors of each Party to safeguard their respective interests with respect to safety, environmental and fiscal matters, the Executive Agencies shall develop specific procedures, subject to national law, for:

   (a) consultation among Inspectors of each Party;
   (b) timely access to information relevant to inspection activities; and
   (c) physical access to Unit Areas for the purpose of inspecting activities therein under a joint inspection regime, including access to metering systems, wherever located.

3. The Inspectors of each Party shall act in cooperation and consult with Inspectors of the other Party to achieve compliance with applicable safety and environmental standards.
4. An Inspector of one Party may, with regard to Facilities located in the Unit Area, request an Inspector of the other Party to exercise his or her powers to ensure compliance with the applicable safety and environmental standards and requirements whenever it appears that circumstances so warrant. In the event of any disagreement between the Inspectors of the Parties, or the refusal of the Inspector of one Party to take action at the request of the Inspector of the other Party, the matter shall be referred to the Executive Agencies.
5. If it appears that it is necessary for the purpose of averting risk to life or serious personal injury or significant damage to the environment, and that circumstances do not permit

[147] *Supra* note 21 in Chap. 4.

[148] Article 19 of the Treaty between US/Mexico, *supra* note 67 in Chap. 2.

> the Inspectors to consult with the Executive Agencies, the Inspector with jurisdiction over the activities giving rise to such risk shall, as authorized under national law, order the immediate cessation of any or all operations upon the request of the other Inspector. Immediately thereafter, but not more than 4 h following the ordered cessation of activity, the Inspectors shall notify the Executive Agencies of such action and the reasons therefore, and the Executive Agencies shall immediately consult regarding actions necessary to address the risk. Nothing in this paragraph shall prevent the right of each Party to authorize the resumption of operations of the relevant Facilities.

However, the fact that most joint development agreements do not refer to the prevention of pollution and the protection of the marine environment does not signify that States are not subject to obligation to preserve and protect the marine environment previously identified. Such obligations ought to be considered when negotiating and implementing a joint development regime.

6.5.8 Applicable Law and Settlement of Disputes with and Between Operators

A significant part of joint development agreements do not include provisions regarding the applicable law or dispute settlement mechanisms, failing to clarify what rules and regulations should be enforced. Most agreements also do not determine which State or joint entities have jurisdiction or control in each case and how the judicial or administrative decisions of one State, or of such entities, should be enforced in the jurisdiction of the other State.

These are fundamental aspects and ought to be considered in joint development regimes, particularly as the lack of specification of these matters may inevitably lead to situations of legal uncertainty. This is the case when trying to determine which law is applicable to employment or compliance of health and safety measures, enforcement of environmental standards, collecting taxes and revenues, granting of rights of exploration and exploitation by States or entities created under the agreement, disputes between operators or between the latter and States or entities regarding the execution of such rights, amongst others. Furthermore, concession agreements signed by States and private entities have an international character because of its many ramifications, such as, the nationality of concessionaires or the applicable law to the concession. As a result, the choice of law applicable to the concession and the performance of the activities is an extremely important aspect of these agreements.

In the Arbitral Award between Saudi Arabia and the Arabian American Oil Company, the Tribunal considered that the applicable law should be that which best corresponded to the nature of the legal relationships between the State and the concessionaire. This was the law of the country with which the contract had the closest natural and effective connection, unless another law could be designated by the conduct of the parties, or as that Tribunal summarized: "*the governing law should coincide with the economic milieu where the operation is to be carried out*."[149] In this case, the Tribunal considered that the applicable law should also be supplemented by the general principles of international law, by custom and industry practice and by doctrine and jurisprudence, in particular whenever certain rights would not be secured by the applicable law.[150] The Arbitral Tribunal also recognized that the concession agreement was the "*fundamental law*" between the State and the concessionaire and that it had the nature of a constitution that had the effect of conferring irretractable or acquired rights to the concessionaire, rights that could not be withdrawn before the end of the concession, except as provided by the same.[151]

In what concerns joint development agreements that do include provisions on the applicable law and the settlement of disputes, these do not follow a common pattern, despite the fact that the majority of agreements determine that conflicts shall be settled by third parties or by joint entities that have the authority to resolve disputes between States or operators, or involving operators and said entities. In these cases, joint entities either apply the law of one of the States, the rules approved under the joint development regime or those referred in the respective licenses or concession or other arrangements signed with operators.[152]

The *Jan Mayen Agreement* provided that the Conciliation Commission would also serve as an instance of appeal should Norway and Iceland fail to reach an understanding on the efficiency of environmental protection measures regarding the exploration and exploitation of hydrocarbon deposits, being that such an appeal will suspend the aforementioned activities.[153]

---

[149] *Saudi Arabia v. Armaco, supra* note 126, pp. 166–167. Also *see* on the arbitration *Petroleum Development (Trucial Coast) Ltd. v. Sheik of Abu Dhabi Award of Lord Asquith of Bishopstone, 1951, Award of Lord Asquith of Bishopstone*, in: 1 *International and Comparative Law Quarterly* (1952), pp. 250–251.

[150] *Saudi Arabia v. Armaco*, *ibid*, pp. 169–172. The Tribunal acknowledged that: "[P]*ublic international law should be applied to the effects of the Concession, when objective reasons lead it to conclude that certain matters cannot be governed by any rule of the municipal law of any State, as is the case in all matters relating to transport by sea, to the sovereignty of the State on its territorial waters and to the responsibility of States for the violation of its international obligations.*"

[151] *Ibid*, p. 168. The Tribunal concluded that: "*Should a new concession contract incompatible with the first, or a subsequent statute, abolish totally or partially that which has been granted by a previous law or concession, this could constitute a clear infringement, by the second contract, of acquired rights or a violation, by the subsequent statute, of the principle of non-retroactivity of laws, with the only exception of rules of public policy. This is because a legal situation acquired by virtue of a previous special statute cannot be abrogated by a subsequent statute—generalia specialibus non derogant—unless the legislator has expressly given retroactive effect to such statute, which the State cannot do in respect of concessions, without engaging its responsibility.*"

[152] On dispute settlement mechanisms commonly used in the oil and gas industry *see* Margaret Ross, "Dispute Management and Resolution", in: *Oil and Gas Law. Current Practice and Emerging Trends*, edited by Greg Gordon and John Paterson (Dundee University Press: Dundee, 2007), pp. 431–460.

[153] *Supra* note 19 in Chap. 4.

Other agreements establish different dispute mechanisms depending on the resources being developed. For example, Article 2 of the agreement signed between Senegal and Guinea-Bissau includes the exploration and exploitation of both living and non-living marine natural resources and provides that all agreements concluded with operators regarding fishing shall include specific regulations stipulating means of settling disputes concerning the interpretation or application of these agreements. This agreement further provides that activities of exploration and exploitation of hydrocarbon deposits shall be subject to the law of Senegal and fisheries to the law of Guinea-Bissau, and that in respect to disputes on the interpretation or application of the Protocol attached to the agreement, States shall resort to arbitration or to the ICJ after attempts to solved through direct negotiations have failed or, in the event of these disputes involving the Agency for Management and Cooperation, States shall try to settle through direct negotiations or by arbitration if a solution could not be reached within 3 months.

Similarly, Article 47 of the agreement signed between Nigeria and Sao Tome and Principe establishes a mechanism for the settlement of disputes between the Joint Authority and operators and between the latter, as well as an additional mechanism applicable to disputes between these States, being that, in this case, there shall be first an attempt to resolve the dispute within the frame of the Joint Authority or the Ministerial Council if the former is not able to settle the matter, which will then have 12 months to find a solution. In the event the Ministerial Council does not succeed in finding a solution, the Heads of State of both countries shall attempt to reach an understanding within 6 months, after which any State may submit the matter to an Arbitral Tribunal in accordance with the terms and conditions set out in the joint development agreement.

Differently, Article 3(5) of the agreement implemented by Colombia and Jamaica establishes that in the joint development area each State shall have jurisdiction over its nationals, referring however that States may implement the measures required in order to achieve that non-nationals comply with the rules and regulations applicable to the activities in the joint development area, and that in these cases jurisdiction shall be determined through consultations between the States. Other agreements establish that the applicable law would be that of the State of the relevant concessionaire, being these resources considered as extracted from the continental shelf of that State, similarly to what is established in Article 19 the agreement signed between Japan and South Korea.

The more intricate a dispute mechanism, the more complex will be its harmonisation between States and their respective legal systems. Amongst the more complex, the regime included in the *Timor Sea Treaty* reveals some of the problems that may occur regarding the use of such procedures and which result from the fact that there are different mechanisms potentially applicable in the joint development area and in areas subject to national jurisdiction. In addition, the lack of harmonisation combined with possible insufficiencies of national legal systems to enforce arbitral decisions may lead an impasse.

Inconsistency of State practice is also evidenced in the fact that very few arrangements, such as the *Timor Gap Treaty* and *Timor Sea Treaty*[154] and the Nigeria and Sao Tome and Principe[155] determine the applicable law to criminal offences in the joint development area. Another joint development agreement that also includes provisions on criminal jurisdiction is the treaty signed by Mauritius and Seychelles. This agreement establishes that both States shall make it an offence under their respective legal orders for any person to conduct resources activities in the joint development area, if not pursuant to the terms of the agreement. It also states that both countries will "*examine different options for addressing offences committed in the* [joint development area]" and that, until then, nationals or residents of one of the States-parties shall be subject to the criminal law of the country of nationality or residence, regarding acts or omissions occurring in the joint development area in respect of seabed activities. The agreement also determines the law applicable to nationals or residents of a third State, as well as that the States-parties shall cooperate on obtaining evidence and information and, eventually, on making arrangements permitting officials of one country to assist in the enforcement of the criminal law of the other country.[156] Other than criminal jurisdiction, the agreement refers that for the purpose of applying the domestic laws of each State, the joint development area shall be considered as being part of its respective territory.[157]

These differences are closely related with the relevant regional context and the circumstances leading to the signing of the agreement, the characteristics of the intervening States and the relation between them, as well as the disparities of the respective legal systems, particularly in what regards offshore development of hydrocarbon deposits.

[154] Article 22 to 29 of the Timor Gap Treaty, *supra* note 113; Article 14 of the Timor Sea Treaty *supra* note 114.

[155] Articles 39 to 44 of the Agreement Nigeria/Sao Tome and Principe, *supra* note 21 in Chap. 4.

[156] Article 3 Agreement Mauritius/Seychelles, *supra* note 66.

[157] Article 7, *ibid.*

Chapter 7
Joint Development Agreements in Areas of Overlapping Claims: State Practice

7.1 Introduction

Despite the fact that there has been a growing awareness of the pragmatic advantages of joint development to forward a functional and legal alternative to overcome deadlock situations caused by opposing or overlapping maritime claims, the number of agreements of offshore hydrocarbon deposits has not grown significantly since the signing of the first agreement more than 50 years ago.[1] In the case of the Asia-Pacific region, States have frequently refused, rejected or ignored attempts to implement provisional arrangements, such as joint development agreements of offshore hydrocarbon deposits, mainly due to the absence of mutual trust or the predisposition to shelve maritime disputes. Instead, States have occasionally chosen to engage in legally-questionable actions, as the granting of drilling or exploration and exploitation rights in disputed maritime areas. They

[1] *Bahrain-Saudi Arabia Frontier Agreement*, done on 22 February 1958 and entered into force on 26 February 1958, published at 1733 U.N.T.S. 3 [online: http://www.un.org/depts/los/LEGISLATIONANDTREATIES/PDFFILES/TREATIES/BHR-SAU1958BA.PDF (accessed on August 2013)], reproduced in: *International Maritime Boundaries*, vol. 2, *supra note 125 in Chap. 3*, pp. 1495–1497. Despite of Iran's sovereignty claims over Bahrain, this agreement determined that the oil resources found in one of the areas of the joint development area would be explored by Saudi Arabia in accordance with the latter's discretion, under the condition that half of the benefits resulting from such exploration would be delivered to Bahrain, without prejudice of Saudi Arabia' sovereignty over this area. *See* Husain M. Albaharna, *The Legal Status of the Arabian Gulf States* (Manchester University Press: Manchester, 1968), pp. 167–195. Ian Townsend-Gault considers that "[t]*he agreement did not resemble joint development in any shape or form but rather privileged the preservation of the unity of the deposit.*" Ian Townsend-Gault, "Maritime cooperation in a functional perspective", in: *Maritime Energy Resources in Asia: Legal Regimes and Cooperation*, edited by Clive Schofield (The National Bureau of Asian Research: Seattle, 2012), p. 17.

V. Becker-Weinberg, *Joint Development of Hydrocarbon Deposits in the Law of the Sea*, Hamburg Studies on Maritime Affairs 30, DOI 10.1007/978-3-662-43570-0_7

have also undertaken military actions in order to reinforce their claims and to secure unrealistic exclusive access to common marine natural resources.[2]

Other regions of the world also offer examples of how the perception of the hydrocarbon potential of certain disputed maritime areas, in addition to territorial disputes, have increased tensions between States, preventing them from reaching an agreement on the delimitation of maritime boundaries. This is the case of the Mediterranean Sea, where the insufficient number of agreements on maritime delimitation has also been the consequence of the perception of the oil and gas potential of its maritime areas, where "*highly developed and highly promising areas alternate with basins of little interest*".[3]

In contradistinction, coastal States of the Baltic Sea have been able to succeed in the delimitation of maritime boundaries,[4] even though there are areas "*with a promising potential for mineral resources*",[5] but generally considered as having "*seemingly limited potential for mineral resources*" and with "*no significant petroleum or mineral resources known*".[6] On the other hand, in the oil and gas-rich North Sea, coastal States have successfully implemented joint development

[2] Mark J. Valencia further identifies the following alternatives: "*1. occupy the area, either expelling or defending against forces of other claimants; 2. unilaterally undertake or license exploration and eventually exploitation in the area; 3. verbally and symbolically maintain the claim while stalling or manoeuvring diplomatically for support; 4. refuse to negotiate in conjunction with any of the above; 5. take no action and ignore the actions and verbiage of other claimants; 6. negotiate directly or through intermediaries to resolve the boundary dispute in conjunction with any of the above; or 7. set aside the boundary dispute and agree to jointly explore and exploit any hydrocarbon resources there.*" Valencia, supra note 70 in Chap. 2, p. 575. Also see Rodman R. Bundy, "State practice in maritime delimitation", in: *World Boundaries. Maritime Boundaries*, vol. 5, edited by Gerald H. Blake (Routledge: London, New York, 1994), pp. 36–40; A. M. Post, *Deepsea Mining and the Law of the Sea*, edited by Shigeru Oda (Martins Nijhoff Publishers: The Hague, Boston, Lancaster, 1983), p. 89.

[3] Leanza, *supra note 125 in Chap. 3*, pp. 374, 376, 378–382, 391–394.

[4] Erik Franckx, "Baltic Sea update", in: *International Maritime Boundaries, vol. 3, supra note 5 in Chap. 5*, pp. 2557–2571.

[5] Erik Franckx, "Baltic Sea maritime boundaries", in: *International Maritime Boundaries*, vol. 1, *supra* note 42 in Chap. 2, p. 366. Also *see* Annick de Marffy, "Recent practice of states in the Baltic Sea", in: *The Baltic Sea: New Developments in National Policies and International Cooperation*, edited by Renate Platzöder and Philomène Verlaan (Martinus Nijhoff Publishers: The Hague, London, Boston, 1996), pp. 91–100.

[6] Hanns J. Buchholz, "The Baltic Sea: lessons learned", in: *Maritime Regime Building. Lessons Learned and Their Relevance for Northeast Asia*, edited by Mark J. Valencia (Martinus Nijhoff Publishers: The Hague, Boston, London, 2001), pp. 19, 21. For an overview of the mineral resource-potential of the Baltic Sea *see* Rempel Hilmar and Michael Schmidt-Thomé, "Hydrocarbon potential of the Baltic Sea region", in: *Mineral Resources of the Baltic Sea-Exploration, Exploitation and Sustainable Development*, coordinated by Emelyan M. Emelyanou, Michael Schmidt-Thomé and Mikhail Spiridonov, 2 *Zeitschrift für Angewandte Geologie* (2004), pp. 17–28.

regimes while achieving a dynamic cooperation on the delimitation of maritime boundaries, which is evidenced by the agreements[7] signed after the decision of the ICJ in the *North Sea Continental Shelf* cases.[8]

Likewise, in the Persian Gulf, where the world's largest oil producers are found, coastal States have managed to agree on the delimitation of maritime boundaries, as

---

[7] *Treaty between the Kingdom of Denmark and the Federal Republic of Germany concerning the Delimitation of the Continental Shelf under the North Sea*, done at Copenhagen on 28 January 1971 and entered into force on 7 December 1972, published at *Bundesgesetzblatt* 1972 II S. 881 [online: http://www.un.org/depts/los/LEGISLATIONANDTREATIES/PDFFILES/TREATIES/DNK-DEU1971CS.PDF (accessed on August 2013], reproduced in: *International Maritime Boundaries*, vol. 2, *supra* note 125 in Chap. 3, pp. 1809–1813; *Agreement between the Government of the Kingdom of Denmark and the Government of the United Kingdom of Great Britain and Northern Ireland relating to the Delimitation of the Continental Shelf between the Two Countries*, done at London on 25 November 1971 and entered into force on 7 December 1972, [online: http://www.un.org/depts/los/LEGISLATIONANDTREATIES/PDFFILES/TREATIES/DNK-GBR1971CS.PDF (accessed on August 2013)], reproduced in: *ibid*, pp. 1833–1834; *Treaty between the Kingdom of the Netherlands and the Federal Republic of Germany on Delimitation of the Continental Shelf under the North Sea*, done at Copenhagen on 28 January 1971 and entered into force on 7 December 1972, published at *Bundesgesetzblatt* 1972 II S. 889 and 857 U.N.T.S. 131 [online: http://www.un.org/depts/los/LEGISLATIONANDTREATIES/PDFFILES/TREATIES/NLD-DEU1971CS.PDF (accessed on August 2013)], reproduced in: *ibid*, pp. 1844–1848; *Protocol to the Treaties of 28 January 1971 between the Federal Republic of Germany and Denmark and the Kingdom of the Netherlands, respectively, concerning the delimitation of the continental shelf under the North Sea*, done at Copenhagen on 28 January 1971, published at 857 U.N.T.S. 155 [online: http://www.un.org/depts/los/LEGISLATIONANDTREATIES/PDFFILES/TREATIES/DEU-DNK-NLD1971CS.PDF (accessed on August 2013)]; *Agreement between Denmark and Norway relating to the delimitation of the continental shelf, 8 December 1965—Exchange of notes constituting an agreement amending the above-mentioned Agreement, 4 June 1974* [online: http://www.un.org/depts/los/LEGISLATIONANDTREATIES/PDFFILES/TREATIES/DNK-NOR1974CS.PDF (accessed on August 2013)], reproduced in: *ibid*, pp. 1821–1823; *Agreement between the United Kingdom of Great Britain and Northern Ireland and the Federal Republic of Germany relating to the Delimitation of the Continental Shelf under the North Sea between the Two Countries*, done at London on 25 November 1971 and entered into force on 7 December 1972, published at 880 U.N.T.S. 185 [online: http://www.un.org/depts/los/LEGISLATIONANDTREATIES/PDFFILES/TREATIES/GBR-DEU1971CS.PDF (accessed on August 2013)], reproduced in: *ibid*, pp. 1856–1858; *Protocol between the Government of the Kingdom of the Netherlands and the Government of the United Kingdom of Great Britain and Northern Ireland amending the Agreement of 6 October 1965 relating to the delimitation of the continental shelf under the North Sea between the two countries*, done at London on 25 November 1971, published at 861 U.N.T.S. 224 [online: http://www.un.org/depts/los/LEGISLATIONANDTREATIES/PDFFILES/TREATIES/NLD-GBR1971CS.PDF (accessed on August 2013)], the Agreement and Protocol are reproduced in: *ibid*, pp. 1866–1869; *Protocol supplementary to the Agreement of 10 March 1965 between the Government of the United Kingdom of Great Britain and Northern Ireland and the Government of the Kingdom of Norway relating to the delimitation of the continental shelf between the two countries*, done at Oslo on 22 December 1978 and entered into force on 20 February 1980 [online: http://www.un.org/depts/los/LEGISLATIONANDTREATIES/PDFFILES/TREATIES/GBR-NOR1978CS.PDF (accessed on August 2013)], the Agreement and Protocol are reproduced in: *ibid*, pp. 1885–1889.

[8] *North Sea Continental Shelf cases, supra note 46 in Chap. 3.*

well as the implementation of cooperative regimes. This has been mostly the result of States' awareness that setting aside political and territorial disputes for the purpose of undertaking joint development would result in securing access to resources and consequently controlling supply, both of which are detrimental to their prosperity.[9]

7.2 The Asia-Pacific Region

7.2.1 A Case-Study

The presence of offshore hydrocarbon deposits in the Asia-Pacific region is a considerable cause of disputes between coastal States and a significant impediment for maritime delimitation. Most disputes in this region regard overlapping maritime claims based on geomorphologic and geological factors, but also territorial conflicts over islands and other features as States attempt to claim the respective adjacent maritime areas, as it happens most notably regarding the Spratly Islands.[10]

---

[9] *See* William T. Onorato, "A case study in joint development: the Saudi Arabia-Kuwait portioned neutral zone", in: *Geology and Hydrocarbon Potential of the South China Sea and Possibilities of Joint Development*, edited by Mark J. Valencia (Pergamon Press: New York, Oxford, Toronto, Sydney, Paris, Frankfurt: 1985), p. 540; Ahmad Razavi, *Continental Shelf Delimitation and Related Maritime Issues in the Persian Gulf*, edited by Shigeru Oda (Martinus Nijhoff Publishers: The Hague, Boston, London, 1997), p. 87; Fesharaki, *supra* note 8 in Chap. 5, p. 1325.

[10] Brian K. Murphy, "Dangerous ground: the Spratly islands and international law", in: 1-1 *Ocean & Coastal L. J.* (1994), pp. 187–212; Jon M. Van Dyke and Dale L. Bennett, "Islands and the delimitation of the Ocean Space in the South China Sea", in: 10 *Ocean Yearbook* (1993), pp. 54–89; Daniel J. Dzurek, "The Spratly Islands dispute: who's on first?", in: 2-1 *Maritime Boundaries* (1996), pp. 1–67; *Ibid*, "Southeast Asian offshore oil disputes", in: 11 *Ocean Yearbook* (1994) 157–164, 172–176; Gerardo M.C. Valero, "Spratly archipelago dispute. Is the question of sovereignty still relevant?", in: 18-4 *Marine Policy* (1994), pp. 314–344; Zou Keyuan, "The Chinese traditional maritime boundary line in the South China Sea and its legal consequences for the resolution of the dispute over the Spratly Islands", in: 14-1 *International Journal of Marine and Coastal Law* (1999), pp. 27–55; Hasjim Djala, "South China Sea island disputes", in: *Security Flashpoints. Oil, Islands, Sea Access and Military Confrontation*, edited by Myron H. Nordquist and John Norton Moore (Martinus Nijhoff Publishers: The Hague, Boston, London, 1998), pp. 109–133; Ted McDorman, "The South China Sea islands dispute in the 1990s—a new multilateral process and continuing friction", in: 8-2 *International Journal of Marine and Coastal Law* (1993), pp. 263–282; Lian A. Mito, "The Timor Gap treaty as a model for joint development in the Spratly islands", in: 13-3 *American University Law Review* (1998), pp. 727–728, 736–742, 749; Christopher C. Joyner, "The Spratly Islands dispute: rethinking the interplay of law, diplomacy, and geopolitics in the South China Sea", in: 13-2 *International Journal of Marine and Coastal Law* (1998), pp. 193–236, *Ibid*, "Toward a Spratly resource development authority: precursor agreements and confidence-building measures", in: *Security Flashpoints, ibid*, pp. 221–224, 229–242; Jon M. Van Dyke, "Disputes over islands and maritime boundaries in East Asia", in: *Maritime Boundary Disputes, Settlement Processes, and the Law of the Sea*, edited by Seoung-Yong Hong and Jon Van Dyke (Martinus Nijhoff Publishers: 2009), pp. 62–75; Alex G. Oude Elferink, "The Islands in the South China Sea: how does their presence

Some States in the Asia-Pacific region have opted for the implementation of joint development regimes as means to surpass similar deadlock situations, particularly when considering that the settlement of these disputes by negotiation and eventually by an agreement on the delimitation of maritime boundaries, or even with the intervention of a third party, would ultimately lead to a long-term procedure incompatible, at least, with States' energy needs. Moreover, in some instances, States were required to establish a legal framework consistent with the fact that operating rights had been granted before the delimitation of maritime boundaries. This was the case of the joint development regimes implemented in the Timor Sea, the Northeast China Sea and the South China Sea.[11]

Considering that the delimitation of the greater part of maritime boundaries in the Asia-Pacific region is a difficult endeavour, joint development agreements provide the necessary legal framework without prejudice of States' sovereignty claims while, at the same time, safeguarding, to a certain extent, the independence and territorial integrity of all States in the region.[12] In this respect, organisations such as the *Association of Southeast Asian Nations* (ASEAN) that have an

(Footnote 10 continued)
limit the extent of the high seas and the Area and the maritime zones of the mainland coasts?, in: 32-2 *Ocean Development and International Law* (2001), pp. 169–191; Schofield/Kuan-Hsiung Wang, *supra* note 54 in Chap. 6.

[11] On the disputes in the South China Sea and the prospects for joint development *see* Becker-Weinberg, *supra* note 1 in Chap. 2; Robert Beckman, "The UN Convention on the Law of the Sea and the Maritime Disputes in the South China Sea", in: 107 *American Journal of International Law* (2013), pp. 142–163; Tara Davenport, Ian Townsend-Gault, Robert Beckman, Clive Schofield, David Ong, Vasco Becker-Weinberg, and Leonardo Bernard, *Conference on Joint Development and the South China Sea—Conference Report* [online: http://cil.nus.edu.sg/wp/wp-content/uploads/2011/06/Report-of-CIL-Conference-on-Joint-Development-and-the-South-China-Sea-2011-04.08.2011.pdf (accessed on August 2013)].

[12] David Ong, "South-east Asian state practice on the joint development of offshore oil and gas deposits", in: *The Peaceful Management of Transboundary Resources*, edited by Gerald H. Blake, William J. Hildesley, Martin A. Pratt, Rebecca J. Ridley and Clive H. Schofield (Graham and Trotman/Martinus Nijhoff: London, Dordrecht, Boston, 1995), pp. 77–96; Park Hee Kwon, *The Law of the Sea and Northeast Asia. A Challenge for Cooperation* (Kluwer Law International: The Hague, London, Boston, 2000), pp. 89–97; R.P. Anand, "South Asia and the law of the sea: problems and prospects", in: *Ocean Governance: Strategies and Approaches for the 21st Century*, edited by Thomas A. Mensah (University of Hawaii: Honolulu, 1996), pp. 331–353; Mark J. Valencia and Jon Van Dyke, "Comprehensive solutions to the South China Sea disputes: some options", in: *Boundaries and Energy: Problems and Prospects*, edited by Gerald Blake, Martin Pratt, Clive Schofield and Janet A. Brown (Kluwer Law International: London, The Hague, Boston, 1998), pp. 85–115; Wu Shicun, ""Joint Development": an ad hoc solution to the South China Sea Dispute", in: 2 *China Oceans Law Review* (2007), pp. 1–10; Steven Kuan-Tsyh Yu, "The law of EEZ/Shelf boundary delimitation: the practice of States in the South China Sea", in: *Proceedings of the International Law Association (ILA) First Asian-Pacific Regional Conference* (Chinese Society of International Law: 1996), pp. 45–48; Donald R. Rothwell, "The law of the sea in the Asian-Pacific region: an overview of trends and developments", in: *ibid*, p. 58; Victor Prescott and Clive Schofield, "Undelimited maritime boundaries of the Asian Rim in the Pacific Ocean", in: 3-1 *Maritime Boundaries* (2001), pp. 1-68; Park, *supra* note 108 in Chap. 6, pp. 175–243, 283.

important role in keeping regional stability, may contribute towards a growing awareness of the benefits resulting from joint development regimes, as well as establishing a high-level interaction between States and consequently encouraging the exercise of self-restraint.[13]

7.2.2 The Timor Sea

The distance between the coasts of East Timor and Australia is approximately 250 nm, allowing for the occurrence of overlapping continental shelves between the two countries in the Timor Sea.[14] Australia always defended the existence of two continental shelves based on the natural prolongation of its submerged land mass until the Timor Trough, at a distance of 150 nm from its northern coast,[15] whereas East Timor has considered this position unacceptable, as did Portugal and Indonesia before, although in different historical periods. East Timor considers that the delimitation in the Timor Sea should be made in accordance with a median line and that the geological characteristics of the Timor Trough should not be considered when confronted with other seabed depressions, given that these depressions do not represent a break or fault of the continental shelf.[16] In addition, the

[13] *See Joint Communiqué 46th ASEAN Foreign Ministers' Meeting Badar Seri Begawan, Brunei Darussalam, 29–30 June 2013*, paras. 90–91 [online: http://www.asean.org/images/2013/news/joint%20communique%20of%20the%2046th%20asean%20foreign%20ministers%20meeting%2046th%20amm%20-%20final%20-%2030%20june%202013.pdf (accessed on August 2013)].

[14] Nuno Marques Antunes, *Towards the Conceptualisation of Maritime Delimitation-Legal and Technical Aspects of Political Process* (Martinus Nijhoff Publishers: Leiven, Boston, 2003), p. 358; Henry Burmester, "Australia and the Law of the Sea", in: *The Law of the Sea in the Asian Pacific Region*, edited by James Crawford and Donald, R. Rothwell (Martinus Nijhoff Publishers: Dordrecht, Boston, London), pp. 51–64.

[15] Malcom Evans, *Relevant Circumstances and Maritime Delimitation* (Claredon Press: Oxford, 1989), pp. 99–118; John Robert Victor Prescott, *The Political Geography of the Oceans. Problems in Modern Geography* (David & Charles Newton Abbot: London, Vancouver, 1975), pp. 191–192; C. Cook, "Filling the gap-delimiting the Australia-Indonesia maritime boundary", in: 10 *Australian Yearbook of International Law* (1981–1983), pp. 170–171; Nuno Marques Antunes, "Spatial allocation of continental shelf rights in the Timor Sea. Reflections on maritime delimitation and joint development", in: *Estudos em Direito Internacional Público* (Almedina: Coimbra, 2004), pp. 274–275, 277; *supra* note 83, 359, 385–389; V. Prescott, *supra* note 50 in Chap. 3, pp. 71–72.

[16] *Continental Shelf (Tunisia/Libya), supra note 10 in Chap. 3*, paras. 62–68, 80; *Continental Shelf (Libya/Malta), supra note 60 in Chap. 3*, paras. 35–41. *See* Masahiro Miyoshi, "Some thoughts on maritime boundary delimitation", in: *Maritime Boundary Disputes, Settlement Processes, and the Law of the Sea*, edited by Seoung-Yong Hong and Jon Van Dyke (Martinus Nijhoff Publishers: 2009), pp. 108–113; Laurent Lucchini, "La délimitation des frontières maritimes dans la jurisprudence internationale: vue d'ensemble", in: *Maritime Délimitation*, edited by Rainer Lagoni and Daniel Vignes (Martinus Nijhoff Publishers: Leiden, Boston, 2006), pp. 4–5; William T. Onorato and Mark J. Valencia, "The new Timor Gap Treaty: legal and political implications", in: 28 *International Centre for Settlement of Investiment Disputes*

adoption of the natural prolongation criterion would not forego the consideration of other criteria, such as equity and the fairness of delimitation and which are not present in Australia's position.[17] In fact, should the latter prevail, the larger part of the known hydrocarbon deposits laying in the seabed and subsoil of the Timor Sea, and which are mostly found closer to the Timorese coast, would be in Australia's continental shelf, whereas the applicability of the median line would place East Timor in a more advantageous position.[18] In this case, Australia would most probably demand the readjustment of the median line, should the issue of delimitation of maritime boundaries be submitted to an International Court or Tribunal, which would almost certainly provoke the intervention of Indonesia.[19]

Australia, Indonesia and East Timor have in different occasions engaged in bilateral negotiations on the delimitation of maritime boundaries in the Timor Sea, having Australia and Indonesia signed two delimitation agreements regarding the maritime boundaries between Papua New Guinea and Indonesia,[20] and in the Arafura Sea, which essentially envisaged settling States' rights regarding the exploration and exploitation of the natural resources of the seabed and subsoil.[21]

(Footnote 16 continued)
Review-Foreign Investment Law Journal (2000), p. 62; Mark J. Valencia and Masahiro Miyoshi, "Southeast Asia seas: joint development of hydrocarbons in overlapping claim areas", in: 16-3 *Ocean Development and International Law Journal* (1986), p. 228; Edward Duncan Brown, "The Tunisia-Libya continental shelf case", in: 7-3 *Marine Policy* (1983), pp. 145–148.

[17] *Continental Shelf (Libya/Malta), supra note 60 in Chap. 3*, paras. 46–50.

[18] Stuart Kaye, "Negotiation and dispute resolution: a case study in international boundary making—the Australia-Indonesia boundary", in: *Oceans Management in the 21st Century: Institutional Frameworks and Responses*, edited by Alex G. Oude Elferink and Donald R. Rothwell (Martinus Nijhoff Publishers: Leiden, Boston, 2004), pp. 146–147; William T. Onorato and Mark J. Valencia, "International cooperation for petroleum development: the Timor Gap Treaty", in: 5-1 *International Centre for Settlement of Investiment Disputes Review—Foreign Investment Law Journal* (1990), pp. 2–3; "The new Timor Gap Treaty: legal and political implications", in: 28 *International Centre for Settlement of Investment Disputes Review—Foreign Investment Law Journal* (2000), p. 61; Charney, *supra* note 37 in Chap. 6, p. 1029; Valencia/Miyoshi, *supra* note 16, p. 230.

[19] Marques Antunes, *supra* note 14, p. 379.

[20] *Agreement between Australia and Indonesia Concerning Certain Boundaries between Papua New Guinea and Indonesia*, done at Jakarta on 12 February 1973 and entered into force 26 November 1974, published at 975 U.N.T.S. 4 [online: http://www.un.org/depts/los/LEGISLATIONANDTREATIES/PDFFILES/TREATIES/AUS-IDN1973PNG.pdf (accessed on August 2013)], reproduced in: *International Maritime Boundaries*, vol. 2, *supra note 125 in Chap. 3*, pp. 1225–1228.

[21] *Agreement between the Government of the Commonwealth of Australia and the Government of the Republic of Indonesia Establishing Certain Seabed Boundaries*, done at Canberra on 18 May 1971 and entered into force on 8 November 1969, published at 974 U.N.T.S. 307 [online: http://www.un.org/depts/los/LEGISLATIONANDTREATIES/PDFFILES/TREATIES/AUS-IDN1971SB.pdf (accessed on August 2013)], reproduced in: *International Maritime Boundaries*, vol. 2, *supra note 125 in Chap. 3*, pp. 1202–1205. Also *see Treaty between the Government of Australia and the Government of the Republic of Indonesia Establishing an Exclusive Economic Zone Boundary and Certain Seabed Boundaries*, done at Perth on 14 March 1997 [online: http://www.un.org/depts/los/LEGISLATIONANDTREATIES/PDFFILES/TREATIES/AUS-IDN1997

The agreement that established the boundaries between Papua New Guinea and Indonesia did not include the continental shelf of East Timor, since at that time the territory of East Timor was under Portuguese rule.[22] Only with the Indonesian invasion and occupation on 7 December 1975, did East Timor begin to be de facto controlled by Indonesia.[23] Australia and Indonesia further agreed on a cooperative regime regarding their mutual interest in the rational management, conservation and optimum utilization of living resources.[24] Only after almost two decades of unsuccessful negotiations did Australia and Indonesia manage to agree on the deferral of the divisive issue of maritime delimitation and engage in the joint development of the non-living natural resources of the Timor Sea, under the legal framework of the *Timor Gap Treaty*.[25]

The *Timor Gap Treaty* was the first wide-ranging joint development regime of offshore hydrocarbon deposits and has been an example for other similar arrangements, such as the joint development regimes established between Senegal

(Footnote 21 continued)
EEZ.pdf (accessed on August 2013)], reproduced in: *International Maritime Boundaries, vol. 4, supra note 8 in Chap. 5*, pp. 2714–2727; *Agreement between the Government of the Commonwealth of Australia and the Government of the Republic of Indonesia Establishing Certain Seabed Boundaries in the Area of the Timor and Arafura Seas, Supplementary to the Agreement of 18 May 1971*, done at Jakarta on 9 October 1972 and entered into force on 8 November 1973, published at 974 U.N.T.S. 319 [online: http://www.un.org/depts/los/LEGISLATIONANDTREATIES/PDFFILES/TREATIES/AUS-IDN1972TA.pdf (accessed on August 2013)], reproduced in: *International Maritime Boundaries*, vol. 2, *supra note 125 in Chap. 3*, pp. 1215–1218.

22 Article 2 of the Agreement Australia/Indonesia, *ibid.*

23 Resolutions (UNSC) n. 384, 22 December 1975; n. 389, 22 April 1976; n. 1236, 7 May 1999; n. 1246, 11 June 1999; n. 1262, 27 August 1999; n. 1264, 15 September 1999; and n. 1272, 25 October 1999. Also *see* Indonesian Law 7, 17 July 1976, which recognizes the integration of East Timor, revoked on 20 October 1999, by an Act of the Indonesian Popular Assembly.

24 *Agreement between the Government of Australia and the Government of the Republic of Indonesia relating to cooperation in fisheries*, done at Jakarta on 22 April 1992, published at 1170 U.N.T.S. 288; *Memorandum of Understanding between the Government of Australia concerning the Implementation of a Provisional Fisheries Surveillance and Enforcement Arrangement*, done at Jakarta 29 October 1981 and entered into force on 1 February 1982, reproduced in: *International Maritime Boundaries*, vol. 2, *supra note 125 in Chap. 3*, pp. 1238–1239; *Memorandum of Understanding between the Government of Australia and the Government of the Republic of Indonesia Regarding the Operations of Indonesian Traditional Fishermen in Areas of the Australian Exclusive Fishing Zone and Continental Shelf*, done at Jakarta on 7 November 1974 and entered into force on 28 February 1975, reproduced in: *ibid*, pp. 1239–1241.

25 Article 33(1) of the Timor Gap Treaty, *supra* note 113. On the Timor Gap Treaty *see* Keith Suter, "Timor Gap treaty. The continuing controversy", in: 17-4 *Marine Policy* (1993), pp. 294–302; Francis M. Auburn and Vivian L. Forbes, "The Timor Gap treaty and the Law of the Sea Convention", in: 10 *Ocean Yearbook* (1993), pp. 40–53; Henry Burmester, "The zone of co-operation between Australia and Indonesia: a preliminary outline with particular reference to applicable law", in: *Joint Development*, vol. 2, *supra* note 70 in Chap. 2, pp. 128–139; Ernst Willheim, "Australia-Indonesia sea-bed boundary negotiations: proposals for a joint development zone in the "Timor Gap" ", in: 29 *Natural Resource Journal* (1989), pp. 821–842; Kusuma-Atmadja, *supra* 111 in Chap. 6, pp. 592–609; Johnston/Valencia, *supra* note 38 in Chap. 4, pp. 70–74. Miyoshi, *supra* note 2 in Chap. 4, pp. 17–21.

and Guinea-Bissau,[26] and Nigeria and Sao Tome and Principe.[27] Its implementation sought to satisfy Australia's growing energy needs, as well as to create the necessary normative regime that allowed for the continuation of the exploration and exploitation activities and the consolidation of operators' rights.[28]

With the end of the Indonesian occupation, the territory of East Timor began to be administrated by the *United Nations Transitional Administration in East Timor* (UNTAET).[29] Its judicial, political and legislative powers included the responsibility for entering into international agreements in representation of the interests and economic viability of the future State of East Timor.[30] UNTAET and Australia began negotiations in view of adapting the *Timor Gap Treaty* to the new *status quo*, *i.e.* an independent East Timor, rinsing the *Timor Gap Treaty* of any potential illegitimacy in light of the internationally recognized right of self-determination of the East Timorese people and its sovereignty over their natural resources.[31] These

---

[26] *Supra note* 116 in Chap. 6.

[27] *Supra* note 2 in Chap. 4. On the agreement *see* Cissé Yacouba, "Les gisements en mer des hydrocarbures transfrontalières: régime juridique en droit de la délimitation maritime", in: 35 *Ottawa L. Rev.* (2003–2004), pp. 64–66; Hurst Groves, "Offshore oil and gas resources: economics, politics and the rule of law in the Nigeria-Sao Tome e Principe joint development zone", in: 59-1 *J. Int'l Affairs* (Fall/Winter 2005), pp. 85–92. Also *see* John Robert Victor Prescott, "Guinea-Bissau-Senegal", in: *International Maritime Boundaries, vol. 3, supra note 5 in Chap. 5*, pp. 2250–2255; Jean-Pierre Queneudec, "L'arbitrage relatif à la détermination de la frontière maritime entre la Guinée-Bissau et le Sénégal 31 Juillet 1989", in: 35 *Annuaire Français de Droit International* (1989), pp. 323–339; Gbenga Oduntan, "The emergent legal regime for exploration of hydrocarbons in the Gulf of Guinea: imperative considerations for participating States and multinationals", in: 57 *Int'l Comp. L. Quarterly* (April 2008), p. 274.

[28] Marques Antunes, *supra* note 14, pp. 393–396; *supra* note 15, p. 287. A similar concern is also present in the agreement between Australia and Papua New Guinea *see* Article 5 of the *Treaty between Australia and the Independent State of Papua New Guinea Concerning Sovereignty and Maritime Boundaries in the area between the Two Countries, including the area known as Torres Strait, and Related Matters*, done at Sydney on 18 December 1978 and entered into force 15 February 1985 [online: www.un.org/depts/los/LEGISLATIONANDTREATIES/PDFFILES/TREATIES/AUS-PNG1978TS.PDF (accessed on August 2013)], reproduced in: *International Maritime Boundaries*, vol. 1, *supra* note 42 in Chap. 2, pp. 937–975. The text of Article 5(1) reads: "*Where prior to 16 September 1975 Australia has granted an exploration permit for petroleum under Australian law in respect of a part of the seabed over which it ceases by virtue of this Treaty to exercise sovereign rights, and a permittee retains rights in respect of that permit immediately prior to the entry into force of this Treaty, Papua New Guinea, upon application by that permittee, shall offer to that permittee a petroleum prospecting licence or licences under Papua New Guinea law in respect of the same part of the seabed on terms that are not less favourable than those provided under Papua New Guinea law to any other holder of a seabed petroleum prospecting licence.*"

[29] Resolution (UNSC) n. 1272, 25 October 1999. UNTAET was replaced by the *United Nations Mission in East Timor*, which was responsible for organizing the referendum on the independence of East Timor that took place on 19 October 1999. *See* Resolution (UNSC) n. 1246, 11 June 1999.

[30] *See* Secretary-General Report S/1999/1024, para. 35, 4 October 1999.

[31] *Supra* note 50 in Chap. 4.

negotiations led to the *Exchange of Notes* and subsequent signing on 10 February 2000 of the MoU on the revised *Timor Gap Treaty*.[32]

The preservation of the joint development model of the *Timor Gap Treaty* implied keeping in force the relevant Indonesian legislation and the decisions and directives issued by the Ministerial Council and the Joint Authority created under the *Timor Gap Treaty*. It also provided for the safeguarding of the rights acquired by operators under product sharing contracts signed with the Joint Authority. UNTAET undertook all rights and obligations of Indonesia under the *Timor Gap Treaty*, with the exception of those contrary to the interests of the people of East Timor. Australia and East Timor would also sign two interim arrangements regarding the development of the *Greater Sunrise gas fields* (*Sunrise* and *Troubadour*) with the purpose of improving the legal and regulatory framework of this unitization regime and achieving the equal sharing of upstream revenues, without prejudice of either States' legal position on or legal rights relating to the delimitation of their respective maritime boundaries, or implying the renunciation, recognition or affirmation of any right or claim relating to the whole or any part of the Timor Sea.[33]

Pursuant to Article 21 of the *Petroleum Mining Code* included in the *Timor Gap Treaty*, whenever a hydrocarbon deposits straddles two or more contract areas, the Joint Authority would require the respective operators to enter into unitization

---

[32] *Supra* note 114 in Chap. 6. On 5 July 2001, UNTAET and Australia signed a second MoU which reaffirmed the joint development model initially adopted, no longer under the auspices of the *Timor Gap Treaty*, but instead of its revised version agreed on the MoU signed on 10 February 2000 and from then onwards referred to as the *Timor Sea Treaty*. These documents are reproduced in: *International Maritime Boundaries, vol. 4, supra note 8 in Chap. 5*, pp. 2763–2795. Also *see* the *Exchange of Notes Constituting an Agreement between the Government of Australia and the Government of the Democratic Republic of Timor-Leste Concerning Arrangements for Exploration and Exploitation of Petroleum in an Area of the Timor Sea between Australia and East Timor*, done at Dili on 20 May 2002 [online: http://www.un.org/depts/los/LEGISLATIONANDTREATIES/PDFFILES/TREATIES/AUS-TLS2002EX.PDF (accessed on August 2013)]; *Treaty between Australia and the Democratic Republic of Timor-Leste on Certain Maritime Arrangements in the Timor Sea*, done at Sydney on 12 January 2006 and entered into force on 23 February 2007 [online: http://www.austlii.edu.au/au/other/dfat/treaties/2007/12.html (accessed on August 2013)]. On the Timor Sea Treaty *see* David Ong, "The new Timor Sea Arrangement 2001: is joint development of common offshore oil and gas deposits mandated under International Law?", in: 17-1 *International Journal of Marine and Coastal Law* (January 2002) 79–122; William T. Onorato, "The Timor Gap Treaty revisted ten years on", in: *International Finance and Development Law. Liber Amicorum Ibrahim F.I. Shihata*, edited by Sabine Schlemmer-Schulte and Ko-Yung Tung (Kluwer Law International: The Hague, London, Boston, New York, 2001), pp. 527–547. In 2013, East Timor instituted proceedings against Australia under lit. b) of Annex B to Article 23 of the *Timor Sea Treaty* on the grounds that Australia seized documents and data from the office of an Australian lawyer representing East Timor regarding the latter's claim that Australia did not negotiate the in good faith the 2006 Treaty on Certain Maritime Arrangements in the Timor Sea [online: http://www.icj-cij.org/docket/files/156/18078.pdf (accessed on August 2014]

[33] *Supra* note 136 in Chap. 6. On the Greater Sunrise agreement *see* Gillian Triggs, "The Timor Sea Treaty and the international agreement for the Greater Sunrise: practical solutions in the Timor Sea", in: 23 *Australian Yearbook of International Law* (2004), pp. 161–176; Marques Antunes, *supra* note 15, p. 289; Ong, *ibid*, p. 80.

agreements subject to its approval and within a reasonable time as determined by this entity, and appoint the operator responsible for production, being that, in the absence of an agreement, the Joint Authority would decide on the applicable unitization regime. Article 9 of the *Timor Sea Treaty* provides that any hydrocarbon deposit that extends across the boundary of the joint development area shall be treated as a single entity, and that both countries shall work to reach an agreement on the manner in which the deposit will be most effectively exploited and on the equitable sharing of the benefits arising from such exploitation.

7.2.3 The Northeast China Sea

The Northeast China Sea is a region with significant precedents of cooperation on the protection of the marine environment, navigation and fishing. However, the existence of offshore hydrocarbon deposits has proven to be a conflicting issue between littoral States, in particular regarding the delimitation of maritime boundaries.[34] The perception of the strong hydrocarbon potential of the East China Sea became generally known with the publication in 1968 of a report prepared by a group of scientists from Japan, South Korea, Taiwan and the United States of America, and sponsored by the *United Nations Economic Commission for Asia and the Far East*.[35] In the year following the publication of this report, Japan, South Korea and Taiwan claimed the greater part of the continental shelf where hydrocarbon deposits were thought to be found and swiftly entered into operating agreements with oil companies in view of its development. In fact, South Korea had already in 1968 proceeded with the division in seven blocks of an area of the East China Sea also claimed by Japan, granting its exploitation to four companies. In return, Japan also granted exploitation rights in areas where claims overlapped.

In 1970, in the absence of an agreement on the delimitation of maritime boundaries in the continental shelf, Japan, South Korea and Taiwan attempted to enter into a joint

[34] *See* Vasco Becker-Weinberg, "Joint development arrangements in Northeast Asia and the Gulf of Tonkin", in: *Beyond Territorial Disputes in the South China Sea: Legal Frameworks for the Joint Development of Hydrocarbon Deposits*, edited by Robert Beckman, Ian Townsend-Gault, Clive Schofield, Tara Davenport and Leonardo Bernard (Edward Elgar: Cheltenham-UK, Northampton-MA,US, 2013), pp. 219–228. Also *see* Mark J. Valencia, "Relevance of lessons learned to Northeast Asia", in: *Maritime Regime Building. Lessons Learned and Their Relevance for Northeast Asia*, edited by Mark J. Valencia (Martinus Nijhoff Publishers: The Hague, Boston, London, 2001), p. 145; Kiyofumi Nakauchi, "Problems of delimitation in the East China Sea and the Sea of Japan", in: 6 *Ocean Development and International Law Journal* (1979), pp. 305-316.

[35] *Economic Commission for Asia and the Far East, Annual Report (18 April 1967—30 April 1968), Official Records of the 45th Session of the Economic and Social Council, 2nd Supplement* (United Nations: New York, 1968), pp. 4–5, 81–86 [online: http://www.unescap.org/EDC/English/AnnualReports/1968(24).pdf (accessed on August 2013)].

[36] Hong Nong and Wu Shicun, "The energy security of China and oil and gas exploitation in the South China Sea", in: *Recent Developments in the Law of the Sean and China*, edited by Myron H. Nordquist,

development agreement of offshore hydrocarbon deposits and defer the maritime delimitation dispute.[36] This attempt met with the objection of China based on its claim over part of the continental shelf that was to be included in said joint development agreement. This in turn lead to the US undertaking efforts to protect the rights of its oil companies who had signed operating agreements with Japan, South Korea and Taiwan. Taiwan eventually withdrew from this trilateral attempt of joint development regime, enabling Japan and South Korea to negotiate a joint development agreement.[37]

In 1974, after nearly 3 years of consultations and more than 20 years of disagreement regarding the relevance of the bay of Okinawa for the purpose of delimitation, Japan and South Korea signed two agreements that ended the maritime dispute on the delimitation of the continental shelf between the two countries in the East China Sea.[38] These two agreements were the result of States' achievement in differentiating between the delimitation of maritime boundaries and the development of marine natural resources, permitting reaching an agreement on delimitation where States' claims did not overlap, and to implement a joint development regime in areas where claims did overlap, despite China's objection to such a cooperative effort.[39] In addition, States agreed not to consider the island of Takeshima/Dokdo in the Sea of Japan for the purpose of delimitation of maritime boundaries as initially claimed by both countries, as well as that the continental shelf delimitation agreement would not affect the legal status of the superjacent waters and air space. This agreement also included a natural resources clause whereby States would endeavour to reach an agreement regarding the development of hydrocarbon deposits that extended across a boundary line or, in

(Footnote 36 continued)
John Norton Moore and Kuen-chen Fu (Martinus Nijhoff Publishers: Leiden, Boston, 2006), p. 145; Jon M. Van Dyke, "The Republic of Korea's maritime boundaries", in: 18-4 *International Journal of Marine and Coastal Law* (2003), pp. 509–540; Jonathan Charney, "Central East Asian maritime boundaries and the Law of the Sea", in: 89-4 *American Journal of International Law* (1995), pp. 746–748; Choon-ho Park, "Oil under troubled waters: the Northeast Asia sea-bed controversy", in: 14 *Harvard International Law Journal* (1973), pp. 212–260; *Ibid*, *supra* note 21 in Chap. 6, p. 1335; Valencia, *supra* note 46 in Chap. 4, pp. 223–247; Miyoshi, *supra* note 108 in Chap. 6, pp. 89–97; Zou, *supra* note 34 in Chap. 6, pp. 161–168.

37 Paul C. Yuan, "China's offshore oil development policy and legislation: an overall analysis", in: 3-2 *International Journal of Estuarine and Coastal Law* (1988), pp. 101–137; *Ibid*, "China's sovereignty over its offshore oil", in: *Geology and Hydrocarbon Potential of the South China Sea and Possibilities of Joint Development*, edited by Mark J. Valencia (Pergamon Press: New York, Oxford, Toronto, Sydney, Paris, Frankfurt, 1985), pp. 520–532.

38 Seo-Hang Lee, "Korea's claims to maritime jurisdiction", in: *Korean Journal of Comparative Law* (1990), p. 70.

39 *Agreement between Japan and the Republic of Korea Concerning the Establishment of Boundary in the Northern Part of the Continental Shelf Adjacent to the Two Countries*, done at Seoul on 30 February 1974 and entered into force on 20 May 1981, published at 1225 U.N.T.S. 104 [online: http://www.un.org/depts/los/LEGISLATIONANDTREATIES/PDFFILES/TREATIES/jap-kor1974north.pdf (accessed on August 2013)], reproduced in: *International Maritime Boundaries*, vol. 1, *supra* note 42 in Chap. 2, pp. 1063–1089; *Agreement between Japan and the Republic of Korea Concerning Joint Development of the Southern Part of the Continental Shelf Adjacent to the Two Countries*, *supra* note 486. On these agreements *see* Miyoshi, *supra* note 108 in Chap. 6, pp. 551–552.

the absence of an agreement, an arbitrator appointed at the request of either State would determine the most effective way to develop said hydrocarbon deposit.

The implementation of the joint development agreement was delayed due to Japan's need to adjust its domestic legislation to the obligations established in the agreement. States were also concerned with the preservation of traditional fishing in the joint development zone and the fact that the latter is located in an area prone to the occurrence of hurricanes and cyclones.[40]

The agreement created a joint development area in the southern part of the continental shelf of both countries, which could be divided into subzones for the purpose of exploration and exploitation activities by Japanese and South Korean concessionaires. Subject to States' approval, concessionaires would enter into operating agreements for the purpose of jointly undertaking such activities, being entitled to an equal stake of the natural resources extracted in the joint development zone and sharing in identical proportions the respective costs. The applicable law would be that of the country of the relevant concessionaire and these resources considered as extracted on the continental shelf of that State.

According to these operating agreements, concessionaires would have to agree on a mechanism for the settlement of disputes, as well as adjust the exploration and exploitation activities with the relevant fisheries interests. In this case, it would have been preferable that States had agreed on a mandatory mechanism in order to reduce the potential impact on exploration and exploitation activities.

Japan and South Korea failed to implement a common regime applicable to the protection of the marine environment, merely establishing in Article 20 that both States shall agree on measures to be taken to prevent collisions and pollution at sea resulting from activities relating to the exploration and exploitation activities in the joint development zone. The only common measures are those referred to in the *Exchange of Notes* following the signing of the agreement. These include, for example, the prohibition of discharge of waste from a ship or a marine facility, the duty for States to promptly undertake measures to prevent the spread of pollution, to prevent the continued discharge of oil and to remove the discharged oil and to promptly provide all available information when a discharge of large quantities of oil from a ship or a marine facility occurs, as well as to undertake measures to seal an abandoned well in order to prevent leakage of mine water or other substances from such well.[41]

This agreement also includes a natural resources clause which establishes that concessionaires shall seek to reach an agreement through consultation and subject to States' approval regarding the most effective means to develop a hydrocarbon deposit that may extend across the lines of the joint development zone. In the event that concessionaires fail to reach an agreement, both States shall present a joint proposal following the required consultations.

---

[40] *See* Miyoshi, *ibid*, pp. 545, 549–550.

[41] *Exchanges of Notes—Agreement between Japan and the Republic of Korea Concerning Joint Development of the Southern Part of the Continental Shelf Adjacent to the two Countries*, done at Seoul on 30 January 1974, *supra* note 108 in Chap. 6.

In what concerns other States of Northeast Asia, there is also report that China and North Korea have agreed on implementing a joint development regime in the Yellow Sea, as well as that China and South Korea have made efforts towards the delimitation of maritime boundaries.[42] China and Japan have yet to agree on a concept of joint development for the East China Sea, despite long acceptance by both countries that only such a cooperative effort may offer a way forward.[43] These countries also have an ongoing dispute regarding the legal status of Okinotorishima.[44]

---

[42] Becker-Weinberg, *supra* note 34. Also *see* Mark J. Valencia, "Conclusions and the way forward", in: 29-2 *Marine Policy* (2005), pp. 185–187; *Ibid*, "Conclusions, regime building and the way forward", in: 28-1 *Marine Policy* (January 2004), pp. 89–96; *Ibid*, "Regime building in the East China Sea", in: 34-1 *Ocean Development and International Law Journal* (2003), p. 199; Yann-huei Song and Zou Keyuan, "Maritime legislation of mainland China and Taiwan: developments, comparison, implications, and potential challenges for the United States", in: 31-4 *Ocean Development and International Law Journal* (2000), pp. 303–345. On the delimitation of maritime boundaries and development of offshore hydrocarbon deposits in the Yellow Sea *see* Mark J. Valencia, "The Yellow Sea. Transnational marine resource management issues", in: 12-4 *Marine Policy* (1988), pp. 383, 385–389, 392–395; *Ibid*, "Northeast Asia: petroleum potential, jurisdictional claims, and international relations", in: 20-1 *Ocean Development and International Law Journal* (1989), pp. 46–48, 57; Yann-Huei Song, "One China, but two sets of maritime legislation: developments, implications, and challenges for the United States", in: *Law of the Sea. Common Heritage and Emerging Challenges*, edited by Harry N. Scheiber (Martinus Nijhoff Publishers: The Hague, London, Boston, 2000), pp. 216–217, 222–225; Jeanette Greenfield, *China's Practice in the Law of the Sea* (Claredon Press: Oxford, 1992), pp. 117–149; Park, *East Asia and the Law of the Sea*, *supra* note 108 in Chap. 6, pp. 15–16, 245–270; Keyuan, *supra* note 66 in Chap. 2, pp. 31–52, 92–95; H. Schulte Nordholt, "Delimitation of the continental shelf in the East China Sea", in: 32 *Netherlands International Law Review* (1985), pp. 123–160; Charney, *supra* note 36, pp. 739–740.

[43] Seokwoo Lee and Young Kil Park, "Maritime delimitation and joint resource development in the East China Sea", in: *Securing the Ocean for the Next Generation*, Papers from the Law of the Sea Institute, UC Berkeley-Korean Institute of Ocean Science and Technology Conference, held in Seoul, Korea, May 2012 [online: http://www.law.berkeley.edu/files/Lee-Park-final.pdf (accessed on August 2013)]; Gao Jianjun, "A note on the 2008 cooperation consensus between China and Japan in the East China Sea", in: 40 *Ocean Development and International Law Journal* (2009), pp. 291–296; Ji Guoxing, "Sino-Japanese jurisdictional delimitation in East China Sea: approaches to dispute settlement", in: *Maritime Boundary Disputes, Settlement Processes, and the Law of the Sea*, edited by Seoung-Yong Hong and Jon Van Dyke (Martinus Nijhoff Publishers: 2009), pp. 77–105; Zou Keyuan, "Implementing the United Nations Convention on the Law of the Sea in East Asia: issues and trends", in: 9 *Singapore Year Book of International Law* (2005), p. 6; Deng Xiaoping, "Speech at the third plenary session of the Central Advisory Commission of the Communist Party of China, October 22, 1984", in: 3 *Selected Works of Deng Xiaoping* (Foreign Languages Press: Beijing, 1984). Also *see China's Path of Peaceful Development and Its View of Regional Security*, 3 November 2008 [online: http://www.fmprc.gov.cn/eng/wjb/zwjg/zwbd/t520658.htm (accessed on August 2013)]; *China-Japan Joint Statement on All-round Promotion of Strategic Relationship of Mutual Benefit*, 22 May 2008 [online: http://www.fmprc.gov.cn/eng/wjdt/2649/t458431.htm (accessed on August 2013)].

[44] Yann-huei Song, "Okinotorishima: A "rock" or and "island"? Recent maritime controversy between Japan and Taiwan/China", in: *Maritime Boundary Disputes, Settlement Processes, and the Law of the Sea*, edited by Seoung-Yong Hong and Jon Van Dyke (Martinus Nijhoff Publishers: 2009), pp. 145–176. Guifang Xue, "How much can a rock get? A reflection from the Okinotorishima Rocks", in: *13-1 China Oceans Law Review* (2011), pp. 1–34.

7.2.4 The Southeast China Sea

In 1979, Thailand and Malaysia signed a MoU establishing a joint authority for the purpose of developing the resources of the continental shelf of the Gulf of Thailand, where these States' sovereignty claims overlapped,[45] without prejudice of ongoing negotiations regarding maritime delimitation.[46] That same year, the two States signed a second MoU identifying the points to be considered for the purpose of delimitating the continental shelf boundary between the two countries in the Gulf of Thailand, while further undertaking to continue negotiations towards final delimitation.[47] This second MoU also included a mineral resources clause establishing that both States would cooperate in seeking an agreement regarding the efficient development of hydrocarbon deposits that straddled a boundary line, even-handedly partaking in all incurred expenses and benefits.[48]

The MoU that established the Joint Authority stipulated that in the event that States achieved an agreement on the delimitation of the continental shelf of the Gulf of Thailand before the end of its term, this entity would be terminated and its gains and assets equally shared by both countries, together with its losses and debts. In the event that States should fail to agree on the delimitation of maritime

---

[45] *Memorandum of Understanding between the Kingdom of Thailand and Malaysia in the Establishment of a Joint Authority for the Exploitation of the Resources of the Sea-bed in a defined Area of the Continental Shelf of the Two Countries in the Gulf of Thailand*, done at Chiang Mai on 21 February 1979 [online: http://cil.nus.edu.sg/rp/il/pdf/1979%20MOU%20between%20Malaysia%20and%20Thailand-pdf.pdf (accessed on August 2013)], reproduced in: *International Maritime Boundaries*, vol. 1, *supra* note 42 in Chap. 2, pp. 1107–1111. *See* Clive Schofield and May Tan-Mullins, "Maritime claims, conflicts and cooperation in the Gulf of Thailand", in: 22 *Ocean Yearbook* (2008), pp. 75–116; David Ong, "Thailand/Malaysia. Joint development agreement 1990", in: 6-1 *International Journal of Estuarine and Coastal Law* (February 1991), pp. 61–63; Ian Townsend-Gault, "The Malaysia/Thailand Joint Development Arrangement", in: *Joint Development*, vol. 2, *supra* note 70 in Chap. 2, pp. 102–107.

[46] These negotiations were pursuant to the guidelines approved in the *Agreed Minutes of the Malaysia-Thailand Official's Meeting on Delimitation of the Continental Shelf Boundary between Malaysia and Thailand in the Gulf of Thailand and in the South China Sea, 27 February—1 March 1978. See* R. Haller-Trost *The Contested Maritime and Territorial Boundaries of Malaysia. An International Law Perspective* (Kluwer Law International: London, The Hague, Boston, 1998), pp. 350–359; Ted McDorman, "Malaysia-Vietnam", in: *International Maritime Boundaries, vol. 3, supra note 5 in Chap. 5*, pp. 2335–2344. Also *see Treaty between the Kingdom of Thailand and Malaysia relating to the Delimitation of the Territorial Sea of the Two Countries*, done at Kuala Lumpur on 24 October 1979 and entered into force on 15 July 1982 [online: http://www.un.org/Depts/los/LEGISLATIONANDTREATIES/PDFFILES/TREATIES/THA-MYS1979TS.PDF (accessed on August 2013)], reproduced in: *International Maritime Boundaries*, vol. 1, *supra* note 42 in Chap. 2, pp. 1096–1098.

[47] Articles 1 and 3 of the *Memorandum of Understanding between the Kingdom of Thailand and Malaysia on the Delimitation of the Continental Shelf Boundary between the Two Countries in the Gulf of Thailand*, done at Kuala Lumpur on 24 October 1979 and entered into force on 15 July 1982 [online: http://www.un.org/Depts/los/LEGISLATIONANDTREATIES/PDFFILES/TREATIES/THA-MYS1979CS.PDF (accessed on August 2013)], reproduced in: *ibid*, pp. 1105–1110.

[48] Article 4 of the 1979 MoU Thailand/Malaysia (Delimitation of the Continental Shelf), *ibid*.

boundaries, the joint development regime would be automatically renewed for the same period of time, as no reference establishing otherwise is included in the MoU.[49] Additionally, this MoU established that States could eventually agree to renegotiate and enter into a new joint development agreement should the common development of resources prove to be more cost-effective.[50] States failed, however, to agree on the extent of the autonomy of the Joint Authority partly due to the inexperience of Thailand in implementing a joint development regime and the difficulty that both countries had in harmonizing the relevant domestic oil legislation, as well as on how exploration rights should be granted to operators in the joint development area.[51] Thailand proposed applying the classic model of concession, whereas Malaysia considered product sharing contracts to be more suitable in light of its experience in offshore development. [52]

These countries would only reach an agreement in 1990 with the implementation of a second joint development regime for the purpose of regulating in detail the powers of the Joint Authority and which actually limited its powers.[53] In fact, while initially the Joint Authority had the autonomy to execute the actions necessary to maximize the revenue of the joint development area, under the second MoU its capacities were subject to acts of parliament of both States. Thailand and Malaysia eventually agreed in this second MoU to use product sharing contracts for the purpose of granting operators exploration and exploitation rights in the joint development zone, as well as to introduce tax and fiscal regulations.[54]

---

[49] Article 6(2) of the 1979 MoU Thailand/Malaysia (Establishment of a Joint Authority), *supra* note 45.

[50] Article 6(1) *ibid.*

[51] Datuk Harun Ariffin, "The Malaysian philosophy of joint development", in: *Geology and Hydrocarbon of the South China Sea and Possibilities of Joint Development, Proceedings of the Second EAPI/CCOP Workshop, East-West Center, Honolulu, Hawaii, 22–26 August 1983*, edited by Mark J. Valencia (Pergamon Press: New York, Oxford, Toronto, Sydney, Paris, Frankfort, 1985), p. 534.

[52] David Ong, "The 1979 and 1990 Malaysia-Thailand joint development agreements: a model for international legal co-operation in common offshore petroleum deposits?", in: 14-2 *International Journal of Marine and Coastal Law* (1999), pp. 228–230; Zhiguo Gao, *International Petroleum Contracts. Current Trends and New Directions* (Grahm&Trotman/Martinus Nijhoff: London, 1994), pp. 23–57.

[53] *Agreement between the Government of Malaysia and the Government of the Kingdom of Thailand on the constitution and Other Matters relating to the Establishment of the Malaysia-Thailand Joint Authority*, done at Kuala Lumpur on 30 May 1990 [online: http://cil.nus.edu.sg/rp/il/pdf/1990%20Agreement%20between%20Malaysia%20and%20Thailand%20on%20the%20MTJA-pdf.pdf (accessed on August 2013)], reproduced in: *International Maritime Boundaries*, vol. 1, *supra* note 42 in Chap. 2, pp. 1111–1123.

[54] Articles 8, 9 to 12, 16, 17 of the 1979 MoU Thailand/Malaysia (Establishment of a Joint Authority), *supra* note 45.

In 1992, Malaysia and Vietnam signed a MoU that established a straightforward joint development regime.[55] Both States agreed that the respective national oil companies would cooperate in undertaking exploration and exploitation activities, being all costs incurred and benefits derived equally borne and shared. This MoU also established that should a petroleum field be located partly in the joint development area and in the continental shelf of Malaysia or Vietnam, these States would arrive at mutually acceptable terms and conditions for the development of such resources.[56]

The joint development area established under this MoU has been accordingly taken into consideration by Thailand and Vietnam when establishing the maritime boundaries between these States in the Gulf of Thailand, since they partly overlap. For that reason, the agreement between Thailand and Vietnam provides that the two countries shall enter into negotiations with Malaysia regarding the reconsideration of its joint development area with Vietnam.[57]

In what concerns the Gulf of Tonkin/Beibu Gulf, China and Vietnam have long disputed access to its considerable potential in marine living and non-living natural resources.[58] These countries established diplomatic relations in 1950 and

[55] *Memorandum of Understanding between Malaysia and the Socialist Republic of Vietnam for the Exploration and Exploitation of Petroleum in a Defined Area of the Continental Shelf Involving the Two Countries*, done at Kuala Lumpur on 5 June 1992 and entered into force on 4 June 1993 [online: http://cil.nus.edu.sg/rp/il/pdf/1992%20MOU%20between%20Malaysia%20and%20Vietnam%20for%20the%20Exploration%20and%20Exploitation%20of%20Petroleum-pdf.pdf (accessed on August 2013)], reproduced in: *International Maritime Boundaries, vol. 3, supra note 5 in Chap. 5*, pp. 2341–2344.

[56] Articles 2, 3 and 8 lit. (d) and (e) *ibid.*

[57] *Agreement between the Government of the Kingdom of Thailand and the Government of the Socialist Republic of Vietnam on the Delimitation of the Maritime Boundary between the Two Countries in the Gulf of Thailand*, done at Bangkok on 9August 1997 and entered into force on 27 February 1998 [online: http://www.un.org/Depts/los/LEGISLATIONANDTREATIES/PDFFILES/TREATIES/THA-VNM1997MB.PDF (accessed on August 2013)], reproduced in: *International Maritime Boundaries, vol. 4, supra note 8 in Chap. 5*, pp. 2692–2694.

[58] *See* Becker-Weinberg, *supra* note 34, pp. 228–233. Also *see* Jianwei Li and Ramses Amer, "Recent practices in dispute management in the South China Sea", in: *Maritime Energy Resources in Asia: Legal Regimes and Cooperation*, edited by Clive Schofield (The National Bureau of Asian Research: Seattle, 2012), p. 79–103; Nguyen Hong Thao and Ramses Amer, "Regional conflict management: challenges of the border disputes of Cambodia, Laos, and Vietnam", in: 2-2 *Austrian Journal of South-East Asian Studies* (2009), pp. 53–79; *Ibid*, "Managing Vietnam's maritime boundary disputes", in: 38 *Ocean Development and International Law* (2007), pp. 305–324; Zou Keyuan, "The Sino-Vietnamese agreement on maritime boundary delimitation in the Gulf of Tonkin", in: 36 *Ocean Development and International Law Journal* (2005), pp. 13–24; *Ibid*, "Maritime boundary delimitation in the Gulf of Tonkin", in: 30 *Ocean Development and International Law* (1999), pp. 235–254; *Ibid*, *supra* note 43, pp. 12–14; Nguyen Hong Thao, "Maritime delimitation and fishery cooperation in the Tonkin Golf", in: 36 *Ocean Development and International Law Journal* (2005), pp. 35–44; Ted L. McDorman, "People's Republic of China-Vietnam", in: *International Maritime Boundaries, vol. 5, supra note 13 in Chap. 4*, pp. 3745–3753; Daniel J. Dzurek, "Maritime agreements and oil exploration in the Gulf of Thailand", in: *Boundaries and Energy: Problems and Prospects*, edited by, Gerald Blake, Martin Prat, Clive Schofield, Janet Allison Brown (Kluwer Law International: London,

negotiations on the delimitation of the Gulf of Tonkin began in 1974. However, due to the deterioration of bilateral ties between both countries until 1992, it has taken over 20 years before an understanding could be reached on the principles applicable to the settlement of the territorial boundary dispute.[59] In the year 2000, China and Vietnam entered into agreement on maritime delimitation, bringing to an end a dispute that had been a major issue in the bilateral relations while simultaneously delimitating China's first maritime boundary.[60] This agreement includes a mineral resources clause according to which China and Vietnam shall reach an agreement on the most effective way to develop any hydrocarbon deposit that may straddle a demarcation line, equally sharing the profits resulting from its development.[61] China and Vietnam continued to hold several negotiation rounds on the delimitation of maritime boundaries outside the mouth of the Gulf of Tonkin, in particular regarding the implementation of a joint development regime in this area. Both States also established cooperation between *China National Offshore Oil Corporation* and *Viet Nam National Oil and Gas Group* regarding oil and natural gas.[62]

Overall, China and Vietnam have agreed to follow the approach of handling easier problems first and strengthening bilateral cooperation in areas such as maritime research, environmental protection or oil and gas exploration and exploitation, while refraining from adopting any measure that could aggravate existing disputes.[63] Yet, China's adherence to the adoption of interim measures

(Footnote 58 continued)
The Hague, Boston, 1998), pp. 117–135; *Ibid*, *supra* note 10, pp. 164–166; Mark J. Valencia, "International conflict over maritime resources in South-East Asia: trends in politicization and militarization", in: *Conflict Over Natural Resources in South-East Asia and the Pacific, Natural Resources of South-east Asia*, edited by Lim Teck Ghee and Mak J. Valencia (Oxford University Press: Singapore, Oxford, New York, 1990), pp. 102, 103, 106–107.

59 *See* Becker-Weinberg, *ibid.*

60 *Agreement on the Delimitation of the Territorial Seas, Exclusive Economic Zones and Continental Shelves in the Beibu Gulf [Gulf of Tonkin] between the People's Republic of China and the Socialist Republic of Vietnam*, done at Beijing, on 25 December 2000 and entered into force on 30 June 2004 [online: http://cil.nus.edu.sg/rp/il/pdf/2000%20Agreement%20between%20China%20and%20Vietnam%20on%20Delimitation%20of%20Territorial%20Seas-pdf.pdf (accessed on August 2013)], reproduced in: *International Maritime Boundaries, vol. 5, supra note 13 in Chap. 4*, pp. 3755–3758.

61 Article 7, *ibid.*

62 *See* Becker-Weinberg, *supra* note 34, p. 232, note 52.

63 *See Agreement on the basic principles guiding the settlement of maritime issues between the Socialist Republic of Viet Nam and the People's Republic of China*, done at Beijing on 11 October 2011 [online: http://www.mofa.gov.vn/en/nr040807104143/nr040807105001/ns111101161013 (accessed on August 2013)]. Also *see* Statements by the Ministry of Foreign Affairs of the People's Republic of China: *Yang Jiechi Meets with Vietnamese Foreign Minister Pham Binh Minh, 11 July 2012* [online: http://www.fmprc.gov.cn/eng/wjb/zzjg/yzs/gjlb/2792/2794/t951393.shtml (accessed on August 2013)]; *Vice President Xi Jinping Meets with Vietnamese Army Chief, 13 April 2012* [online: http://www.fmprc.gov.cn/eng/wjb/zzjg/yzs/gjlb/2792/2794/t923082.shtml (accessed on August 2013)]; *Vice Premier Li Keqiang Meets with Vietnamese Deputy Prime Minister Hoang Trung Hai, 31 March 2012* [online: http://www.fmprc.gov.cn/eng/wjb/zzjg/yzs/gjlb/2792/2794/t921052.shtml (accessed on

and efforts to implement joint development regimes with Vietnam in the Gulf of Tonkin notwithstanding,[64] China continues to grant rights in maritime areas of the South China Sea pending the delimitation of boundaries and with disregard for the claims of other States.[65] On the other hand, China has strongly protested at the passing by the National Assembly of Vietnam of the *Vietnamese Law of the Sea*, which includes maritime areas claimed by China, extending the country's sovereignty and jurisdiction to the Paracel Islands (Xisha in Chinese, Hoang Sa in Vietnamese) and the aforementioned Spratly Islands (Nansha in Chinese, Truong Sa in Vietnamese).[66]

The controversy between the two States is also reflected in the joint submission of Malaysia and Vietnam to the CLCS in respect of the southern part of the South China Sea, as well as, for that matter, of the partial submission made by Vietnam in respect of the northern part of the East Sea.[67] In both cases, China maintains the validity of its commonly referred to as *nine-dash line*, *nine-dotted lines* or *U-shaped line* map of the South China Sea that includes most part of the areas and islands and other features disputed in this region,[68] while the other relevant littoral countries

(Footnote 63 continued)
August 2013)]; and Statement by the Ministry of Foreign Affairs of Viet Nam: *Viet Nam, China negotiate cooperation at sea, 30 May 2012* [online: http://www.mofa.gov.vn/en/nr040807104143/nr040807105001/ns120531093517 (accessed on August 2013)].

64 Statement by the Ministry of Foreign Affairs of Viet Nam: *Viet Nam, China discuss sea area off Tonkin Gulf, 24 May 2012* [online: http://www.mofa.gov.vn/en/nr040807104143/nr040807105001/ns120524140255 (accessed on August 2013). Both countries have reiterated in 2012 their agreement to work together to speed up the negotiations and discuss cooperation for joint development in the area according to the principles stipulated in the aforementioned agreement on basic principles guiding the settlement of maritime issues.

65 Statement by the Ministry of Foreign Affairs of Viet Nam: *Viet Nam protests Tan Bao Hao's activities which violate sovereignty and jurisdictional rights of Viet Nam,* 7 November 2011 [online: http://www.mofa.gov.vn/en/tt_baochi/pbnfn/ns111107161135 (accessed on August 2013)]. Recently the two countries made statements at the 24th Meeting of States Parties to UNCLOS (9–13 June 2014) regarding the placing of a drilling rig (Haiyang Shiyou 981) by China in a disputed maritime area, which sparked strong reactions in the South China Sea. China eventually removed the drilling rig.

66 Statement by the Ministry of Foreign Affairs of the People's Republic of China: *Vice Foreign Minister Zhang Zhijun Lodges a Solemn Representation Concerning the Passing of the Maritime Law of Vietnam by the Vietnamese Parliament*, 21 June 2012 [online: http://www.fmprc.gov.cn/eng/wjb/zzjg/yzs/gjlb/2792/2794/t945207.shtml (accessed on August 2013)]; Statement by the Ministry of Foreign Affairs of Viet Nam: *Foreign Minister Pham Binh Minh clarifies content of Viet Nam Maritime Law*, 26 June 2012 [online: http://www.mofa.gov.vn/en/nr040807104143/nr111027144142/ns120625222640/view (accessed on August 2013)].

67 [online: http://www.un.org/depts/los/clcs_new/submissions_files/mysvnm33_09/mys_vnm2009excutivesummary.pdf (accessed on August 2013)]. On the extended continental shelf submission in East and Southeast Asia *see* I Made Andi Arsana and Clive Schofield, "Adding further complexity? Extended continental shelf submissions in East and Southeast Asia", in: *Maritime Energy Resources in Asia: Legal Regimes and Cooperation*, edited by Clive Schofield (The National Bureau of Asian Research: Seattle, 2012), pp. 49–58.

68 [online: http://www.un.org/depts/los/clcs_new/submissions_files/mysvnm33_09/chn_2009re_mys_vnm_e.pdf (accessed on August 2013)]. On the drawing of the U-shaped line and the

oppose the international legal basis of this map and the sovereignty claims made by the aforementioned submitting coastal States.[69]

Likewise, despite the efforts between the Philippines and China to implement interim measures regarding their dispute in the South China Sea,[70] and the importance given by both countries in maintaining "*healthy and stable development of the bilateral relations*",[71] in early 2013, the Philippines conveyed to China a notification to submit the said dispute to arbitration under Article 287 and Article 1 of Annex VII of UNCLOS.[72] Accordingly, the Philippines consider that China's *nine-dash line* interferes with its EEZ and continental shelf. The Philippines also maintain that China's claim and occupation within the *nine-dash line* of submerged banks, reefs and low tide elevations hinder its rights towards these features. In addition, the Philippines consider that China's claim of maritime zones greater than 12 nm surrounding features occupied by China and that it deems as rocks under Article 121(3) of UNCLOS, result in the encroachment of these zones on the Philippines' EEZ.

With the submission to arbitration of the dispute on maritime jurisdiction in the West Philippine Sea, the Philippines seek to obtain a declaration of the inconsistency of China's *nine-dash line*, as well as on the legal status of certain maritime features, particularly if these are capable of generating entitlement to maritime zones greater than 12 nm. The Philippines do not, however, intend to establish sovereignty over the islands claimed by both countries proceeded with the delimitation of maritime boundaries. These would be impossible considering China's declaration made after ratifying UNCLOS.[73]

(Footnote 68 continued)
Chinese perception of the South China Sea disputes, see Kuen-Chen Fu, "Safeguarding China's national interests in the South China Sea: rectification, services, leadership, and maritime delimitation", in: *17-1 China Oceans Law Review* (2013), pp. 12–19; Wu Schicun, *Solving Disputes for Regional Cooperation and Development in the South China Sea: A Chinese Perspective* (Chandos Publishing: Oxford, Cambridge, Philadelphia, New Delhi, 2013), pp. 34–39, 77–83.

[69] [online: http://www.un.org/depts/los/clcs_new/submissions_files/submission_mysvnm_33_2009.htm (accessed on August 2013)].

[70] Communiqué by the Ministry of Foreign Affairs of the People's Republic of China: *Ambassador Liu Jianchao pays Courtesy call on Philippine Foreign Affairs Secretary Romulo, 13 March 2009* [online: http://www.fmprc.gov.cn/eng/wjb/zwjg/zwbd/t542281.htm (accessed on August 2013)].

[71] *China and the Philippines held 19th Foreign Ministry Consultations, 14 June 2013* [online: http://www.fmprc.gov.cn/eng/zxxx/t1050565.shtml (accessed on June 2013)].

[72] *Notification and Statement of Claim on the West Philippine Sea, 22 January 2013* [online: https://www.dfa.gov.ph/index.php/2013-06-27-21-50-36/unclos (accessed on August 2013)].

[73] The declaration of China made under Article 298 of UNCLOS, made on 25 August 2006, after ratification of the Convention, reads: "*The Government of the People's Republic of China does not accept any of the procedures provided for in Sect. 2 of Part XV of the Convention with respect to all the categories of disputes referred to in paragraph 1(a) (b) and (c) of Article 298 of the Convention.*" [online: http://www.un.org/Depts/los/convention_agreements/convention_declarations.htm (accessed on August 2013)]. See Haiwen Zhang and Chenxi Mi, "Jurisprudential tenability of the Philippines v

The Philippines primary concern in bringing this arbitration against China is to challenge the latter's "*claims to areas of the South China Sea and the underlying seabed as far as 870 nautical miles from the nearest Chinese coast, to which China has no entitlement under* [UNCLOS], *and which, under the Convention, constitute the Philippines'* [EEZ] *and continental shelf*".[74]

In response, the Chinese Ambassador to the Philippines sent a *Note Verbale* expressing China's rejection of the said notification and statement of claim, reiterating that China has indisputable sovereignty over the entire South China Sea, as encompassed by its *9-dash line* claim. China considers that "*its sovereignty over the Nansha Islands and their adjacent waters is supported by abundant historical and legal evidence*" and that the notification and statement of claim "*contains many grave errors both in fact and in law, and includes many false accusations against China*". China also stated that "*bearing in mind the larger interest of China-Philippines relations and regional peace and stability*" the Philippines should return to and uphold the established consensus between China and ASEAN countries of resolving the disputes through bilateral negotiation stipulated in the *Declaration on the Conduct of Parties in the South China Sea*. China believes it has made every effort to maintain stability and to promote regional cooperation in the South China Sea, requesting the Philippines to respond "*positively to China's proposals on establishing a bilateral regular consultation mechanism on maritime issues*" and resuming "*the operation of the Confidence Building Measures Mechanism established between the two countries*".[75]

The Philippines remain, nonetheless, committed to pursuing the arbitration procedure under Annex VII of UNCLOS, underlying that "*the 5-member arbitration panel will be formed with or without China.*"[76] Indeed, on 24 June 2013, ITLOS announced the appointment of the new arbitrator and president in the

(Footnote 73 continued)
China Arbitration on South China Sea Disputes?", in: *The South China Sea Arbitration: A Chinese Perspective*, edited by Stefan Talmon and Bing Bing Jia (Hart Publishing: Oxford and Portland, Oregon, 2014), pp. 137–158.

[74] *Supra* note 72, p. 1.

[75] *Chinese Spokesperson Hong Lei's remarks on China returned the Philippines' Notification on the submission of South China Sea issue to international Arbitration, 19 February 2013* [online: http://ph.china-embassy.org/eng/xwfb/t1014903.htm (accessed on August 2013)]. Also *see Foreign Ministry Spokesperson Hua Chunying's Remarks on the Philippines' Efforts in the Pushing for the Establishment of the Arbitral Tribunal in Relation to the Disputes between China and the Philippines in the South China Sea, 26 April 2013* [online: http://www.fmprc.gov.cn/eng/xwfw/s2510/2535/t1035577.shtml (accessed on August 2013)]. Recently the two countries made statements at the 24th Meeting of States Parties to UNCLOS (9–13 June 2014) regarding the arbitration.

[76] *The Department of Foreign Affairs' Statement on China's Response to the Philippines' Arbitration Case, 19 February 2013*, [online: http://www.gov.ph/2013/02/19/dfa-statement-on-chinas-response-to-the-ph-arbitration-case-before-unclos/ (accessed on August 2013)].

arbitral proceedings, thus completing the composition of the five-member Annex VII Arbitral Tribunal.[77]

Also in the South China Sea, on 16 March 2009, Malaysia and Brunei signed an *Exchange of Letters* marking the successful conclusion of 20 years of negotiations on outstanding bilateral issues between the two countries. The "*mutually-beneficial formula*" incorporated in the *Exchange of Letters* marked a historic and momentous occasion for Brunei and Malaysia, resulting in a "*win-win solution for both countries*".[78] However, as this document has not been made public, it is not possible to verify the exact content of the same. Still, based on official press statements issued by both ministries of foreign affairs, it is possible to conclude that Brunei and Malaysia have agreed on the final delimitation of maritime boundaries of the respective territorial seas, the continental shelf and the EEZ, the establishment of a "*Commercial Arrangement Area*" regarding oil and gas, the modalities for the final demarcation of the land boundary, "*unsuspendable*" rights of maritime access for nationals and residents of Malaysia across Brunei's maritime zones en route to and from their destination in Sarawak (Malaysia).[79]

According to the "*Commercial Arrangement Area*", Malaysia's oil concession Blocks L (Blocks CA1) and M (Blocks CA2), which coincided with Brunei's Blocks J and K, are recognised as being situated within Brunei's maritime areas, being Malaysia allowed to participate in "*joint development of oil and gas on commercial basis in the two areas for a period of 40 years*". In turn, as part of "*extended gestures of friendship*", both States would share the revenues from the exploitation of oil and gas, even though there is no mention of any applicable *ratio*.[80]

---

[77] I.T.L.O.S. Press Release n. 191, 25 April 2013, *Arbitrators appointed in the arbitral proceedings instituted by the Republic of the Philippines against the People's Republic of China* [online: http://www.itlos.org/fileadmin/itlos/documents/press_releases_english/PR_191_E.pdf (accessed on August 2013)]; I.T.L.O.S. Press Release n. 197, 24 June 2013, New arbitrate and president appointed in the *arbitral proceedings instituted by the Republic of the Philippines against the People's Republic of China* [online: http://www.itlos.org/fileadmin/itlos/documents/press_releases_english/PR_197_E.pdf (accessed on August 2013)]

[78] [online: http://www.mofat.gov.bn/index.php/bilateral-relations/item/76-malaysia (accessed on August 2013)], [online: http://www.bt.com.bn/business-national/2010/05/03/najib-sees-win-win-solution-both-countries (accessed on August 2013)]; *Joint Statement on the 15th Annual Leaders' Consultation between Brunei Darussalam and Malaysia*, 12 September 2011 [online: http://www.mofat.gov.bn/index.php/news/news-highlights/item/195-joint-statement-on-the-15th-annual-leaders-consultation-between-brunei-darussalam-and-malaysia (accessed on August 2013)].

[79] *Joint press statement by leaders on the occasion of the working visit of Yab Dato' Seri Abdullah Haji Ahmad Badawi, Prime Minister of Malaysia to Brunei Darussalam on 15–16 March 2009* [online: http://bn.chineseembassy.org/eng/wlxw/t542877.htm (accessed on August 2013)].

[80] Ministry of Foreign Affairs of Malaysia: *Brunei Will Share Oil Revenue on Disputed Oil Block—Anifah*, 7 June 2010 [online: https://www.kln.gov.my/archive/content.php?t=4&articleId=695563 (accessed on August 2013)];

M'sia-Brunei agreement will benefit both countries, Borneo Bulletin: 3 May 2010 [online: http://www.kln.gov.my/web/brn_begawan/n2009/-/asset_publisher/ME2g/blog/m-sia-brunei-agreement-

In 2001, Cambodia and Thailand signed a MoU regarding the maritime area where claims overlapped in the continental shelf of the Gulf of Thailand, whereby both countries agreed to conclude an agreement for the joint development of offshore hydrocarbon deposits found therein, as well as the delimitation of the territorial sea, continental shelf and EEZ. These States further agreed to create a Joint Technical Committee for the purpose of preparing the joint development regime to be implemented, as well as to settle their maritime boundaries.[81]

Lastly, it should be mentioned that Cambodia and Vietnam have signed an agreement on the creation of joint historic waters that allows for the exploitation of the natural resources found in the joint development historical waters zone.[82]

7.3 Lessons Learned

The delimitation of maritime boundaries between two or more States by means of an agreement or compulsory settlement mechanism is not currently an attainable undertaking in the Asia-Pacific region, partly because coastal States have taken refuge behind their nationalistic views which are often fuelled by the presence of hydrocarbon deposits in areas of overlapping claims or that are believed to have a significant hydrocarbon potential. The lack of an accurate estimate of the mineral resource-potential of disputed maritime areas and the general perception of the existence of abundant reserves heightens the tension between States. In this respect, the undertaking of joint surveys of the seabed and subsoil could provide a real assessment of the mineral resource-potential of the areas of overlapping claims and constitute, as a result, a first step towards cooperation and ultimately making the delimitation of maritime boundaries or the sharing of resources achievable endeavours.

Another key aspect for the durable character of the disputes in the Asia-Pacific region is the fact that certain coastal States do not clarify their claims as required under international law and in the particular the law of the sea, thus preventing the assessment of overlapping entitlements and eventually the possible identification of maritime areas for joint development. These are, evidently, indispensable elements for the negotiation of an agreement on the delimitation of maritime

---

(Footnote 80 continued)
will-benefit-both-countries;jsessionid=3474BF022A92915154B3B74A6234E55A?redirect=%2Fweb%2Fbrn_begawan%2Fn2009 (accessed on August 2013)].

[81] *Memorandum of Understanding between the Royal Government of Cambodia and the Royal Thai Government regarding the Area of their Overlapping Maritime Claims to the Continental Shelf*, done at Phnom Penh 18 on June 2001, reproduced in: *International Maritime Boundaries, vol. 5, supra note 13 in Chap. 4*, pp. 3745–3746.

[82] *Agreement on Historic Waters of Vietnam and Kampuchea, made in Ho Chi Minh City*, done on 7 July 1982, reproduced in: *International Maritime Boundaries, vol. 3, supra note 5 in Chap. 5*, pp. 2364–2365.

boundaries or joint development. The clarification of States' claims and their respective legal tile is, for that reason, a vital component of any solution for the Asia-Pacific region. The absence of clarification and/or of an agreement has often resulted in competitive drilling, jeopardizing the stability of the region. It may also lead to the waste of resources and is a serious threat to the protection and preservation of the marine environment in disputed maritime areas.[83]

The joint development agreements implemented in the Asia-Pacific region were the result of States' creative and pragmatic approach towards hurdling deadlock situations and seeking a legal solution for the complex problems associated with the development of common offshore hydrocarbon deposits, without prejudice to their sovereignty claims. This pragmatism, however, was the consequence of an economic drive, rather than a legal one. States' willingness to enter into joint development agreements was primarily conditioned by the need to satisfy the demand for energy sources and not the conviction that States were legally bound to share common mineral resources.[84] These agreements also provided the indispensable legal assurance that was required for operators to undertake the complex, time consuming and highly costly offshore exploration and exploitation activities.[85]

Despite some possible similarities, the joint development agreements of the Asia-Pacific region do not have identical legal structure or content as a result of the discretionary nature of these agreements and the specific circumstances in which each agreement was negotiated. It is also a consequence of the political, social and economic circumstances, such as energy dependence, diversity of States' capabilities regarding exploration and exploitation of offshore hydrocarbon deposits or compatibility of national legal systems.

Joint development agreements in the Asia-Pacific region have also been more or less complex depending on the scope and extent of their functional character.

---

[83] Vasco Becker-Weinberg, "Seabed activities and the protection and preservation of the marine environment in disputed maritime areas of the Asia-Pacific region", in: *Securing the Ocean for the Next Generation*, Papers from the Law of the Sea Institute, UC Berkeley-Korean Institute of Ocean Science and Technology Conference, held in Seoul, Korea, May 2012 [online: http://www.law.berkeley.edu/files/Becker-Weinberg-final.pdf (accessed on August 2013)].

[84] David Ong accepted the hypothesis of the formation or the existence of an emergent rule in a regional context that could offer a good basis for the applicability of a joint development model and consequently solve the problem of delimitation of maritime boundaries in respect of the development of non-living marine natural resources, although rejecting the idea of a legal customary foundation of joint development agreements. Accordingly, these regions would include the North Sea, the Persian Gulf and the East and South China Seas, as these are regions with precedents in implementing cooperative regimes and the respective geomorphologic characteristics, particularly the fact that they are enclosed or semi-enclosed seas. *See* Ong, *supra* note 61 in Chap. 3, p. 795.

[85] William T. Onorato, "Promoting foreign investment through international petroleum joint development regimes", in: 1-1 *International Centre for Settlement of Investiment Disputes Review—Foreign Investment Law Journal* (Spring 1986), pp. 81–88; Valencia, *supra* note 49 in Chap. 6, p. 683; Lagoni, *supra* note 70 in Chap. 2, pp. 360–362; Wälde, *supra* note 70 in Chap. 2, p. 156.

From including the development of both living and non-living marine natural resources, to creating more than one joint development area to be subject to the jurisdiction of one or more States or, instead, to be managed by an entity created under the agreement for the purpose of undertaking States' rights and obligations, including granting access to exploration and exploitation activities.

Although there are no legal conditions for joint development, the existence of precedents of cooperation between States or within the relevant region has favoured reaching an agreement on the exploration and exploitation of common resources, even if there have also been agreements signed between States that did not share a relevant pattern of cooperation, or in regions where there was no prior awareness of the problems represented by transboundary or shared resources. In both situations, the existing reliability between States determined, to a certain extent, the successful implementation of the agreements. It should be noted that coastal States in the Asia-Pacific region have seemed to prefer bilateral negotiations to regional efforts or compulsory settlement of disputes regarding common mineral resources. States' accomplishment in implementing cooperative regimes regarding the development of common offshore hydrocarbon deposits in the Asia-Pacific region may eventually ease the delimitation of maritime boundaries, since States no longer have to deal with the difficult matter concerning the exploration and exploitation of these resources. It also provides States with a valuable framework for the exchange of information, which is also an important step towards effective and efficient cooperation.

Joint development agreements in the Asia-Pacific region have contributed towards confidence building, cooperation and stability at the bilateral level between the relevant coastal States. In those cases where States have not succeed in negotiating or implementing joint development agreements, they are, nonetheless, required by international law to make their claims and respective legal title known and to act in accordance with their obligations of cooperation and good faith previously mentioned, as well as to refrain from the threat or use of force and the peaceful settlement of disputes.

Chapter 8
The Role of Mineral Resources in Maritime Delimitation

8.1 The EEZ and Continental Shelf Boundaries

The principle that has been widely accepted and acknowledged is that maritime delimitation shall be determined by agreement, as opposed to unilateral delimitation.[1] Yet, for nearly half a century, judges, lawyers and scholars have debated on the rules and methods applicable to the delimitation of maritime boundaries of the EEZ and the continental shelf between neighbouring coastal States when States fail to reach an agreement.

Over the years, international jurisprudence has attempted to clarify the principles applicable to the delimitation of maritime boundaries, even though it tried to combine, on the one hand, the certainty of the law when it comes to establishing and defining the principles applicable to the delimitation of maritime boundaries and, on the other hand, the ability of adjusting to the specificities of each case or situation. In fact, in 1952, Gilbert Gidel had already questioned the usefulness of pre-defined delimitation rules instead of allowing States to reach an agreement, especially when considering the specificities of each continental shelf.[2]

The second sentence of Article 6(1) of the 1958 Convention on the Continental Shelf reads that:

> In the absence of agreement, and unless another boundary line is justified by special circumstances, the boundary is the median line, (…).

This provision sets forth what has been commonly referred to as the *triple rule* or *three-point rule* of delimitation, i.e. delimitation in the absence of delimitation by

---

[1] Article 6(1) of the 1958 Convention on the Continental Shelf and Articles 74(1) and 83(1) of UNCLOS. Also *see Delimitation of the Maritime Boundary in the Gulf of Maine Area, supra note 60 in Chap.* 3, para. 112.

[2] Gidel, *supra* note 52 in Chap. 3, pp. 16–17. Also *see* Georges Scelle, *Plateau Continental et Droit International* (A. Pedone: Paris, 1955), pp. 12–18, 30–32.

V. Becker-Weinberg, *Joint Development of Hydrocarbon Deposits in the Law of the Sea*, Hamburg Studies on Maritime Affairs 30, DOI 10.1007/978-3-662-43570-0_8

agreement making use of the median line corrected by special circumstances. The median line is an equidistance line applied if the coasts of two or more States are opposite to each other. However, the reference to equidistance and special circumstances without any clarification as to the relation between these two concepts and their scope resulted in different positions on the applicability of this rule.

To further underline the difficulty of interpreting this rule, the ICJ in the *North Sea Continental Shelf cases* (which in 1964 was the first dispute on the delimitation of maritime boundaries after the entry into force of the 1958 Convention on the Continental Shelf) made an attempt to determine the applicable maritime delimitation rules under international customary law. The Court did not refer to the abovementioned Article 6 but to existing customary law because Germany had not ratified the Convention. It concluded that this provision did not reflect the existing international law and that States were only subject to the principle of maritime delimitation determined by agreement. In this respect, the ICJ further considered that given the unique character of each border and the fact that the use of the equidistance method of delimitation was not part of *jus cogens*, States were only subject to the obligation to the principle of delimitation by agreement in accordance with equitable principles taking into account all "*relevant circumstances*."[3] The ICJ noted that:

> The Court has already stated why it considers that the international law of continental shelf delimitation does not involve any imperative rule and permits resort to various principles or methods, as may be appropriate, or a combination of them, provided that, by the application of equitable principles, a reasonable result is arrived at.[4]

The Court's *dictum* that the continental shelf must be the natural prolongation of its land territory was followed by the Arbitral Tribunal in the *Anglo-French Continental Shelf* case, even if not forming a basis for the delimitation of

---

[3] *North Sea Continental Shelf cases, supra note 46 in Chap.* 3, paras. 101(C)(1), 47, 69–72, 77, 81, 82–83. Also *see Delimitation of the Maritime Boundary in the Gulf of Maine Area, supra note 60 in Chap.* 3, paras. 113–115, 124–125, 159; and *Continental Shelf (Libya/Malta), supra note 60 in Chap.* 3, para. 46. On the judgment of the ICJ in the North Sea cases *see* L.F.E Goldie, "The North Sea continental shelf case. A ray of hope for the International Court?", in: 6–2 *New York Law Forum* (1970), pp. 327–377; Charles Vallée, *Le Plateau Continental dans le Droit Positif Actuel* (A. Pedone: Paris, 1971), pp. 253–292; Prosper Weil, *Perspectives du Droit de la délimitation maritime* (Pedone: Paris, 1988), pp. 155–169; Douglas M. Johnston, *The Theory and History of Ocean Boundary-Making* (McGill-Queen's University Press: Kingston, Montreal, 1988), pp. 165–178; Donat Pharan, "International decisions. Affaire du plateau continental de la Mer du Nord (1969)", in: *The Continental Shelf and the Exclusive Economic Zone/Le Plateau Continental et la Zone Économique Exclusive* edited by Donat Pharand and Umbero Leanza (Martinus Nijhoff Publishers: Dordrecht, Boston, London, 1993), pp. 95–104; Tanja, *supra* note 60 in *Chap.* 3, pp. 62–80, 119–153; Oude Elferink, *supra* note 60 in Chap. 3, pp. 22–25, 47–55, 171; Brown, *supra* note 26 in Chap. 3, pp. 164–165, 172–184; *Ibid, supra* note 60 in Chap. 3, pp. 49–87; Nelson, *supra* note 60 in Chap. 3, p. 838; Kwiatkowska, *supra* note 59 in Chap. 3, pp. 254–255; Evans, *supra* note 15 in Chap. 7, pp. 13–27. A. G. Oude Elferink, *The Delimitation of the Continental Shelf between Denmark, Germany and the Netherlands. Arguing Law, Practicing Politics?* (Cambridge University Press, Cambridge, 2013), pp. 161–341.

[4] *North Sea Continental Shelf cases, supra note 46 in Chap.* 3, para. 90.

boundaries between neighbouring States.[5] In this case, the Arbitral Tribunal also had the opportunity to develop the rules on maritime delimitation between neighbouring States seeing that, despite both States involved in this dispute had signed the 1958 Convention on the Continental Shelf, France presented a reservation to Article 6 to which the United Kingdom objected and therefore, at least, this provision would not be applicable.[6]

Similar to the *North Sea Continental Shelf* cases, the Arbitral Tribunal only referred to general and also imprecise rules of customary law, making it difficult to assess how the established boundary line derived from the principles of customary law.[7] The Arbitral Tribunal referred to a single rule, a "*combined equidistance-special circumstances rule*,"[8], adding that the "*Court does not overlook that under Article 6 the equidistance principle ultimately possess an obligatory force which it does not have in the same measure under the rules of customary law.*"[9] In effect,

---

[5] *Ibid*, paras. 43–44, 46; *Anglo-French Continental Shelf Arbitration* (1977–1978), in: 18 *International Legal Materials* 397, paras. 75–80, 100.

[6] *Anglo-French Continental Shelf Arbitration, ibid*, para. 63: "*The fact that Article 6 is not applicable as between the Parties to the extent that it is excluded by the French reservations does not mean that there are no legal rules to govern the delimitation of the boundary in areas where the reservation operates. On the contrary, as the International Court of Justice observed in the North Continental Shelf cases, "there are still rules and principles of law to be applied" (I.C.J. Reports 1969, paragraph 83); and these are the rules and principles governing delimitation of the continental shelf in general international law.*"

[7] *Ibid*, para. 67: "*Thus, in invoking the application of Article 6, paragraph 1, the United Kingdom claims that this paragraph places an onus of proof upon the French Republic to show the existence of any "special circumstance" on which it relies and to show that these circumstances justify a boundary other than the median line as defined by the paragraph. The French Republic, on the other hand, in contesting the applicability of Article 6 and invoking the rules of customary law, claims the governing principle to be that the delimitation must be equitable and the equidistance principle to be merely one of numerous "methods" which may in certain circumstances be used to produce an equitable delimitation. Neither of these views of equidistance "principle" or "method" however, appears to the Court to place it in its true perspective.*" Also *see* Churchill/Lowe, *supra* note 68 in Chap. 2, p. 185.

[8] *Anglo-French Continental Shelf Arbitration supra* note 5, para. 68.

[9] *Ibid*, para. 70. The Arbitral Tribunal also referred that it "*took the view that the equidistance principle itself is not inherent in the doctrine of the continental shelf nor a logical necessity of that doctrine derived from any fundamental principle of proximity or adjacency; and it also emphasised the doubts voiced in the International Law Commission as to the possibility that, in certain cases, the geographical configuration of the coast would render a boundary drawn on the basis of equidistance inequitable. So it was that the Court was led to conclude that in customary law the basic principle of delimitation is that, failing agreement, the boundary must be determined in accordance with equitable principles.*" (para. 82) "*These observations, to which the present Court of Arbitration in general subscribes, indicate that the validity of the equidistance method, or of any other method, as a means of achieving an equitable delimitation of the continental shelf is always relative to the particular geographical situation. In short, whether under customary law or Article 6, it is never a question either of complete or of no freedom of choice as to method; for the appropriateness—the equitable character—of the method is always a function of the particular geographical situation.*" (para. 84) "*In short, this Court considers that the appropriateness of the equidistance method or any other method for the*

the Arbitral Tribunal used the equidistance method, which considered being predictable and later took into account equity considerations to adjust the rule to the specificities of the case[10] In this respect the Arbitral Tribunal observed that:

> Just as it is not the function of equity in the delimitation of the continental shelf completely to refashion geography, so it is also not the function of equity to create a situation of complete equity where nature and geography have established an inequity.[11]

In the *Continental Shelf* (*Tunisia/Libyan Arab Jamahiriya*) case, the ICJ also had the opportunity to identify with more accuracy the principles and rules applicable to the delimitation of maritime boundaries, particularly as neither of the States was party to the 1958 Convention on the Continental Shelf. Indeed, as Edward Duncan Brown emphasized, some clarification would have been useful at this point, seeing that drafting and approval process of UNCLOS was in its final stage and that Article 83(1) had already been included in the final text.[12] Still, the Court reiterated the ruling of the *North Sea Continental Shelf* cases and although rejecting the equidistance method as being prescribed by a mandatory rule of customary law, it considered the use of a median line in accordance with equitable principles[13]:

> The equitableness of a principle must be assessed in light of its usefulness for the purpose of arriving at an equitable result. It is not every such principle which is in itself equitable; it may acquire this quality by reference to the equitableness of the solution. The principles to be indicated by the Court have to be selected according to their appropriateness for

(Footnote 9 continued)
purpose of effecting an equitable delimitation is a function or reflection of the geographical and other relevant circumstances of each particular case" (para. 97).

10 *Ibid*, para. 238–245.

11 *Ibid*, para. 249.

12 Brown, *supra* note 26 in Chap. 3, p. 179.

13 *Continental Shelf* (*Tunisia/Libya*), *supra note 10 in Chap.* 3, paras. 45–48, 70–71. On the Court's decision *see* Robert Kolb, *Case Law on Equitable Maritime Delimitation/Jurisprudence sur les Délimitations Maritimes Selon L'Équeté. Digest and Commentaries/Répertoire et commentaries*, edited by Vaughan Lowe (Martinus Nijhoff Publishers: The Hague, London, New York, 2003), pp. 40–41, 166–170; Jean-Pierre Queneudec, "Note sur l'arret de la cour internationale de justice relatif a la délimitation du plateau continental entre la Tunisie et la Libye", in: 27 *Annuaire Français de Droit International* (1981), pp. 203–212; Mark B. Feldman, "The Tunisian-Libya continental shelf case: geographic justice or judicial compromise?", in: 77-2 *American Journal of International Law* (April: 1983), pp. 219–238; Emmanuel Decaux, "L'arrêt de la Cour Internationale de Justice dans l'affaire du plateau continental (Tunisie/Libye)", in: 28 *Annuaire Français de Droit International* (1982), pp. 357–391; *Ibid*, "L'arrêt de la C.I.J. sur la requête a fin d'intervention de Malte, dans l'affaire du plateau continental entre la Tunisie et la Libye (14 avril 1981)", in: 27 *Annuaire Français de Droit International* (1981), pp. 177–202; Keith Highet, "Whatever became of natural prolongation?", in: *Rights to Oceanic Resources. Deciding and Drawing Maritime Boundaries*, edited by Dorinda G. Dallmeyer and Louis Devorsey, Jr. (Martinus Nijhoff Publishers: Dordrecht, Boston, London, 1989), pp. 87–100; Nelson, *supra* note 60 in Chap. 3, pp. 843–846. Tanja, *supra* note 60 in Chap. 3, pp. 180–206; Oude Elferink, *supra* note 60 in Chap. 3, pp. 64–73; Brown, *supra* note 60 in Chap. 3, pp. 138–165. Also *see Continental Shelf* (*Libya/Malta*), *supra note 60 in Chap.* 3, paras. 45–46, 63.

> reaching an equitable result. (...) Equity as a legal concept is a direct emanation of the idea of justice."[14] Adding that, "each continental shelf case in dispute should be considered and judged on its own merits, having regard to its peculiar circumstances; therefore, no attempt should be made to overconceptualize the application of the principles and rules relating to the continental shelf.[15]

Edward Duncan Brown concluded that it was legitimate to question the practical guidance and insight that the Court's ruling offered "*into judicial thinking on delimitation problems.*"[16]

In the *Gulf of Maine* case, the ICJ considered that the equitable criteria should be determined in a case-by-case basis and therefore could not be regulated by law. The Court pointed out that the rule prescribed by international law in maritime delimitation between neighbouring States in the absence of an agreement, is that:

> [D]elimitation is to be effected by the application of equitable criteria and by the use of practical methods capable of ensuring, with regard to the geographic configuration of the area and other relevant circumstances, an equitable result.[17]

Thus, the Court placed the focus in reaching an equitable result, excluding any considerations on the existence of legal rules applicable to defining equitable criteria or methods.[18] Likewise, the ICJ in the *Libya/Malta* case, as well as the Arbitral Tribunals in the arbitrations between *Guinea/Guinea-Bissau*[19] and *St. Pierre et Miquelon,*[20] endorsed the said focus in finding an equitable result, even

---

[14] *Continental Shelf (Tunisia/Libya), supra note 10 in Chap.* 3, paras. 70–72. Also *see North Sea Continental Shelf cases, supra note 46 in Chap.* 3, paras. 91–93; *Continental Shelf (Libya/Malta), supra note 60 in Chap.* 3, paras. 45, 48, 63; *Maritime Delimitation in the Area between Greenland and Jan Mayen, supra note 33 in Chap.* 4, para. 57; *Case Concerning Maritime Delimitation in the Black Sea, supra note 53 in Chap.* 3, paras. 111, 155.

[15] *Continental Shelf* (Tunisia/Libya), *ibid*, para. 132. Also *see Delimitation of the Maritime Boundary in the Gulf of Maine Area, supra note 60 in Chap.* 3, para. 59; *Continental Shelf* (Libya/Malta), *ibid*, para. 65; *Maritime Delimitation in the Area between Greenland and Jan Mayen, ibid*, paras. 55–56.

[16] Brown, *supra* note 26 in Chap. 3, p. 181.

[17] *See Delimitation of the Maritime Boundary in the Gulf of Maine Area, supra note 60 in Chap.* 3, para. 112(2). Also *see Land and Maritime Boundary between Cameroon and Nigeria, supra note 83 in Chap.* 3, para. 294: "*The Court is bound to stress in this connection that delimiting with a concern to achieving an equitable result, as required by current international law, is not the same as delimiting in equity. The Court's jurisprudence shows that, in disputes relating to maritime delimitation, equity is not a method of delimitation, but solely an aim that should be borne in mind in effecting the delimitation.*"

[18] *Delimitation of the Maritime Boundary in the Gulf of Maine Area, ibid*, paras. 124, 157–158, 162–163.

[19] *Case concerning the delimitation of the maritime boundary between Guinea and Guinea-Bissau, Arbitral Award of 14 February 1985*, para. 89.

[20] *Delimitation of Maritime Areas between Canada und the French Republic* (*St. Pierre et Miquelon*), 31 *International Legal Materials* (1992), para. 38.

though the ICJ did also adopt the corrective approach used by in the *Anglo-French Continental Shelf* arbitration previously referred.[21]

In the *Greenland/Jan Mayen* case, where the ICJ had to determine the applicable rules on delimitation of maritime boundaries between the continental shelf and the EEZ (fishery zone) without the relevant States having agreed on a single maritime boundary, the Court decided that, in what the continental shelf was concerned, to proceed with the delimitation of a median line and then subsequently adjust or modify this line in light of special circumstances. The novel fact of this case is that the ICJ recognized that the principles of maritime delimitation enshrined in Articles 74 and 83 reflected customary international law.

> The Court however notes that Article 74, paragraph 1, and Article 83, paragraph 1, of that Convention provide for the delimitation of the continental shelf and the exclusive economic zone between States with opposite or adjacent coasts to be effected "by agreement on the basis of international law, as referred to in Article 38 of the Statute of the International Court of Justice, in order to achieve an equitable solution". That statement of an "equitable solution" as the aim of any delimitation process reflects the requirements of customary law as regards the delimitation both of continental shelf and of exclusive economic zones.[22]

This acknowledgement would also be followed in subsequent cases.[23] In fact, similarly to the *North Sea Continental Shelf cases*, the ICJ in the case concerning the territorial and maritime dispute between Nicaragua and Colombia had to determine the law applicable to the delimitation of maritime boundaries because Colombia was not party to UNCLOS.[24] In this case, Nicaragua and Colombia agreed that the applicable law was customary international law and that several provisions of UNCLOS reflected customary international law, particularly the provisions regarding the delimitation of the EEZ and the continental shelf, as well as, for that matter, the legal regime of islands.[25] The Court had already in the *Nicaragua v. Honduras* case concluded that:

---

[21] *Continental Shelf (Libya/Malta), supra note 60 in Chap.* 3, paras. 45, 73.

[22] *Maritime Delimitation in the Area between Greenland and Jan Mayen, supra note 33 in Chap.* 4, para. 48. For a critical view of the Court's position *see* R.R. Churchill, "The Greenland—Jan Mayen case and its significance for the international law of maritime boundary delimitation", in: 9-1 *International Journal of Marine and Coastal Law* (1994), p. 17.

[23] *Territorial and Maritime Dispute (Nicaragua v. Colombia) 2012, supra note 53 in Chap.* 3, para. 139; *Maritime Delimitation and Territorial Questions between Qatar and Bahrain, supra note 26 in Chap.* 6, paras. 167, 230; *Case Concerning Maritime Delimitation in the Black Sea, supra note 53 in Chap.* 3, para. 115–122. Also *see The Eritrea—Yemen Arbitration (second phase), supra* note 23 in Chap. 4, paras. 131–132; *Barbados/Trinidad and Tobago, Award of the Arbitral Tribunal, 11 April 2006*, paras. 242, 306 [online: http://www.pca-cpa.org/upload/files/Final%20Award.pdf (accessed on August 2013)]; *Guyana/Suriname, supra note 25 in Chap.* 2, paras. 342, 335.

[24] *Territorial and Maritime Dispute (Nicaragua v. Colombia) 2012, ibid*, para. 137.

[25] Articles 74, 83 and 121 of UNCLOS. See *Territorial and Maritime Dispute (Nicaragua v. Colombia) 2012, ibid*, para. 138.

> For the exclusive economic zone and the continental shelf, Articles 74, paragraph 1, and 83, paragraph 1, of UNCLOS provide that they are to be delimited by "agreement on the basis of international law" to "achieve an equitable solution". As to the plotting of a single maritime boundary the Court has on various occasions made it clear that, when a line covering several zones of coincident jurisdictions is to be determined, the so-called equitable principles/relevant circumstances method may usefully be applied, as in these maritime zones this method is also suited to achieving an equitable result (...) The jurisprudence of the Court sets out the reasons why the equidistance method is widely used in the practice of maritime delimitation: it has a certain intrinsic value because of its scientific character and the relative ease with which it can be applied. However, the equidistance method does not automatically have priority over other methods of delimitation and, in particular circumstances, there may be factors which make the application of the equidistance method inappropriate.[26]

The judgement of the *Cameroon/Nigeria case* related this understanding with Articles 74(1) and 83(1) of UNCLOS. The Court identified these provisions as being the relevant provisions of UNCLOS and that the applicable criteria, principles and rules of delimitation were included in the equitable principles/relevant circumstances method. The Court went on to clarify that this method involved drawing an equidistant line and then considering the factors that may require the adjustment of that line in order to achieve an equitable result.[27] In so doing, the Court applied the same method as in the case *Maritime Delimitation and Territorial Questions between Qatar and Bahrain* (*Qatar* v. *Bahrain*),[28] which consisted of drawing an equidistance line and then assessing whether there are relevant circumstances that might make it necessary to adjust that line.

As previously referred, Articles 74(1) and 83(1) of UNCLOS have the same wording as a result of the compromise formula adopted during the negotiation of the Convention so as not to exclude any of the different positions present during the negotiations and drafting of these provisions, namely the supporters of the *equidistance/special circumstances* and of the *equitable principle/special*

---

[26] *Territorial and Maritime Dispute between Nicaragua and Honduras in the Caribbean Sea (Nicaragua v. Honduras), Judgment, I.C.J. Reports 2007*, paras. 270–272.

[27] *Land and Maritime Boundary between Cameroon and Nigeria, supra note 83 in Chap*. 3, paras. 285, 288, 290.

[28] *Maritime Delimitation in the Area between Greenland and Jan Mayen, supra note 33 in Chap*. 4, para. 55: "*It is thus apparent that special circumstances are those circumstances which might modify the result produced by an unqualified application of the equidistance principle. General international law, as it has developed through the case-law of the Court and arbitral jurisprudence, arid through the work of the third United Nations Conference on the Law of the Sea, has employed the concept of 'relevant circumstances. This concept can be described as a fact necessary to be taken into account in the delimitation process.*"

circumstances method.[29] Consequently, these provisions do not include any reference to a delimitation method but rather only to mention that the aim is to achieve a delimitation of maritime boundaries that is equitable,[30] which prompted Robin Churchill and Lowe Vaughn to describe the said compromise formula as not being very "*meaningful*."[31] Nonetheless, through interpretation, this formula has become clearer and its legal sense refined.[32] Only recently ITLOS, in the case between *Bangladesh/Myanmar* made an overview of the international law applicable to delimitation of maritime boundaries under Articles 74(1) and 83(1) of UNCLOS, referring that[33]:

> The Tribunal observes that the issue of which method should be followed in drawing the maritime delimitation line should be considered in light of the circumstances of each case. The goal of achieving an equitable result must be the paramount consideration guiding the action of the Tribunal in this connection. Therefore the method to be followed should be one that, under the prevailing geographic realities and the particular circumstances of each case, can lead to an equitable result.[34]

In another recent case the ICJ began by determining what were the relevant coasts of the States and their extent, namely, those coasts the projections of which overlapped, observing that the task of delimitation consists in resolving the

---

[29] *See Continental Shelf* (*Libya/Malta*), *supra note 60 in Chap.* 3, paras. 41–43, 56–57; *Territorial and Maritime Dispute between Nicaragua and Honduras in the Caribbean Sea*, *supra* note 26, para. 272. Also *see* Jean-Pierre Quéneudec, "L'affaire de la délimitation du plateau continental entre la France et le Royaume-Uni", in: 83 *Revue Générale de Droit International Public* (1979), pp. 53–103; Edward Duncan Brown, "The Anglo-French continental shelf case", in: 19 *San Diego Law Review* (1979), pp. 461–530; D.M. McRae, "Delimitation of the continental shelf between the United Kingdom and France: the channel arbitration", in: 15 *Canadian Yearbook of International Law* (1977), pp. 173–197; Maurice Mendelson, "On the quasi normative effect of maritime boundary agreements", in: *Liber Amicorum Judge Shigeru Oda*, vol. 2, edited by Nisuke Ando, Edward McWhinney and Rüdiger Wolfrum (Kluwer Law International: The Hague, London, New York, 2002), pp. 1069–1086; Bernard H. Oxman, "Political, strategic, and historical considerations", in: *International Maritime Boundaries*, vol. 1, *supra* note 42 in Chap. 2, pp. 3–40; Tanaka, *supra* note 23 in Chap. 3, p. 191; Blecher, *supra* note 59 in Chap. 3, pp. 65–68; Weil, *supra* note 13, pp. 112–114; Charney, *supra* note 136, p. 585; Brown, *supra* note 60 in Chap. 3, pp. 262–278; Johnston, *supra* note 13, pp. 200–210; Andrassy, *supra* note 46 in Chap. 3, pp. 91–107.

[30] "*The delimitation of* [the exclusive economic zone/the continental shelf] *between States with opposite and adjacent coasts shall be effected by agreement on the basis of international law, as referred to in Article 38 of the Statute of the International Court of Justice, in order to achieve an equitable solution.*"

[31] Churchill/Lowe, *supra* note 68 in Chap. 2, p. 191.

[32] For a critical view *see* Malcom D. Evans, "Maritime boundary delimitation: where do we go from here?", in: *The Law of the Sea: Progress and Prospects*, edited by David Freestone, Richard Barnes and David Ong (Oxford University Press: Oxford, 2006), pp. 137–160.

[33] Bangladesh/Myanmar, *supra* note 71 in Chap. 3, paras. 225–234.

[34] *Ibid*, para 235. Also *see Territorial and Maritime Dispute between Nicaragua and Honduras in the Caribbean Sea*, *supra* note 26, paras. 280–281. Bangladesh/India, supra note 69 in Chap. 6, paras. 339–346.

overlapping claims by drawing a line of separation between the maritime areas concerned. The Court than turned to the methodology to be employed in effecting the delimitation, having identified three stages. In the first stage, the Court established a provisional delimitation line using methods that were geometrically objective and appropriate for the geography of the area. This would consist of an 'equidistance line' (adjacent coasts) or a 'median line' (opposite coasts), unless the establishment of such a line is not possible, although the Court acknowledged that there was no legal consequences resulting from the use of either one of these terms, seeing that the method of delimitation in each case involved constructing a line each point on which is an equal distance from the nearest points on the two relevant coasts, using the most appropriate base points on the coasts.[35]

The Court further noted that subsequent to the drawing of the provisional delimitation line, it took into account the relevant circumstances that would require the adjustment of the equidistance/median line for the purpose of achieving an equitable result. Lastly, the Court assessed the proportionality effect of the line to the relevant coasts in order to achieve an equitable result. However, as the Court noted, this process may not "*be applied in a mechanical fashion.*"[36]

8.2 Consideration of Special or Relevant Circumstances

8.2.1 Concept

The compromise formula included in Article 83(1) of UNCLOS did not provide much clarity regarding the method to be used in the delimitation of maritime boundaries between States with opposite or adjacent coasts. Likewise, Article 6 of the 1958 Convention on the Continental Shelf merely referred that, in the absence of an agreement, the boundary line between States with opposite or adjacent coasts should be taken into consideration as special circumstances, without adding any reference to what these special circumstances were and how they should be considered when drawing a boundary line.

International jurisprudence has established that the appropriate delimitation method is that which considers the circumstances of each case in order to achieve an equitable result, even though it remains to be evaluated in every situation the weight to be attributed to the different circumstance for the purpose of adjusting a provisional delimitation line. Such circumstances include the configuration of the

[35] *Territorial and Maritime Dispute (Nicaragua v. Colombia) 2012, supra note 53* in Chap. 3, paras. 141, 155, 167, 177, 184, 190, 191.

[36] *Ibid*, paras. 192–194, 242, 247.

coasts,[37] the presence of islands[38] or the differences in the lengths of the coastlines.[39]

The concept of 'special circumstance' was firstly used in reference to the territorial sea and included in Article 12(1) of the 1958 Convention on the Territorial Sea and the Contiguous Zone, and later in Article 15 of UNCLOS. These provisions adopted almost identical wording and remained connected with the equidistance method first referred in Article 12(1).[40]

In addition to the reference to 'special circumstances' made in the first boundary delimitation cases submitted to International Courts and Arbitral Tribunals, international jurisprudence has also made reference to 'relevant circumstances'. In fact, while the 1958 Convention on the Continental Shelf refers to 'special circumstances', general international law based on international jurisprudence mentions 'relevant circumstances'. Despite having different sources and

---

[37] *North Sea Continental Shelf cases, supra note 46 in Chap.* 3, paras. 82, 91, 95, 96, 101(D)(1); *Delimitation of the Maritime Boundary in the Gulf of Maine Area, supra note 60 in Chap.* 3, paras. 205–206, 211–213, 215; Bangladesh/Myanmar, *supra* note 71 in Chap. 3, paras. 292, 297; *Land and Maritime Boundary between Cameroon and Nigeria, supra note 83 in Chap.* 3, paras. 295–297; *Territorial and Maritime Dispute (Nicaragua v. Colombia) 2012, supra note 53 in Chap.* 3, para. 229.

[38] *Anglo-French Continental Shelf Arbitration, supra* note 5, paras. 243–251; *Continental Shelf (Tunisia/Libya), supra note 10 in Chap.* 3, paras. 120, 128–129; *Continental Shelf (Libya/Malta), supra note 60 in Chap.* 3, paras. 64, 68–69, *Case Concerning Maritime Delimitation in the Black Sea, supra note 53 in Chap.* 3, para. 149, 187–188; *The Eritrea—Yemen Arbitration (second phase), supra* note 23 in *Chap.* 4, paras. 139, 147, 150–151, 165–168; *Delimitation of the Maritime Boundary in the Gulf of Maine Area, supra note 60 in* Chap. 3, para. 222; *Land and Maritime Boundary between Cameroon and Nigeria, supra note 83 in Chap.* 3, para. 299; *Maritime Delimitation and Territorial Questions between Qatar and Bahrain, supra note 26 in Chap.* 6, paras. 219, 222; *Territorial and Maritime Dispute between Nicaragua and Honduras in the Caribbean Sea, supra* note 26, paras. 304–305; Bangladesh/Myanmar, *supra* note 71 in Chap. 3, paras. 316–319.

[39] *North Sea Continental Shelf cases, supra note 46 in Chap.* 3, para. 98, 101(D)(3); *Continental Shelf (Tunisia/Libya), supra note 10 in Chap.* 3, paras. 103–105, 108, 130–131, 133B(5); *Continental Shelf (Libya/Malta), supra note 60 in Chap.* 3, paras. 56–58, 74–75, 78; *Maritime Delimitation in the Area between Greenland and Jan Mayen, supra note 33 in Chap.* 4, paras. 65–69; *Delimitation of the Maritime Boundary in the Gulf of Maine Area, supra note 60 in Chap.* 3, para. 218, *Delimitation of Maritime Areas between Canada und the French Republic (St. Pierre et Miquelon), supra* note 20, paras. 33, 93; *Land and Maritime Boundary between Cameroon and Nigeria, supra note 83 in Chap.* 3, para. 301; *Barbados/Trinidad and Tobago, supra* 23, paras. 337–338, 376–379; *Guyana/Suriname, supra note 25 in Chap.* 2, para. 393; *Case Concerning Maritime Delimitation in the Black Sea, supra note 53 in Chap.* 3, para. 213.

[40] The Article 12(1) of the 1958 Convention on the Territorial Sea and Contiguous Zone reads: "*Where the coasts of two States are opposite or adjacent to each other, neither of the two States is entitled, failing agreement between them to the contrary, to extend its territorial sea beyond the median line every point of which is equidistant from the nearest points on the baselines from which the breadth of the territorial seas of each of the two States is measured. The above provision does not apply, however, where it is necessary by reason of historic title or other special circumstances to delimit the territorial seas of the two States in a way which is at variance therewith.*"

wordings, the two concepts are similar in substance, seeing that they both seek to achieve an equitable result. There is in fact a tendency towards the integration of the two concepts.[41]

8.2.2 *Mineral Resources*

8.2.2.1 Conservation of the Unity of the Deposit

The *Memorandum on the Regime of the High Seas* published by the United Nations established that the unity of hydrocarbon deposits should be taken into account for the purpose of delimitation of maritime boundaries, particularly when considering the well-established principle of unity applicable to rivers that provided for a certain area of a river being considered as a unity regarding its use and share of resources. However, the Memorandum made only but a brief and general note to the principle of unity of hydrocarbon deposits in relation to the delimitation of maritime boundaries, without elaborating on the terms and conditions under which this principle should be applied and what consequences should result thereof.[42] Moreover, the principle of unity applicable to rivers is connected with the use and share of resources and not with the delimitation of boundaries. The said Memorandum did not present a reasonable substantiation to support the existence of an international binding rule requiring States to consider the unity of hydrocarbon deposits for the purpose of delimitation of maritime boundaries.[43]

Also, as examined in Chap. 3, the ILC decided not to continue the work on shared natural resources regarding oil and gas, seeing that the specific and complex issues relating to transboundary reserves, particularly when pending the delimitation of maritime boundaries, are best settled between States through cooperation.

---

[41] *Maritime Delimitation in the Area between Greenland and Jan Mayen, supra note 33 in Chap.* 4, paras. 54–56. Also *see Anglo-French Continental Shelf Arbitration*, *supra* note 5, para. 148.

[42] *The Memorandum on the Regime of the High Seas,* prepared by the Secretariat of the United Nations (A/CN.4/32), done on 14 July 1950, published in *Yearbook of the International Law Commission, 1950, vol.* 2, 112, paras. 337–338 [online: http://untreaty.un.org/ilc/documentation/french/a_cn4_32.pdf (accessed on August 2013)]. On customary law applicable to rivers *see* Joseph W. Dellapenna, "The customary international law of transboundary fresh waters", in: *Exploitation of Natural Resources in the 21st Century*, edited by Malagosia Fitzmaurice and Milena Szuniewicz (Kluwer Law International: The Hague, London, New York, 2003), pp. 153–188; Eyal Benvenisti, "Collective action in the utilization of shared freshwater: the challenges of international water resources law", in: 90 *American Journal of International Law* (1996), pp. 384–415; Ann Berkley Rodgers and Albert E. Utton, "The Ixtapa draft agreement relating to the use of transboundary groundwaters", in: 25 *Natural Resources Journal* (July 1985), pp. 713–722.

[43] Lagoni, *supra* note 63 in Chap. 2, p. 240; Barberis, *supra* note 63 in Chap. 2, pp. 49–50; Utton, *supra* note 70 in Chap. 2, p. 68. Also *see* William E. Kenworthy, "Joint development of international waters", in: 54-3 *American Journal of International Law* (July 1960), pp. 592–602.

Differently, M.W. Mouton adopted the concept advanced by the said Memorandum and developed a legal notion of the principle of unitization that considered the "*essential unity of a deposit*"[44] in relation with the delimitation of maritime boundaries, or what Jacques Vuilleimier described as States' and operators' obligation to develop hydrocarbon deposits as *bon pater familiae*, seeking to establish and keep the balance between maximizing production and avoiding waste.[45] M.W. Mouton believed that the presence of a hydrocarbon deposit was a special circumstance that should be considered as an exception to the applicability of the median line in case of countries with opposite coasts, and that States should have in mind the protective perimeter of hydrocarbon deposits in order to avoid that drilling immediately outside the perimeter could affect the development of the deposit. Adding, however, that States' interests would be better served by means of an agreement due to the difficulty of developing a general rule that could be applied to every situation where a common deposit would straddle a maritime boundary.[46] In fact, the relevance of the discretionary element remains a key element of provisional arrangements pending the delimitation of maritime boundaries, and is also essential for the implementation of joint development arrangements of offshore hydrocarbon deposits.

8.2.2.2 Relevance for the Delimitation of Maritime Boundaries

International Courts and Tribunals have had the opportunity of examining the role of mineral resources for the delimitation of maritime boundaries. In several cases, States have unsuccessfully requested that the (potential) presence of mineral resources or oil concession areas, in which operators have acquired rights to explore and exploit these resources, be considered special circumstances for the purpose of maritime delimitation, even if in most cases States were not prepared to base their arguments solely on the consideration of these factors.[47] In this manner, States attempted to reinforce their claims regarding the disputed maritime area and eventually the recognition of a de facto boundary line, which would also safeguard the rights granted to operators for the exploration and exploitation of mineral resources in disputed maritime areas.

The ICJ considered in the *North Sea Continental Shelf* cases that, "*so far as known or readily ascertainable, the physical and geological structure, and natural resources, of the continental shelf areas involved*" should be taken into account during negotiations towards the delimitation of maritime boundaries, in addition to other relevant factors, namely, the general configuration of the coasts and the

---

[44] Mouton, *supra* note 21 in Chap. 2, p. 421.

[45] Vuilleimier, *supra* note 62 in Chap. 2, p. 283a.

[46] Mouton, *supra* note 21 in Chap. 2, p. 421.

[47] *Separate Opinion of Judge Jessup*, paras. 67, 72, *North Sea Continental Shelf cases, supra note 46 in Chap.* 3.

presence of any special or unusual features, and the element of a reasonable degree of proportionality.[48] The Court would again consider the importance of these resources in the *Libya/Malta* case, referring that they "*are the essential objective envisaged by States when they put forward claims to sea-bed areas containing them.*"[49]

The ICJ in the *North Sea Continental Shelf* cases further determined that there was no limit for the circumstances that could be taken into consideration and that would be the "*balancing-up*" of all such considerations that will produce an equitable result. The Court added that the importance of each circumstance will depend on the specificities of each case:

> In fact, there is no legal limit to the considerations which States may take account of for the purpose of making sure that they apply equitable procedures, and more often than not it is the balancing-up of al1 such considerations that will produce this result rather than reliance on one to the exclusion of al1 others. The problem of the relative weight to be accorded to different considerations naturally varies with the circumstances of the case.[50]

In what concerns mineral resources in particular, the Court recognized the factual value of hydrocarbon deposits and the importance of preserving the unity of these resources to ensure their efficient development. It did not, however, take them into account for the purpose of delimitation, despite their relevance in light of all factors.

> Another factor to be taken into consideration in the delimitation of areas of continental shelf as between adjacent States is the unity of any deposits. The natural resources of the subsoil of the sea in those parts which consist of continental shelf are the very object of the legal regime established subsequent to the Truman Proclamation. Yet it frequently occurs that the same deposit lies on both sides of the line dividing a continental shelf between two States, and since it is possible to exploit such a deposit from either side, a problem immediately arises on account of the risk of prejudicial or wasteful exploitation by one or other of the States concerned. To look no farther than the North Sea, the practice of States shows how this problem has been dealt with, and al1 that is needed is to refer to the undertakings entered into by the coastal States of that sea with a view to ensuring the most efficient exploitation or the apportionment of the products extracted (…) The Court does not consider that unity of deposit constitutes anything more than a factual element which it is reasonable to take into consideration in the course of the negotiations for a delimitation. The Parties are fully aware of the existence of the problem as also of the possible ways of solving it.[51]

In the same case, Judge Jessup observed in his *Separate Opinion* that there is no rule of international law requiring States to proceed with the fair apportionment of mineral resources[52] and that the aim was to achieve a delimitation that was equitable and not an equal apportionment of maritime areas.[53]

---

48 *North Sea Continental Shelf cases, ibid*, para. 101(D).

49 *Continental Shelf (Libya/Malta), supra note 60 in Chap.* 3, para. 50.

50 *North Sea Continental Shelf cases, supra note 46 in Chap.* 3, para. 93.

51 *Ibid*, para. 97.

52 *Separate Opinion of Judge Jessup*, paras. 78, 83, *ibid.*

53 *Ibid*, paras. 17–18. Also *see Case Concerning Maritime Delimitation in the Black Sea, supra note 53 in Chap.* 3, para. 111.

In what oil and gas concessions are concerned, the ruling of the ICJ in the case concerning the *Continental Shelf* (*Tunisia/Libyan*), which has been subsequently upheld, including by Arbitral Tribunals, is that, despite States' practice revealing the existence of a *modus vivendi*, these are not relevant factors in the context of maritime delimitation, namely when representing or adjusting a provisional delimitation line, except if such oil concessions are based on express or tacit agreement between the States.[54] More recently, in fact, the ICJ found in the *Cameroon/Nigeria* case that:

> Overall, it follows from the jurisprudence that, although the existence of an express or tacit agreement between the parties on the siting of their respective oil concessions may indicate a consensus on the maritime areas to which they are entitled, oil concessions and oil wells are not in themselves to be considered as relevant circumstances justifying the adjustment or shifting of the provisional delimitation line. Only if they are based on express or tacit agreement between the parties may they be taken into account. In the present case there is no agreement between the Parties regarding oil concessions. The Court is therefore of the opinion that the oil practice of the Parties is not a factor to be taken into account in the maritime delimitation in the present case.[55]

In the case concerning *Delimitation of the Maritime Boundary in the Gulf of Maine Area* (*Canada/United States of America*), the Chamber of the ICJ also took into account the conduct of the States but it did not consider that it amounted to the existence of a *modus vivendi* and therefore the circumstances regarding granting of oil concessions was different from the *Tunisia/Libya* case. The Chamber considered that in the *Tunisia/Libya* case, the ICJ took into account the conduct of France and Italy that was continued after the independence of Tunisia and Libya when they granted oil concessions. The Chamber also considered that the period claimed by Canada to establish a *modus vivendi* was too brief to have produced the intended legal effect.[56] Likewise, in the case concerning the *Continental Shelf* (*Libyan Arab Jamahiriya/Malta*), the ICJ did not accept as evidence of acquiescence the claim made by Malta that, through State practice regarding the limits of certain oil concessions, States had indicated that the median line was relevant to the final delimitation of a maritime boundary, having concluded that:

> [The Court was not able] to discern any pattern of conduct on either side sufficiently unequivocal to constitute either acquiescence or any helpful indication of any view of either Party as to what would be equitable differing in any way from the view advanced by

---

[54] *Case Concerning Maritime Delimitation in the Black Sea, supra note 53 in Chap.* 3, paras. 189–198; *Delimitation of the Maritime Boundary in the Gulf of Maine Area, supra note 60 in Chap.* 3, paras. 148–150; *Continental Shelf (Tunisia/Libya), supra note 10 in Chap.* 3, paras. 107, 117–119. Also *see Guyana/Suriname, supra note 25 in Chap.* 2, para. 390; *Barbados/Trinidad and Tobago, supra* 23, para. 364. The Arbitral Tribunal in the Guinea/Guinea Bissau case declined to take into consideration an oil concession granted by Portugal *supra* note 19, p. 281, para. 63.

[55] *Land and Maritime Boundary between Cameroon and Nigeria, supra note 83 in Chap.* 3, paras. 303–304.

[56] *Delimitation of the Maritime Boundary in the Gulf of Maine Area, supra note 60 in Chap.* 3, paras. 126, 149–152.

that Party before the Court. Its decision must accordingly be based upon the application to the submissions made before it of principles and rules of international law.[57]

Also, in the case concerning the *Delimitation of Maritime Areas between Canada und the French Republic* (*St. Pierre et Miquelon*), the Tribunal concluded that it did not have a reason to consider the potential mineral resources as having bearing on the delimitation.[58]

In the case concerning territorial sovereignty between Eritrea and Yemen, the Arbitral Tribunal considered that offshore petroleum contracts entered into by both countries could to a certain extent lend support to a median line between the opposite coasts and in so doing divide the jurisdiction of the States, but they did not strengthen States' sovereignty claims over the disputed islands.[59]

The aforementioned can also be said regarding joint development areas. In effect, the Arbitral Tribunal in the award that opposed Barbados to Trinidad and Tobago rejected the former's request that the joint development area created under the agreement signed by Barbados and Guyana should be considered for the purpose of establishing the maritime boundary between Barbados and Trinidad and Tobago,[60] since this agreement was *res inter alios acta* in respect of Trinidad and Tobago and for this reason could not influence the delimitation between the latter and Barbados, except in so far as it would reflect the limits of Barbados'

[57] *Continental Shelf (Libya/Malta), supra note 60 in Chap.* 3, paras. 24–25.

[58] *Supra* note 20, para. 89.

[59] *The Eritrea—Yemen Arbitration (first phase), supra* note 103 in Chap. 3, paras. 437–439. Also *see The Eritrea—Yemen Arbitration (second phase), supra* note 226, paras. 86, 132; *Separate Opinion of Judge* Jessup, para. 81, *North Sea Continental Shelf cases, supra note 46 in Chap.* 3; *Continental Shelf (Tunisia/Libya), supra note 10 in Chap.* 3, paras. 117–118, 125. On the *Eritrea/ Yemen Arbitration see* Barbara Kwiatkowska, "The Eritrea-Yemen Arbitration: landmark progress in the acquisition of territorial sovereignty and equitable maritime boundary delimitation", in: 32-1 *Ocean Development and International Law* (2001), pp. 1–25; *Ibid*,"The contribution of the 2002 (Cameroon v. Nigeria: Equatorial Guinea intervening) land and maritime boundary between Cameroon and Nigeria judgment of the International Court of justice to equitable maritime boundary delimitation", in: 17 *The Hague Yearbook of International Law* (2004), pp. 95–135; Yoshifumi Tanaka, "Reflections on the Eritrea/Yemen Arbitration of 17 December 1999 (second phase: maritime delimitation)", in: 48 *Netherlands International Law Review* (2001), pp. 197–225; *Ibid*, "Reflections on maritime delimitation in the Cameroon/Nigeria case", in: 53 *International and Comparative Law Quarterly* (2004), pp. 369–406; Nuno Marques Antunes, "The 1999 Eritrea-Yemen maritime delimitation award and the development of international law", in: *Estudos em Direito Internacional Público* (Almedina: Coimbra, 2004), pp. 225–268; *Ibid*, "The pending maritime delimitation in the Cameroon v. Nigeria case: a piece in the jigsaw puzzle of the Gulf of Guinea", in: 15-2 *International Journal of Marine and Coastal Law* (2000), pp. 163–192; Maurice Mendelson, "The Cameroon-Nigeria case in the ICJ: some territorial sovereignty and boundary delimitation issues", in: 75 *British Yearbook of International Law* (2004), pp. 223–247; Prescott/Schofield, *supra* note 51 in Chap. 6, pp. 234–235.

[60] *Exclusive Economic Zone Co-operation Treaty signed between the Republic of Guyana and the State of Barbados*, done at London on 2 December 2003, reproduced in: *International Maritime Boundaries, vol. 5, supra note 13 in Chap.* 4, pp. 3587–3594.

maritime claim.[61] The possibility of coastal States and third States claiming in the future any effect of lines establishing a joint development area for the purpose of establishing a de facto boundary line should follow the same reasoning.

Similarly, the ICJ in the territorial and maritime dispute between Nicaragua and Colombia did not consider the joint development area created by the agreement between Jamaica and Colombia as being part of the maritime area to be delimited between Nicaragua and Colombia. The Court considered that this maritime area could not extend beyond the area in which the entitlements of Nicaragua and Colombia overlapped, or where one of the States had no entitlement, whether because of an agreement it had concluded with a third State or because that area lied beyond a judicially determined boundary with a third State. In the specific case of the joint development area between Jamaica and Colombia, the Court noted that more than half of it was located beyond 200 nm from Nicaragua and, therefore, could not constitute part of the relevant area in any event, and that neither Nicaragua nor Colombia sought to include it.[62]

The ICJ, in *Case Concerning Maritime Delimitation in the Black Sea* (*Romania v. Ukraine*) concluded that Ukraine's oil concessions practice offered no support in order to prove a tacit agreement or *modus vivendi* between the States regarding the line which would separate their respective EEZ and continental shelves, and therefore such activities should not constitute a relevant circumstance.[63]

Despite the reservations of International Courts and Tribunals in granting full weight to offshore hydrocarbon deposits for the purpose of delimitation, it does not imply that the problems represented by transboundary resources, or the advantages of joint development agreements in this regard, have been ignored. As early as the *North Sea Continental Shelf* cases, the ICJ considered that it would be within States' powers to determine how to proceed in the event of transboundary resources, appointing as a possible outcome the "*joint jurisdiction*" of all or part of the overlapping maritime areas, should States fail to reach an agreement or not intend to equally divide such areas.[64] Judge Jessup further referred in his *Separate Opinion* that the reference made to the cooperative development of offshore hydrocarbon deposits constituted an attempt to contribute towards a better understanding of the principles of equity.[65] Likewise, the recommendations made by Judge Evensen in his *Dissenting Opinion* in the *Case Concerning the*

---

[61] *Barbados/Trinidad and Tobago, supra* 23, para. 349. Also *see* Article 34 of the Vienna Convention.

[62] *Territorial and Maritime Dispute (Nicaragua v. Colombia) 2012, supra note 53 in Chap.* 3, para. 163.

[63] *Case Concerning Maritime Delimitation in the Black Sea, supra note 53 in Chap.* 3, paras. 189–198.

[64] *North Sea Continental Shelf cases, supra note 46 in Chap.* 3, paras. 94, 97, 99, 101(C)(2), (D)(2).

[65] *Separate Opinion of Judge Jessup*, paras. 78, 83, *North Sea Continental Shelf cases, supra note 46 in Chap.* 3. Also *see Dissenting Opinion of Vice-President Weeramantry*, paras. 108–113, *KasikililSedudu Island (Botswana/Namibia) Judgment, I.C.J. Report 1999,* p. 1045.

Continental Shelf between Tunisia and Libya had also the same intent and reportedly did contribute towards the implementation of a joint development regime between both countries.[66] Judge Evensen essentially considered that cooperative regimes, such as joint development agreements, could be a corollary for other equity considerations that had already been referred by the ICJ in the *North Sea Continental Shelf* cases.[67] The Arbitral Tribunal also considered in the second stage of proceedings concerning maritime delimitation between Eritrea and Yemen that these countries should give every consideration to cooperative exploitation of resources that straddle maritime boundaries, particularly in light of improving relations between both countries and the existence of State practice.[68] In fact, in the case between Guyana and Suriname, the Arbitral Tribunal went as far as acknowledging that joint development agreements have been "*particularly encouraged by international courts and tribunals*", although it did not elaborate on the grounds for such a solution being legally required under international law.[69] In this case, the Tribunal also considered that oil practice between States could not be taken into account in the delimitation of the maritime boundary, except if such practice would have been based on express or tacit agreement between States.[70]

In the maritime delimitation case between Myanmar and Bangladesh, ITLOS noted that it was not "*unusual*" for States to enter into agreements to address the "*complex legal and practical problems*" resulting from the delimitation of maritime boundaries, such as those involving transboundary resources. ITLOS specifically referred to an area that was beyond 200 nm from the coast of Bangladesh but within 200 nm from the coast of Myanmar, thus accepting the possibility of such agreements taking place in areas of overlapping claims beyond 200 nm, in the outer continental shelf.[71]

In what living marine natural resources are concerned, the decision of the *Grisbådarna Case* recognized the importance of preserving the unity of these resources, as well as, for that matter, of protecting acquired fishing rights. However, the Court eventually awarded the sovereignty of the Grisbådarna banks to Sweden based on its historic claim and not in consideration of the need to ensure

---

66 *Agreement between the Great Libyan Arab Socialist People's Jamahariya and the Republic of Tunisia to Implement the Judgment of the International Court of Justice in the Tunisia/Libya Continental Shelf Case*, done at Benghazi on 8 August 1988, reproduced in: *International Maritime Boundaries*, vol. 2, *supra note 125 in Chap.* 3, pp. 1679–1680. On the case before the ICJ *see* Jonathan I. Charney, "The delimitation of ocean boundaries", in: *Rights to Oceanic Resources. Deciding and Drawing Maritime Boundaries*, edited by Dorinda G. Dallmeyer and Louis Devorsey, Jr. (Martinus Nijhoff Publishers: Dordrecht, Boston, London, 1989), pp. 40–41. On the agreement *see* Miyoshi, *supra* note 2 in Chap. 4, pp. 35–36.

67 *Dissenting Opinion of Judge Evensen*, pp. 320–321, *Continental Shelf (Tunisia/Libya), supra note 10 in Chap.* 3.

68 *The Eritrea—Yemen Arbitration* (*second phase*), *supra* note 23 in Chap. 4, paras 84–86.

69 *Guyana/Suriname, supra note 25 in Chap.* 2, paras. 462–463.

70 *Ibid*, para. 390.

71 Bangladesh/Myanmar, *supra* note 71 in Chap. 3, para. 472.

the preservation and efficient management of living resource.[72] Just as the ICJ in *Gulf of Maine* case also did not consider that delimitation should take into account the conservation and efficient management of living resources or fishing rights or the fact that fishing was an essential activity. Instead, the Court concluded that it would be up to the relevant States to settle these issues through an agreement that could best serve their interests.[73]

The Arbitral Tribunal in the *St. Pierre et Miquelon* case recognized that the criteria governing delimitation is to be found primarily in geographical facts, despite the economic relevance of fishery resources and fishing activities, as well as of the need to safeguard the legitimate interests of fishing communities, particularly when, in this case, the solution reached did not have a radical impact on fishing rights and on the economic dependence of those most affected by delimitation. The Arbitral Tribunal also considered that States' hydrocarbon interests in overlapping areas were not relevant for the purpose of delimitation.[74]

---

[72] *Grisbådarna Case (Norway v. Sweden), P.C.A. Award 23 October 1909*, 111, 129, reproduced in: *United Nations Reports of International Arbitral Awards, vol. XI*, pp. 147–166. Also *see* Moritaka Hayashi, "The management of transboundary fish stocks under the LOS Convention", in: 8-2 *International Journal of Marine and Coastal Law* (1993), pp. 245–261; Shigeru Oda, "Sharing of ocean resources-unresolved issues in the Law of the Sea", in: *La Gestion des Ressources pour l'Humanité: Le Droit de la Mer/The Management of Humanity's Resources: The Law of the Sea*, edited by René-Jean Dupuy (Martinus Nijhoff Publishers: The Hague, Boston, London, 1982), pp. 49–62; *Ibid*, "Fisheries under the United Nations Convention on the Law of the Sea", in: *Law of the Sea*, edited by Hugo Caminos (Ashgate/Dartmouth: Aldershot, Burlington US, Singapore, Sydney, 2001), pp. 283–299.

[73] *Delimitation of the Maritime Boundary in the Gulf of Maine Area, supra note 60 in Chap.* 3, paras. 231–238, 240. Also *see Maritime Delimitation in the Area between Greenland and Jan Mayen, supra note 33 in Chap.* 4, paras. 75, 76. *See* Jan Schneider, "The Gulf of Maine Case: the nature of an equitable result", in: 79-3 *American Journal of International Law* (July 1985), p. 575; David Vanderzwagg, "Transboundary challenges and cooperation in the Gulf of Maine region: riding a restless sea toward misty shores", in: *Law of the Sea. Common Heritage and Emerging Challenges*, edited by Harry N. Scheiber (Martinus Nijhoff Publishers: The Hague, London, Boston, 2000), pp. 265–285; Davis R. Robinson, David A. Colson and Bruce C. Rashkow, "Some perspectives on adjudicating before the world court: the Gulf of Maine case", in: 79-3 *American Journal of International Law* (July 1985), pp. 578–597; Levi E. Clain, "Gulf of Maine—a disappointing first in the delimitation of a single maritime boundary", in: 25 *Virginia Journal of International Law* (1984–1985), pp. 538–543; Sang-Myon Rhee, "Equitable solutions to the maritime boundary dispute between the United States and Canada in the Gulf of Maine", in: 75 *American Journal of International Law* (1981), pp. 590–628. On the status of the offshore boundaries of the United States of America and Canada *see* Jonathan I. Charney, "The offshore jurisdiction of the States of the United States and the Provinces of Canada-a comparison", in: 12-2 *Ocean Development and International Law* (1982), pp. 301–335; Albert W. Koers, "Implications of alternative regimes for the Northeast Atlantic treaty system", in: *Fisheries Conflicts in the Northeast Atlantic: Problems of Management and Jurisdiction*, edited by Giulio Pontecorvo and Norma Hench Hagist (Ballinger Publishing Company: Cambridge Massachusetts, 1974), pp. 53–70; Johnston, *supra* note 3, pp. 178–191.

[74] *Supra* note 20, paras. 83–91. On this case *see* Ted L. McDorman, "The Canada–France maritime boundary case: drawing a line around St. Pierre and Miquelon", in: 84 *American Journal of International Law* (1990), pp. 157–189.

Nonetheless, in the *Greeland/Jan Mayen* case, the ICJ did in fact take into account the presence of natural resources and divided the area so that the States could share the resources.

> The southernmost zone, zone 1, corresponds essentially to the principal fishing area referred to in paragraph 73 above. In the view of the Court, the two Parties should enjoy equitable access to the fishing resources of this zone. (…) The Court considers that an equal division of the whole area of overlapping claims would give too great a weight to this circumstance. Taking into account the equal division of zone 1, it considers that the requirements of equity would be met by the following division of the remainder of the area of overlapping claims (…).[75]

Furthermore, without prejudice of general consensus among certain States as to the critical relevance of hydrocarbon deposits for the purpose of delimitation,[76] the acceptance by International Courts or Arbitral Tribunals of their decisive importance would certainly result in the use of delimitation lines that would not provide the lasting stability which is required of a maritime boundary. Accordingly, when considering all relevant circumstances, attention should also be given to the enduring characteristics of the relevant maritime area, seeing that the delimitation of boundaries is a process that permanently alters the legal, political, social and economic context between neighbouring coastal States, and inevitably has consequences at the regional and global level.

---

[75] *Maritime Delimitation in the Area between Greenland and Jan Mayen, supra note 33 in Chap.* 4, para. 92. Also *see Ibid*, paras. 75–76.

[76] This is the case, for example, of the Persian Gulf. *See* Ian Townsend-Gault, "Maritime Boundaries in the Arabian Gulf", in: *The Razor's Edge: International Boundaries and Political Geography*, edited by Clive Schofield, David Newman, Alasdair Drysdale and Janet Alison Brown (Kluwer Law International: London, The Hague, New York, 2002), pp. 223–236; *Ibid*, "Offshore boundary delimitation in the Arabian/Persian Gulf", in: *Ocean Boundary Making: Regional Issues and Developments*, edited by, Douglas M. Johnston and Phillip M. Saunders (Croom Helm: London, New York, Sydney, 1988), pp. 217–218; Richard Young, "The Persian Golf", in: 3 *New Directions in the Law of the Sea*, edited by Robin Churchill, Myron Nordquist, S. Houston Lay (Oceana Publications Inc.: New York, 1973), p. 233; *Ibid*, "Equitable solutions for offshore boundaries: the 1968 Saudi Arabia-Iran Agreement", in: 64 *American Journal of International Law* (January 1970), pp. 152–157; S.H. Amin, "Law of the continental shelf delimitation: the Gulf example", in: 27 *Netherlands International Law Review* (1980), pp. 335–346.

[illegible] in the *Greenland/Jan Mayen* case the ICJ did in fact take into account the presence of natural resources in [illegible] and the area [illegible] States could share the resources.

[illegible]

Chapter 9
Resources that Lie Between the Continental Shelf and the Area (Article 142 of UNCLOS)

9.1 The Common Heritage of Mankind

The improvement of the technology that allowed for the exploitation of mineral resources at greater depths caused a general concern that States with such technology or the means to develop or acquire it, would progressively be able to claim these maritime areas as being part of their respective continental shelf.[1]

Within this context, the Truman Proclamation took into consideration the global need for mineral resources and encouraged efforts towards the discovery of new sources, having in mind the prospective mineral potential of the north-American continental shelf and the technological improvements in progress at the time that would allow for their recovery. The Truman Proclamation further recognized that jurisdiction over these resources was required in the interest of their conservation and prudent utilization.

The ILC in its commentary to the provisions on the continental shelf noted that the idea had been argued for the internationalization of the resources of the seabed, whereby their exploitation would be entrusted to international bodies, instead of coastal States. Yet, at the time, the ILC did acknowledge that the circumstances presented "*insurmountable practical difficulties*" and "*would not ensure the effective exploitation of natural resources necessary to meet the needs of mankind.*"[2] The ILC also recognized the possible conflict between the exploration and exploitation of the seabed and subsoil and the freedom of the seas, although concluding that this was not "*a sufficient reason for obstructing a development which (...) can be to the benefit of mankind.*" In this respect, the ILC considered that "[t]*he necessary steps must be taken to ensure that this development affects*

---

[1] Resolution (UNGA) n. 2574 A (XXII), 15 December 1969, recognized that the developing technology is making the entire seabed and ocean floor progressively accessible and exploitable for scientific, economic, military and other purposes.

[2] *Yearbook of the International Law Commission, 1956, supra note 59 in Chap.* 5, p. 296.

V. Becker-Weinberg, *Joint Development of Hydrocarbon Deposits in the Law of the Sea*, Hamburg Studies on Maritime Affairs 30, DOI 10.1007/978-3-662-43570-0_9

the freedom of the seas no more than is absolutely unavoidable, since that freedom is of paramount importance to the international community."[3]

An overwhelming number of States, particularly developing countries and those that had recently gained independence from colonial rule, were especially concerned that the awareness of the immense potential of the deep seabed might result in only the States that had, or could acquire, the know-how and the technological capability would have access to, and benefit from, the development of the mineral resources found therein.[4] Such circumstances could cause a race to seize these resources, inevitably leading to their competitive development and waste, causing international instability and aggravating even further the gap between the developing and the developed countries. This was also the time where the international community was influenced by the ideas supporting the establishment of a *New International Economic Order* based on equity, sovereign equality, interdependence, common interest and cooperation among all States, including, for example, promoting the transfer of technology.[5]

The United Nations acknowledged the common interest of mankind in the seabed and ocean floor and that its exploitation and use should be conducted in view of maintaining international peace and security and for the benefit of all mankind, taking into account the special interests and needs of the developing countries.[6] The United Nations further envisaged international cooperation and coordination for the peaceful exploitation and use of the seabed and ocean floor.[7] As a result, the *Committee on the Peaceful Uses of the Sea-Bed and the Ocean Floor beyond the Limits of National Jurisdiction* was established for the purpose

---

[3] *Ibid.*

[4] Mark W. Zacher and James G. McConnell, "Down to the sea with stakes: the evolving law of the sea and the future of the deep seabed regime", in: 21-1 *Ocean Development and International Law* (1990), pp. 79–81.

[5] Resolution (UNGA) n. 3201(S-VI), 1 May 1974, *Declaration on the establishment of a New International Economic Order*. Also *see* Resolution (UNGA) n. 3202 (S-VI), 1 May 1974, *Programme of action on the establishment of a New International Economic Order*; Resolution (UNGA) n. 1515 (XV), 15 December 1960, *Concerted action for economic development of economically less developed countries.*

[6] On the principle of the common heritage of mankind applicable to the Area and its resources *see* Alexandre-Charles Kiss, "The common heritage of mankind: utopia or reality?", in: 40 *International Journal* (1985), pp. 423–441; *Ibid*, "La notion de patrimoine commun de l'humanité", in: 175-2, *Recueil des Cours de l'Académie de Droit International* (1982), pp. 109–119, 196–224; Rüdiger Wolfrum, "Hohe See und Tiefseeboden (Gebiet)", in: *Handbuch des Seerechts*, edited by Wolfgang Graf Vitzthum (Verlag C.H. Beck: Munich 2006), pp. 338–340; *Die Internationalisierung staatsfreier Räume: die Entwicklung einer internationalen Verwaltung für Antarktis, Weltraum, Hohe See und Meeresboden* (Berlin: Springer, 1984), 331–396; *Ibid*, "The principle of the common heritage of mankind", in: 43 *Zeitschrift für Ausländisches Öffentliches Recht und Völkerrecht* (1983), pp. 312–337.

[7] Resolution (UNGA) n. 2340 (XXII), 18 December 1967, *Examination of the question of the reservation exclusively for peaceful purposes of the sea-bed and the ocean floor, and the subsoil thereof, underlying the high seas beyond the limits of present national jurisdiction, and the use of their resources in the interests of mankind.*

of, inter alia, studying the legal principles and norms that could promote international cooperation in the exploration and use of the seabed and ocean floor and to ensure that their resources would be exploited for the benefit of mankind.[8]

In his prominent address at the United Nations, the Maltese ambassador Avi Pardo referred to the problem that unrestricted national exploitation of certain resources of the ocean floor could have on world economies regarding setting of prices and curtailment of markets, adding that, the strategic importance of such resources could result in a great unbalance between the developed and the developing countries. Ambassador Pardo drew attention to the peril that the military use of the ocean floor could represent for the escalation of the arms race and the increase of instability, and also of the need to safeguard the marine environment from permanent impairment. He called for an international regime applicable to the seabed and the ocean floor beyond national jurisdiction as the only alternative that could provide assurance that its resources would be exploited for the benefit of all countries. Ambassador Pardo concluded referring that "*the seabed and the ocean floor are a common heritage of mankind and should be used and exploited for peaceful purposes and for the exclusive benefit of mankind.*"[9] In this regard, Lewis Alexander wrote that, concerning the problem of the development of mineral resources of the seabed beyond coastal States' sovereignty, "[t]*he principle goals of legal regimes in the sea should be to maximize net benefits received from the marine environment and to provide a means for the equitable resolution of conflicts.*"[10]

Following the Maltese ambassador's address, the United Nations adopted several of its recommendations and, most importantly, the applicability of the concept of 'common heritage of mankind' to the seabed and ocean limits of national jurisdiction.[11] The United Nations prepared the relevant guiding principles and called for the establishment of the legal regime to safeguard the common interest of mankind in the seabed and ocean floor, while, at the same time, imposing a moratorium so that States and persons should refrain from all activities of exploitation of its resources and no claim to any of its part or its resources

[8] Resolution (UNGA) n. 2467 A (XXIII), 21 December 1968, *ibid.*

[9] First Committee, General Assembly of the United Nations (Twenty-second session), United Nations Doc A/C.1/PV 1515, 1 November 1967; United Nations Doc A/C.1/PV 1516, 1 November 1967.

[10] Lewis Alexander, "National jurisdiction and the use of the sea", in: 8-3 *National Resources Journal* (1968), p. 398.

[11] The reference to 'heritage of mankind' was initially made regarding the protection of marine living resources. It was included in a proposal of 1927 by José León Suárez to the *Experts Committee for the Progressive Codification of International Law of the League of Nations*. This proposal forwarded the idea that these resources, particularly whales, should be considered part of the heritage of mankind in order to protect them from extinction: "*Les richesses de la mer, en particulier les richesses immenses de la région antarctique, constituent un patrimoine de l'humanité, et notre Commission, constituée par la Société des Nations, est tout indiquée pour proposer au Gouvernement un moyen d'action avant qu'il ne soit trop tard.*" *See* Scovazzi, *supra* note 1 in Chap. 6, p. 93.

should be recognized.[12] In referring to the legal content of the concept of the common heritage of mankind, Alexandre-Charles Kiss concluded in his seminal work that:

> [L]es principes proclamés par l'Assemblée générale des Nations Unies à l'unanimité et à plusieurs reprises combinés avec le vote de la grand majorité des Etats du monde á l'issue de la troisième Conférence sur le droit de la mer et la mise en œuvre effective de la convention dans plusieurs de ses parties avant même son adoption formelle permettraient de reconnaître, erga omnes, l'opposabilité du principe instituant aux grands fonds marins un patrimoine commun de l'humanité.[13]

9.2 Delimitation and Use of the Area

The Area corresponds to the seabed and ocean floor and subsoil thereof, beyond the limits of national jurisdiction.[14] It can be located 200 nm from the territorial sea baselines or more in the case of an extended continental shelf.[15] In this case, the final determination of the Area will depend on the outcome of States' submissions to the CLCS regarding the extension of their continental shelves beyond 200 nm.[16] It would be very difficult for the Authority to dispute a boundary line, established either unilaterally, or by means of an agreement between two or more States, or even as a result of extended continental shelf. In this case, although the recommendations of the CLCS are final and binding, it is the coastal State that has the legal capacity to determine the outer limit of the continental shelf.[17] Indeed, ITLOS in the *Bangladesh/Myanmar* case noted:

---

[12] Resolution (UNGA) n. 2749 (XXV), 17 December 1970, *Declaration of principles governing the sea-bed and the ocean floor, and the subsoil thereof, beyond the limits of national jurisdiction*; Section D of Resolution (UNGA) n. 2574 (XXIV), *supra* note 56 in Chap. 3.

[13] "La notion de patrimoine commun de l'humanité", *supra* note 6, p. 235.

[14] Article 1(1) of UNCLOS.

[15] Article 76 of UNCLOS.

[16] Article 76(8) and Annex II of UNCLOS. *See* Erik Franckx, "The International Seabed Authority and the common heritage of mankind: the need for states to establish the outer limits of their continental shelf", in: 25 *International Journal of Marine and Coastal Law* (2010), pp. 552–562.

[17] Articles 76(8), 84(2) and 134(3) and (4) of UNCLOS. *See* Ted L. McDorman, "The role of the Commission on the Limits of the Continental Shelf: a technical body in a political world", in: 17-3 *International Journal of Marine and Coastal Law* (2002), p. 306; Edward Duncan Brown, *Sea-bed Energy and Mineral resources and the Law of the Sea—Volume II: The Area Beyond the Limits of National Jurisdiction* (Graham and Trotman: London, 1986), p. II.6.23–24. Also *see* International Law Association, *Report of the International Committee on Legal Issues of the Outer Continental Shelf to the Biennial International Law Association Conference in Berlin, Germany, August 2004*, by Alex Oude Elferink (Seventy-First Conference, London: 2004), pp. 773–819; *Ibid*, *Second Report of the International Committee on Legal Issues of the Outer*

> It is clear from Article 76, paragraph 8, of the Convention that the limits of the continental shelf beyond 200 nm can be established only by the coastal State. Although this is a unilateral act, the opposability with regard to other States of the limits thus established depends upon satisfaction of the requirements specified in Article 76, in particular compliance by the coastal State with the obligation to submit to the Commission information on the limits." Adding, however, that "[t]he forgoing does not imply that entitlement to the continental shelf depends on any procedural requirements. A coastal State's entitlement to the continental shelf exists by the sole fact that the basis of entitlement, namely, sovereignty over the land territory, is present. It does not require the establishment of outer limits. Article 77, paragraph 3, of the Convention, conforms that the existence of entitlement does not depend on the establishment of the outer limits of the continental shelf by the coastal State.[18]

Until the introduction of the Area with UNCLOS, the seabed outside the area of the territorial sea to a depth of 200 m or, beyond that limit, to where the depth of the superjacent waters allowed for the exploitation of mineral resources, could be integrated in the continental shelf.[19] Likewise, all parts of the sea that were not included in the territorial sea or in the internal waters of a coastal State would be considered as being the high seas.[20] On the one hand, all States enjoyed the freedoms of the high seas beyond the territorial sea,[21] while, on the other hand, coastal States had sovereign rights for the purpose of exploring and exploiting the mineral resources of their respective continental shelf.[22]

The entering into force of UNCLOS has undoubtedly been a great achievement and has contributed towards the expansion of the scope of the international legal regime applicable to the seas and oceans. The Area regime was one of the most novel parts of UNCLOS. Differently from the EEZ which was a foreseeable evolution of the law of the sea, the Area regime had an innovative character in the 'new law of the sea', due to the underlying concept of common heritage of mankind. This significant alteration is rightfully acknowledged in the Preamble of UNCLOS, despite the fact that the Convention does not provide a definition of the concept.[23]

The unique aspects of the Area regime, which derive from the concept of common heritage of mankind, and the fact that the development of its resources is for the benefit of mankind,[24] exclude any State from making sovereignty claims over the Area or developing its resources as part of the freedom of the high seas.[25]

(Footnote 17 continued)
Continental Shelf to the Biennial International Law Association Conference in Toronto, Canada, June 2006, by Alex Oude Elferink (Seventy-Second Conference: London, 2006), pp. 215–253.

18 Bangladesh/Myanmar, *supra* note 71 in Chap. 3, paras. 407–409.

19 Article 1 of the 1958 Convention on the Continental Shelf.

20 Article 1 of the 1958 Convention on the High Seas.

21 Article 2 of the 1958 Convention on the High Seas.

22 Article 2(1) and (2) of the 1958 Convention on the Continental Shelf.

23 On the reference made in the *Preamble* and its legal significance *see* Brown, *supra* note 17, p. II.3.4-6.

24 Article 140(1) of UNCLOS.

25 Articles 137(1) and 87(2) of UNCLOS. On the relation between the regime of the high seas and Part XI of UNCLOS *see* Alex G. Oude Elferink, "The regime of the Area: delineating the

9.3 Cooperation Between Coastal States and the Authority

UNCLOS provides that the Area shall be exclusively used for peaceful purpose and that States shall act in relation to the Area in accordance with the applicable regime, the UN Charter and other rules of international law, in the interests of maintaining peace and security and promoting international cooperation and mutual understanding.[26] Accordingly, regarding the Area, States are subject to the obligation to cooperate and to act in good faith, which are implicit in their duty to share information on the existence and intention to develop shared or transboundary resources, when these are found straddling a boundary line or are susceptible of being exploited wholly or partly from either side of that line.

In the case of the Area regime, the applicability of these obligations should also be taken into account towards the fulfilment of the general purpose of the Area. This is evident, for example, in the fact that UNCLOS established that States on becoming parties to the Convention agree that there shall be no amendment to the basic principle relating to the common heritage of mankind and that States shall not be party to any agreement in derogation thereof.[27] UNCLOS also employs distinct terms in certain provisions regarding the principles governing the Area, either referring to 'States parties'[28] or 'States',[29] recognizing, as a result, the existence of obligations that bind all States and not just those that are party to UNCLOS. In consideration of the above, the duty for States to cooperate and act in good faith should also be extended to any dealings between States and the Authority,[30] such as regarding marine scientific research or transfer of technology.[31]

In what concerns specifically mineral resources that lie between the continental shelf and the Area, the extent of States' obligation of cooperation is different when taking into account the regimes applicable to the activities of exploration and exploitation of the resources of the Area that lie across limits of national jurisdiction, and those same activities when referring to transboundary resources straddling the continental shelf of two or more countries. UNCLOS provides in Article 142 that the development of deposits found in the Area that straddle the limits of national jurisdiction shall be conducted with due regard for the rights and legitimate interests of the relevant coastal States and that consultations shall be

(Footnote 25 continued)
scope of application of the common heritage principle and freedom of the high seas", in: 22-1 *International Journal of Marine and Coastal Law* (2007), pp. 144–147. Also *see* Brown, *supra* note 17, p. II.2.1–42.

[26] Articles 138 and 141 of UNCLOS.

[27] Article 311(6) of UNCLOS.

[28] Articles 139 and 143 of UNCLOS.

[29] Articles 137, 138, 140 and 141 of UNCLOS.

[30] Articles 157(4) and 300 of UNCLOS.

[31] Articles 143, 144 and 256 of UNCLOS.

maintained in order to avoid trespassing on such rights and interests.[32] The Convention also provides that in the event that such activities in the Area result in the exploitation of resources found within national jurisdiction, the prior consent of the coastal State shall be required.[33] However, UNCLOS did not include the same rule for transboundary resources found in the continental shelf that may extend across delimitated areas or maritime areas where claims overlap.[34] In this respect, UNCLOS did not introduce any changes to the 1958 Convention on the Continental Shelf. In fact, during the drafting of this Convention, although the ILC issued a report that made clear that it was not willing to accept the sovereignty of coastal States over the seabed and subsoil of the continental shelf, it considered, nonetheless, that the rights of the coastal State in the continental shelf are connected with and concern the exploration and exploitation of the resources of the continental shelf, and that these rights are exclusive in the sense that, if the coastal State does not undertake the exploration and exploitation of the resources of the continental shelf, no other State may do so with the latter's consent.[35] Consequently, States do not have to cooperate regarding such resources when only found in the continental shelf since they are considered as belonging exclusively to the relevant coastal State.

The fundamental distinction between these two regimes is the fact that when mineral resources are partially located in the Area they are also considered as being included in the concept of the common heritage of mankind[36] and therefore should too be appropriately developed for the benefit of mankind.[37] In this respect and as rightfully noted, the reference to this concept should be considered as an

---

[32] On development of resources beyond States' jurisdiction *see* Jonathan I. Charney, "The equitable sharing of revenues from seabed mining", in: *Policy Issues in Ocean Law: Studies of the Working Group on Technical Issues of the Law of the Sea*, edited by Herbert G. Knight (*et al.*) (St. Paul Minnesota 1975), pp. 53–120; Jon L. Jacobson, "The United States and the law of the sea conference: interim resource options", *ibid*, pp. 121–189; Bernard Oxman, "Law of the sea forum: the 1994 Agreement on implementation of the seabed provisions of the Convention on the Law of the Sea. The 1994 Agreement and the Convention", in: 88-4 *American Journal of International* Law (1995), pp. 687–696.

[33] Article 142 of UNCLOS.

[34] International Law Association, *supra* note 15 in Chap. 2, p. 22.

[35] *Yearbook of the International Law Commission, 1956*, *supra* note 59 in Chap. 5, p. 297: "*The Commission desired to avoid language lending itself to interpretations alien to an object which the Commission considers to be of decisive importance, namely, the safeguarding of the principle of the full freedom of the superjacent sea and the air space above it. Hence it was unwilling to accept the sovereignty of the coastal State over the seabed and subsoil of the continental shelf. On the other hand, the text as now adopted leaves no doubt that the rights conferred upon the coastal State cover all rights necessary for and connected with the exploration and exploitation of the natural resources of the continental shelf. Such rights include jurisdiction in connexion with the prevention and punishment of violations of the law. The rights of the coastal State are exclusive in the sense that, if it does not exploit the continental shelf, it is only with its consent that anyone else may do so.*"

[36] Article 136 of UNCLOS.

[37] Article 140 of UNCLOS.

"*underlying philosophy*", seeing that it "*does not give indications concerning the formal or substantive contents of a legal regime in a general sense.*"[38] The terms and conditions applicable to the development of these resources must be, as a result, determined between the Authority and the respective coastal State.

There are, nonetheless, certain implications that derive from the consideration of the concept of the common heritage of mankind that ought to be taken into account. To begin with, differently from the development of the resources of the continental shelf, which are subject to the discretionary powers of the relevant coastal States, UNCLOS provides that the exploration and exploitation of the resources of the Area shall be carried out "*to foster healthy development of the world economy and balanced growth of international trade, and to promote international cooperation for the over-all development of all countries, especially developing States.*"[39] Also coastal States are not subject to an obligation of conservation or optimum utilization of the marine natural resources of the continental shelf, either living or non-living. In the continental shelf, the development of mineral resources is predominantly subject to economic and financial considerations and principally guided by their efficient and profitable recovery. In contradistinction, the Area regime provides that its resources shall be orderly, safe and rationally managed in accordance with sound principles of conservation and the avoidance of unnecessary waste, and to ensure supply to consumers, the promotion of just and stable prices and of long-term equilibrium between supply and demand, and the conditions of access to markets for the imports of minerals produced from the resources of the Area and for imports of commodities produced from such minerals.[40] Therefore, any cooperative effort between coastal States and the Authority regarding mineral resources partly found in the Area and within national jurisdiction, must necessarily take into consideration the limitations that result from the Area regime.

9.4 Due Regard and the Duty of Information

UNCLOS provides that there should be reasonable and reciprocal regard between the activities in the Area and other activities in the marine environment.[41] The Convention defines activities in the Area as meaning "*all activities of exploration*

---

[38] Nele Matz-Lück, "The concept of the common heritage of mankind: its viability as a management tool for deep-sea genetic resources", in: *The International Legal Regime of Areas beyond National Jurisdiction: Current and Future Developments*, edited by Alex G. Oude Elferink and Erik J. Molenaar (Martinus Nijhoff Publishers: Leiden, Boston, 2010), p. 65. Matz-Lück concludes that "[i]*t is a matter of methodology whether the common heritage concept is a legal principle from which one can deduct the necessary legal consequences or whether it is an idea that is reflected in different treaty provisions.*" (p. 66).

[39] Article 150 of UNCLOS.

[40] Article 150 lit. (b), (e), (f) and (j) and Article 13(1) of Annex III of UNCLOS.

[41] Article 147(1), (3) of UNCLOS.

for, and exploitation of, the resources of the Area,"[42] being these considered as "*all solid, liquid or gaseous mineral resources in situ in the Area at or beneath the seabed, including polymetallic nodules.*"[43]

In particular, the Convention takes into consideration the possibility of mineral resources straddling the Area and into the continental shelf of a coastal State and being wholly or partly developed from either side of the delimitation line between these two maritime areas. In these cases, Article 142(1) of UNCLOS provides that activities in the Area shall be conducted with due regard for the rights and legitimate interests of the relevant coastal State. This provision has been referred to as emphasizing "[t]*he primacy of coastal State rights and jurisdiction over the resources of the continental shelf.*"[44]

Although Article 142(1) does not make any express reference to the obligation of mutual-restraint regarding "*resource deposits in the Area which lie across limits of national jurisdiction,*" it would seem that this obligation would be implied in the duty of due regard in order to avoid causing damage to the relevant coastal State, seeing that the duty of due regard for the rights and interests of third States is connected with States' obligation to prevent and not to cause harm. In the event that either the coastal State or the Authority would cause damage, the relevant rules on liability would be applicable.[45] In this respect, the Seabed Disputes Chamber considered that carrying out an environmental impact assessment was a 'direct obligation' under UNCLOS and a general obligation under customary international law and, consequently, "*should be included in the system of consultations and prior notifications set out in Article 142.*"[46] As a result, coastal States and the Authority must exchange information regarding mineral resources

---

[42] Article 1(3) of UNCLOS.

[43] Article 133 lit. (a) of UNCLOS. The Seabed Disputed Chamber considered that the activities of the Area to ensure effective protection for the marine environment from harmful effects—which may arise from such activities—indicates the activities in respect of which the Authority should adopt rules, regulations and procedures. The Seabed Disputed Chamber concluded that the activities referred in Article 145 of UNCLOS are included in the concept of activities in the Area, namely, drilling, dredging, excavation, disposal of waste, construction and operation or maintenance of installations, pipelines and other devices related to such activities. *See Request for Advisory Opinion submitted to the Seabed Disputes Chamber, supra note 77 in Chap.* 2, para. 85.

[44] *UNCLOS: A Commentary, supra note 36 in Chap.* 5, p. 947.

[45] Article 187 lit. (b) of UNCLOS.

[46] *Request for Advisory Opinion submitted to the Seabed Disputes Chamber, supra note 77 in Chap.* 2, paras. 145, 148: "*Although aimed at the specific situation under discussion by the Court, the language used seems broad enough to cover activities in the Area even beyond the scope of the Regulations. The Court's reasoning in a transboundary context may also apply to activities with an impact on the environment in an area beyond the limits of national jurisdiction; and the Court's references to "shared resources" may also apply to resources that are the common heritage of mankind. Thus, in light of the customary rule mentioned by the ICJ, it may be considered that environmental impact assessments should be included in the system of consultations and prior notifications set out in article 142 of the Convention with respect to "resource deposits in the Area which lie across limits of national jurisdiction.*"

that may be found to straddle the demarcation line between the continental shelf and the Area, as well as adopt restraint regarding any action that may potentially cause harm to their legitimate interests. The recognition of a duty of information and an obligation of due regard does not imply, however, the obligation for coastal States and the Authority to cooperate in sharing these resources.

9.5 Sharing the Common Resources of the Area and the Continental Shelf

Without foregoing the observations made in Chap. 6 on revenue-sharing regarding mineral resources exploited by two or more coastal States in an area of overlapping claims in the continental shelf beyond 200 nm, Part XI of UNCLOS does not include or exclude the possibility of coastal States and the Authority—which derives its legitimacy from the consent of the States-parties to UNCLOS[47]—implementing a cooperative arrangement for the common development of the straddling resources of the Area.[48] Consequently, UNCLOS does not disregard the possible applicability of the rules on the system of exploration and exploitation of the resources of the Area to the development of these resources when they lie across limits of national jurisdiction, as well as, for that matter, the rules, regulations and procedures of the Authority applicable to operations.[49] In this case, it would be the Enterprise entering into such cooperative arrangements as the organ of the Authority through which it takes part in deep seabed activities, as well as the transporting, processing and marketing of the minerals recovered from the Area.[50]

However, the applicability of this system of exploration and exploitation would greatly limit the discretionary powers of coastal States regarding the scope and framework of similar cooperative arrangements. These would include, for example, carrying out the activities in the Area in association with the Enterprise and being subject to regular inspections of offshore installations in the Area.[51]

---

[47] Rüdiger Wolfrum, "Legitimacy of international law and the exercise of administrative functions: the example of the International Seabed Authority, the International Maritime Organization (IMO) and the International Fisheries Organizations", in: 9-11 *German Law Journal* (2004), pp. 2054, 2059.

[48] Articles 157(2) and 170, and Annex IV of UNCLOS.

[49] Article 17 of Annex III of UNCLOS. *See* Michael C. Wood, "The International Seabed Authority: fifth to twelfth sessions (1999–2006)", in: 11 *Max Planck Yearbook of United Nations Law* (2007), p. 56: "*The possibility of joint exploitation of deposits straddling the boundary between the Area and the continental shelf of a coastal state is anticipated in article 142 of the Convention (which deals with the rights and legitimate interests of coastal states). If such deposits were to be found and joint exploitation seemed desirable, some practical ad hoc arrangement would be needed (for which there are precedents in agreements between states).*"

[50] Article 158(2) of UNCLOS.

[51] Article 153(2) lit. (b),(5) of UNCLOS. Also *see* Articles 147(2) and 187 lit. (c), (d) and e) of UNCLOS.

UNCLOS also provides that activities in the Area must be carried out in accordance with a formal written plan of work,[52] which has to be approved by the Council, after review by the Legal and Technical Commission.[53] Additionally, the Authority would have control of the activities in the Area and confer the operator the exclusive right to explore and exploit the resources in the area covered by the plan of work,[54] as well as the right to apply penalties to contractors for violation of contract.[55] Furthermore, the plan of work, except for those presented by the Enterprise, shall be in the form of a contract concluded between the Authority and the applicant or applicants, taking into consideration the objectives set out in UNCLOS for the financial terms of the contracts.[56] The Convention also establishes that these contracts may provide for joint arrangements between the contractor and the Authority through the Enterprise, in the form of joint ventures or production sharing, as well as any other form of joint arrangement, which shall have the same protection against revision, suspension or termination as contracts with the Authority.[57]

In what concerns liability, UNCLOS provides that States-parties shall assist the Authority for the purpose of securing compliance with the relevant provisions of its regime and of the plans of work for the exploration and exploitation of its resources. To this effect, States-parties shall undertake all measures necessary to ensure compliance by the sponsored contractor, being liable for the damages caused by the failure to carry out their responsibility,[58] without prejudice of the applicability of other rules of international law on liability.[59]

The liability of the State-party takes place with the occurrence of damage that results from act or omission to carry out its own obligations under Part XI of UNCLOS, as well as its responsibility towards the sponsored contractor, namely regarding ensuring compliance.

> 121. The obligations of sponsoring States are not limited to the due diligence “obligation to ensure”. Under the Convention and related instruments, sponsoring States also have obligations with which they have to comply independently of their obligation to ensure a certain behaviour by the sponsored contractor. These obligations may be characterized as direct obligations.
>
> 122. Among the most important of these direct obligations incumbent on sponsoring States are: the obligation to assist the Authority in the exercise of control over activities in the Area; the obligation to apply a precautionary approach; the obligation to apply best

---

52 Article 153(3) and Articles 3 and 6(3) of Annex III of UNCLOS.

53 Articles 162(2) lit. (j) and 165(2) lit. (b) of UNCLOS.

54 Article 153(4) and Article 3(4) lit. (b) of Annex III of UNCLOS.

55 Article 18 of Annex III of UNCLOS.

56 Articles 3(5) and 13 of Annex III of UNCLOS.

57 Article 11 of Annex III of UNCLOS.

58 Articles 139 and 153 of UNCLOS. *See Request for Advisory Opinion submitted to the Seabed Disputes Chamber, supra note 77 in Chap.* 2, paras. 213–217.

59 Articles 139(2) and 304 and Article 4(4) of Annex III of UNCLOS.

> environmental practices; the obligation to take measures to ensure the provision of guarantees in the event of an emergency order by the Authority for protection of the marine environment; the obligation to ensure the availability of recourse for compensation in respect of damage caused by pollution; and the obligation to conduct environmental impact assessments (...).[60]

However, the State-party is exempt of liability for the failure of the sponsored contractor to meet its own obligations.[61] Accordingly, the obligation of a State-party constitutes an obligation of means or conduct, rather than an obligation of result.[62]

UNCLOS did not include the possibility of a State that is not a party to the Convention to be able to put forward a claim against the Authority in the Seabed Disputes Chamber, or for that matter in a municipal court, due to the Authority's immunity.[63] Unless the Authority should waive its immunity, it would be very difficult for the coastal State to prosecute the latter for damages caused by the development of common resources of the Area and the continental shelf. In these cases, the relevant coastal State would only be able to act, in the appropriate judicial forum, against the entity that, in association with the Enterprise,[64] carried out the activities in the Area and was responsible for the damages caused as a result.[65]

---

[60] *See Request for Advisory Opinion submitted to the Seabed Disputes Chamber, supra note 77 in Chap.* 2, paras. 121–122.

[61] *Ibid*, paras. 121, 172–177, 182, 204, 217. The text of Article 4(4) of Annex III of UNCLOS reads: "*The sponsoring State or States shall, pursuant to article 139, have the responsibility to ensure, within their legal systems, that a contractor so sponsored shall carry out activities in the Area in conformity with the terms of its contract and its obligations under this Convention. A sponsoring State shall not, however, be liable for damage caused by any failure of a contractor sponsored by it to comply with its obligations if that State Party has adopted laws and regulations and taken administrative measures which are, within the framework of its legal system, reasonably appropriate for securing compliance by persons under its jurisdiction.*" The text of Article 22 of Annex III of UNCLOS reads: "*The contractor shall have responsibility or liability for any damage arising out of wrongful acts in the conduct of its operations, account being taken of contributory acts or omissions by the Authority. Similarly, the Authority shall have responsibility or liability for any damage arising out of wrongful acts in the exercise of its powers and functions, including violations under article 168, paragraph 2, account being taken of contributory acts or omissions by the contractor. Liability in every case shall be for the actual amount of damage.*"

[62] *Request for Advisory Opinion submitted to the Seabed Disputes Chamber, supra note 77 in Chap.* 2, para. 110.

[63] Articles 178 and 187 and article 22 of Annex III and article 37 of Annex VI of UNCLOS. *See* Brown, *supra* note 17, p. II.4.53.

[64] Article 2(3) of Annex IV of UNCLOS reads: "*Nothing in this Convention shall make the Enterprise liable for the acts or obligations of the Authority, or make the Authority liable for the acts or obligations of the Enterprise.*"

[65] Article 139(2) of UNCLOS.

9.6 Protection of the Marine Environment from Pollution Caused by Activities in the Area

The legal regime applicable to the protection and preservation of the marine environment is included in Part XII of UNCLOS. In accordance, States are bound to the general obligation to protect and preserve the marine environment,[66] and in particular to take all measures necessary to ensure that activities under their jurisdiction or control are undertaken so as not to cause damage by pollution to other States. States are further obliged to prevent pollution from spreading beyond the areas where they exercise sovereign rights.[67] Part XII also includes provisions applicable to different sources of pollution, including seabed activities under the jurisdiction of coastal States[68] and activities in the Area. In the case of the latter, Part XII provides that international rules, regulations and procedures shall be established in order to prevent, reduce and control pollution of the marine environment from activities in the Area. States shall also adopt laws and regulations to prevent, reduce and control pollution of the marine environment from activities in the Area undertaken by vessels, installations, structures and other devices flying their flag or of their registry or operating under their authority, as the case may be.[69] Part XII of UNCLOS also refers that the enforcement of the said rules, regulations and procedures shall be subject to the Area regime.[70] Accordingly, Article 145 lit. (a) of UNCLOS provides that the Authority shall adopt such rules, regulations and procedures for, amongst other, the prevention, reduction and control of pollution, in particular from the effects of drilling, dredging, excavation, disposal of waste, construction and operation or maintenance of installations, pipelines and other devices related to such activities.[71] In this respect, despite the definition of activities in the Area provided by UNCLOS only referring to the activities of exploration and exploitation of resources of the Area,[72] the Seabed Disputes Chamber has considered that the aforementioned activities should also be included in the notion of activities in the Area.[73]

---

[66] Articles 192 and 194(1) of UNCLOS.

[67] Article 194(2) of UNCLOS.

[68] Articles 194(3) lit. (c), 208 and 214 of UNCLOS.

[69] Article 209 of UNCLOS.

[70] Article 215 of UNCLOS.

[71] Also *see* Article 17 lit. (f) of Annex III of UNCLOS. On the protection and preservation of the marine environment in the Area *see* Brown, *supra* note 17, p. II.9.3–18. On the role of the Authority in implementing the Area regime *see* James Harrison, *Making the Law of the Sea: A Study in the Development of International Law* (Cambridge University Press: Cambridge, 2011), pp. 115–153.

[72] Article 1(1)3) of UNCLOS.

[73] *Request for Advisory Opinion submitted to the Seabed Disputes Chamber, supra note 77 in Chap.* 2, para. 87: "*The provisions considered in the preceding paragraphs confirm that processing and transporting as mentioned in Annex IV, article 1, paragraph 1, of the Convention are excluded from the notion of "activities in the Area". They set out lists of activities whose harmful*

The Area regime also provides that the Legal and Technical Commission (one of the organs of the Council), has the responsibility to make assessments of the environmental implications of activities in the Area, as well as to present proposals and recommendations to the Council on matters concerning the protection of the marine environment.[74] On the other hand, the Council may disapprove areas for exploitation by contractors or the Enterprise in the case of risk of serious harm to the marine environment.[75]

Article 142(3) of UNCLOS determines that the Area regime shall not affect the right of coastal States to take the relevant measures, consistent with Part XII, that are necessary to prevent, mitigate or eliminate grave and imminent danger to their coastline or related interests from pollution or threat thereof or from hazardous occurrences resulting from or caused by any activities in the Area. Accordingly, Article 194 of UNCLOS establishes that States shall take all measures that are necessary towards that end, but refrain from unjustifiable interfering with activities carried out by other States, or in this case, the activities in the Area. On the other hand, States should cooperate by notifying other States that are likely to be affected by damage by pollution, as well as regarding the preparation of contingency plans.[76] States are also to endeavour to monitor the risks and effects of pollution of the marine environment and share the respective results.[77]

(Footnote 73 continued)
effects are indicated as directly resulting from such activities. These lists may be seen as an indication of what the Convention considers as included in the notion of "activities in the Area". These activities include: drilling, dredging, coring, and excavation; disposal, dumping and discharge into the marine environment of sediment, wastes or other effluents; and construction and operation or maintenance of installations, pipelines and other devices related to such activities."

[74] Articles 163(1) lit. (b) and 165(2) lit. (d), (e), (f), (g), (h), (k), (l) of UNCLOS.

[75] Article 162(2) lit. (x) of UNCLOS.5.

[76] Articles 198 and 199 of UNCLOS.

[77] Articles 204 and 205 of UNCLOS.

Chapter 10
Conclusions and Outlook

10.1 Main Conclusions

The dissertation was set out to explore the concept of joint development in its two main distinctions: as provisional arrangements pursuant to Article 83(3) of UNCLOS when pending the delimitation of maritime boundaries and as cooperative efforts regarding the exploration and exploitation of transboundary mineral resources. Within these cooperative efforts the dissertation also examined the relation between joint development and natural resources clauses included in maritime delimitation treaties, as well as, the possibility of joint development of mineral resources that lie between the continental shelf and the Area.

The dissertation focused on determining whether under international law and specifically the law of the sea there is an obligation for States to enter into provisional arrangements or cooperative efforts, and whether these must necessarily result in joint development. It sought to define States' obligations regarding mineral resources found in areas of overlapping claims or extending across a boundary line, specifically concerning the undertaking of exploration and exploitation activities and safeguarding the rights of neighboring coastal States. The identification of the legal regime applicable to disputed maritime areas and transboundary mineral resources provided the basis to evaluate the relevant legal and functional aspects of joint development, but also to identify States' rights and obligations in the absence of an agreement concerning seabed activities in these situations. In addition, the research concentrated on analyzing State practice, international jurisprudence and the relevant international legal instruments.

The main finding of the dissertation was that there is no obligation in international law for States to cooperate in the development of the mineral resources found in areas of overlapping claims or that straddle a boundary line. Although cooperation is undoubtedly indispensable to achieve resource-efficiency and will ultimately benefit the interests of the relevant States, joint development agreements are essentially driven by economic purposes. They have a pragmatic

V. Becker-Weinberg, *Joint Development of Hydrocarbon Deposits in the Law of the Sea*,
Hamburg Studies on Maritime Affairs 30, DOI 10.1007/978-3-662-43570-0_10

character and take into consideration, first and foremost, the interests of the respective States and the specificities of each case.

State practice on joint development has not been founded on the convention that these regimes are required by international law. There is no consistent practice regarding joint development agreements implemented thus far, nor is there substantial or sufficient proof of State practice, or any evidence that States have perceived or actively pursued joint development as the appropriate conduct to undertake in situations of shared mineral resources. States have also not made substantial efforts for the progress of international law in that respect. The general practice of States on joint development over the past five decades has showed that the relevant States and those most affected have unambiguously not supported the notion of a joint development customary rule, neither have they made any declarations to that effect. Likewise, State practice did not suffer considerable changes during that time. International jurisprudence and other relevant international documents, particularly those issued by international legal bodies, have also not recognized that such a customary rule exists, nor have they managed to clarify the rules applicable to the sharing of common mineral resources as has occurred, for example, in relation to the law applicable to maritime delimitation or transboundary aquifers.

A customary rule regarding joint development would depend on practice evidencing States' conviction that joint development was required by international law. The lack of legal consistency of certain practices is manifested in the absence of the belief that they are required and would consequently not qualify them as customary, but as a mere usage that takes into account best practices applicable to seabed activities. Although such a belief would stand on its own, its existence is dependent on an appropriate level of generality foremost manifested in State practice. Yet, the more impressive the evidence that such a belief exists, more examples of State practice would be likely to occur. The relevance of such a circular thinking is particularly felt in respect of a new practice contrary to existing customary law. In these cases, the understanding that a new practice is legal cannot result independently from general State practice to support it. Indeed, although it may be arguable that the presence of the subjective element (*opinio juris sive necessitasis*) is not always necessary or relevant, or that it may be difficult to establish its separate existence from State practice. In the case of the constitutive process of a joint development rule State practice could not do without it.

The practice of States on joint development reveals that they would not consider the enforcement of a joint development rule based on usage or that practice had developed without consideration for *opinio juris*. The overwhelming number of joint development agreements signed before the delimitation of maritime boundaries acknowledges their interim character and that they do not have any relation with States' claims over the disputed maritime areas. Their implementation is exclusively the outcome of States willingness to cooperate, rather than the recognition of an obligation to develop common resources. Likewise, joint development agreements signed after the delimitation of maritime boundaries are in most cases not the result of an obligation pursuant to natural resources clauses inserted in delimitation treaties. These arrangements are strictly connected with the specificities of every situation upon which States draw their decision to negotiate and implement a joint development regime. In the Asia-Pacific region, for example, where the problems of overlapping maritime areas and territorial disputes and the consequential sharing of common mineral resources are particularly felt and where, as a result, joint development may have an important role, one cannot conclude that a regional practice has developed or is evolving towards a joint development rule, or that it may constitute a potential case of regional international law-making, which, in any event, would not be considered applicable to all States as a general customary law.

The underlying principle of cooperation in joint development agreements does not go any further than the scope and content of the general rule of cooperation. This can be observed in the obligation to make every effort to enter into provisional arrangements referred in Article 83(3) of UNCLOS. This provision merely reinforces the general obligation to negotiate in good faith and of mutual-restraint, without requiring or establishing the obligation to enter into an agreement regarding the development of common mineral resources. It does not constitute a special rule of cooperation that applies a general standard existing under international law, taking into account the concrete circumstances of each case, particularly as there would be no situation that would justify the specialization of the general rule of cooperation in the form of joint development, nor has such a rule been crystallized in international law. In fact, should a special rule of joint development exist in relation to a general rule of cooperation, it would alter the compromise reached in Articles 74(3) and 83(3) of UNCLOS that interim measures should not be mandatory. This would constitute an evident breach of the Convention and an outcome not expressed or envisaged by any of the States that has entered into joint development agreements before the delimitation of maritime boundaries. It would also narrow the applicability of these provisions to the delimitation of maritime boundaries by agreement or the implementation of provisional arrangements.

A special rule of joint development would also be likely to disrupt the established balance in the law of the sea in respect to the exercise of sovereignty, since the legal titles upon which two or more States derive their sovereign, exclusive and inherent rights to explore and exploit the mineral resources of the continental shelf are identical in their nature. The prescription to share common mineral resources or an obligation of joint development would result in an unjustified limitation to these rights for the sake of reaching an equitable result. It would have the effect of conceptualizing an obligation of joint development that could be triggered by any of the relevant States that intended to develop the mineral resources that straddle a boundary line or that are found in areas of overlapping claims, overriding the right of the coastal State not to engage in exploration and exploitation activities and merely agreeing to disagree, as well as not having consideration for the State's capabilities to undertake seabed activities. It would also be nearly impossible to ensure the effectiveness of the enforceability of such an obligation or to underline the terms and conditions that could eventually fulfil its content. This arbitrary result would simply not be based on the applicability of an international legal rule determining a certain outcome. It would consist of adopting a solution that relinquished the applicability of existing international law regarding common mineral resources for the sake of arriving at a consensus.

The balancing of different sovereign rights and the finding of a solution for situations where these rights are in conflict is only achieved through the applicability of international rules. These are fundamentally, the exercise of due regard and mutual-restraint and the adoption of procedural obligations, such as the sharing of information. Indeed, as international binding agreements between States, these regimes are subject to international law and in particular the law of the sea. The principal consequence being that States may not undertake exploration and exploitation activities of common mineral resources without the consent of other legally entitled States. Coastal States are further subject to inform neighboring States of the presence of common mineral resources and of their intention to initiate exploration and exploitation activities.

Concurrent with the obligation of States to make every effort to enter into provisional arrangements, States must continue to negotiate, in good faith, towards the settlement of maritime boundaries, even though there is no connection between joint development and delimitation, seeing that, as provisional arrangements, they are without prejudice to States' claims or the final delimitation of maritime boundaries. After the delimitation of maritime boundaries, joint development agreements do not alter the legal status of the relevant maritime area or the respective rights and obligations of coastal States and of other States, such as the freedom of navigation, the protection and preservation of the marine environment from pollution caused by seabed activities, or those related with the operation and decommissioning of offshore installations and structures.

The dissertation intends to contribute towards a better understanding of the concept of joint development and the overall conclusion of the adequacy of the legal regime included in UNCLOS to deal with the situations examined in this work. The balance achieved during the negotiation and drafting of the Convention provides a suitable consideration of the different interests at hand concerning the exploration and exploitation of mineral resources found in maritime areas of overlapping claims or that straddle a maritime boundary. Although UNCLOS requires limited cooperation regarding mineral resources when compared with other matters included in the Convention, this framework engages States to deal with the exploration and exploitation of these resources in accordance with their best interests. Indeed, UNCLOS acknowledges States' sovereignty and jurisdiction, while also requiring States to undertake their obligations. These include the obligations for States to negotiate in good faith the peaceful settlement of disputes, to clarify their claims and to identify the disputed maritime area, and the obligations to exercise self-restraint and to refrain from the use or threat of the use of force.

10.2 Outlook for the Progress of Joint Development in the Law of the Sea

The demand for fossil fuels is expected to carry on growing and with it the need to secure access to energy sources and their control. This remains a dominant aspect that will continue structuring relations between States, particularly as technology allows for the development of offshore hydrocarbon deposits at increasingly greater depths and in more extreme conditions.

States may increasingly resort to International Courts and Tribunals for the settlement of maritime disputes, but it is unlikely that this will be the case of the conflicts analyzed in this dissertation regarding common mineral resources. States do not seem willing to renounce to their claims to maritime areas with great hydrocarbon potential, including those adjacent to islands or other features, nor are they prepared to accept the possibility of settlement by means of compulsory mechanisms due to the length of procedures and the uncertainty of its outcome. Joint development remains a valuable alternative in situations where mineral resources straddle a maritime boundary or are found in maritime areas of overlapping claims, including beyond 200 nm. There are also new challenges for joint development, such as regarding mineral resources in the Area which lie across limits of national jurisdiction.

Despite the growing awareness of the pragmatic and legal aspects of joint development agreements and the possible rise of its numbers, it is doubtful that this may lead to the materialization of a customary rule of joint development in the foreseeable future or the acknowledgment of the existence of a special obligation to cooperate sharing offshore mineral resources. It is also improbable that States will make declarations supporting the notion of a joint development customary rule. These circumstances notwithstanding, the fulfillment by States of their duties may seriously contribute towards the successful de-escalation and prevention of the continuance of conflicts regarding mineral resources found in areas of overlapping claims or that straddle a maritime boundary, allowing States to engage in meaningful negotiations and eventually consider the possibility of joint development.

List of Documents

International Jurisprudence

Permanent Court of International Justice

- *Case of the Mavrommatis Palestine Concessions, Judgment, 1924 P.C.I.J., Series A, No. 2*
- *S.S. Lotus (France v. Turkey), 1927 P.C.I.J., Series A, No. 10*
- *The Factory at Chorzów (Claim for Indemnity), The Merits, 1928 P.C.I.J., Series A, No. 17*
- *Free Zones of the Upper Savoy and the District of Gex (France v. Switzerland), 1932 P.C.I.J., Series A/B, No. 46*
- *Legal status of Eastern Greenland, Judgment, 1933 P.C.I.J. Series A/B, No. 53*
- *Phosphates in Morocco (Italy v. Fr.), 1938 P.C.I.J., Series A/B, No. 74.*

International Court of Justice

- *Corfu Channel case, Judgment of April 9th, 1949, I.C.J. Reports 1949*
- *Interpretation of Peace Treaties, Advisory Opinion: I.C.J. Reports 1950*
- *Competence of Assembly regarding admission of the United Nations, Advisory Opinion: I.C.J. Reports 1950*
- *Anglo-Iranian Oil Co. case (jurisdiction), Judgment of July 22nd, 1952: I.C.J. Reports 1952*
- *Case concerning Right of Passage over Indian Territory (Merits), Judgement of 12 April 1960, I.C.J. Reports 1960*
- *Case concerning the Temple of Preah Vihear (Cambodia v. Thailand), Merits, Judgment of 15 June 1962, I.C.J. Reports 1962*
- *South West Africa Cases (Ethiopia v. South Africa; Liberia v. South Africa), Preliminary Objections, Judgment of 21 December 1962, I.C.J. Report 1962*

V. Becker-Weinberg, *Joint Development of Hydrocarbon Deposits in the Law of the Sea*, Hamburg Studies on Maritime Affairs 30, DOI 10.1007/978-3-662-43570-0

- *Case concerning Northern Cameroons (Cameroon v. United Kingdom), Preliminary Objections, Judgment of 2 December 1963: I.C.J. Reports 1963*
- *South West Africa, Second Phase, Judgment, I.C.J. Reports 1966*
- *North Sea Continental Shelf cases (Federal Republic of Germany/Denmark; Federal Republic of Germany/Netherlands), Judgment, I.C.J. Reports 1969*
- *Barcelona Traction, Light and Power Company, Limited, Judgment, I.C.J. Reports 1970*
- *Nuclear Tests (Australia v. France), Judgment, I.C.J. Reports 1974*
- *Fisheries Jurisdiction (United Kingdom v. Iceland), Merits, Judgment, I.C.J. Reports 1974*
- *Aegean Sea Continental Shelf, Judgment, I.C.J. Reports 1978*
- *Continental Shelf case (Tunisia/Libyan Arab Jamahiriya), Judgment, I.C.J. Reports 1982*
- *Delimitation of the Maritime Boundary in the Gulf of Maine Area, Judgment, I.C.J. Reports 1984*
- *Continental Shelf (Libyan Arab Jamahiriya/Malta), Judgment, I.C.J. Reports 1985*
- *Frontier Dispute, Judgment, I.C.J. Reports 1986*
- *Military and Paramilitary Activities in und against Nicaragua (Nicaragua v. United States of America), Merits, Judgment, I.C.J. Reports 1986*
- *Applicability of the Obligation to Arbitrate under Section 21 of the United Nations Headquarters Agreement of 26 June 1947, Advisory Opinion, I.C.J. Reports 1988*
- *Arbitral Award of 31 July 1989, Judgment, I.C.J. Reports 1991*
- *Case concerning the Land, Island and Maritime Dispute (El Salvador/ Honduras: Nicaragua intervening), Judgment, I.C.J. Reports 1992*
- *Maritime Delimitation in the Area between Greenland and Jan Mayen, Judgment, I.C.J. Reports 1993*
- *Territorial Dispute Case (Libyan Arab Jamahiriya v Chad), Judgment, I.C.J. Reports 1994*
- *East Timor (Portugal v. Australia), Judgment, I.C.J. Reports 1995*
- *Request for an Examination of the Situation in Accordance with paragraph 63 of the Court's Judgment of 20 December 1974 in the Nuclear Tests (New Zealand v. France) Case, I.C.J. Reports 1995*
- *Legality of the Threat or Use of Nuclear Weapons, Advisory Opinion, I.C.J. Reports 1996*
- *Gabčíkovo-Nagymaros Project (Hungary/Slovakia), Judgment, I.C.J. Reports 1997*
- *KasikiliSedudu Island (Botswana/Namibia) Judgment, I.C.J. Report 1999*
- *Maritime Delimitation and Territorial Questions between Qatar and Bahrain (Qatar v. Bahrain), Merits, Judgment, I.C.J. Reports 2001*
- *Land and Maritime Boundary between Cameroon and Nigeria (Cameroon v. Nigeria: Equatorial Guinea intervening), Judgment, I.C.J. Reports 2002*
- *Sovereignty over Pulau Ligitan and Pulau Sipadan (Indonesia/Malaysia), Judgment, I.C.J. Reports 2002*

- *Territorial and Maritime Dispute between Nicaragua and Honduras in the Caribbean Sea (Nicaragua v. Honduras), Judgment, I.C.J. Reports 2007*
- *Territorial and Maritime Dispute between Nicaragua and Honduras in the Caribbean Sea (Nicaragua v. Honduras), Judgment, I.C.J. Reports 2007 (II)*
- *Case Concerning Maritime Delimitation in the Black Sea (Romania v. Ukraine), Judgment, I.C.J. Reports 2009*
- *Pulp Mills on the River Uruguay (Argentina v. Uruguay), Judgment, I.C.J. Reports 2010*
- *Territorial and Maritime Dispute (Nicaragua v. Colombia), Judgement, I.C.J. Reports 2012.*

International Tribunal for the Law of the Sea

- *Southern Bluefin Tuna Cases (New Zealand v. Japan; Australia v. Japan), Requests for Provisional Measures, Order, I.T.L.O.S. Nos. 3 and 4, 1999*
- *The MOX Plant Case (Ireland v. United Kingdom), Request for provisional measures, I.T.L.O.S. No.10, 2001*
- *Case Concerning Land Reclamation by Singapore in and around the Straits of Johor (Malaysia v. Singapore), Provisional Measures, I.T.L.O.S. No.12, 2003*
- *Dispute concerning Delimitation of the Maritime Boundary between Bangladesh and Myanmar in the Bay of Bengal, Judgment of 14 March 2012, Case n. 16*
- *Responsibilities and obligations of States sponsoring persons and entities with respect to activities in the Area, Advisory Opinion (Request for Advisory Opinion submitted to the Seabed Disputes Chamber), 1 February 2011, Case n. 17.*

Arbitral awards

(I)

- *Award between the United States and the United Kingdom relating to the rights of jurisdiction of United States in the Bering's sea and the preservation of fur seals, decision of 15 August 1893*
- *Grisbådarna Case (Norway v. Sweden), P.C.A. Award 23 October 1909*
- *The Island of Palmas Case (or Miangas) (US v. Netherlands), Award of the Tribunal, P.C.A., 4 April 1928*
- *Dickson Car Wheel Co. Case (1931) (US v. Mexico)*
- *Lac Lanoux Arbitration (France v. Spain), 1957*
- *Trail Smelter Case (United States, Canada) 16 April 1938 and 11 March 1941*
- *Anglo-French Continental Shelf Arbitration (1977–1978)*
- *Arbitration between the United Kingdom of Great Britain and Northern Ireland and the French Republic on the Delimitation of the Continental Shelf, Decision of the Court of Arbitration dated 30 June 1977*

- *Case concerning the delimitation of the maritime boundary between Guinea and Guinea-Bissau, Arbitral Award of 14 February 1985*
- *Delimitation of Maritime Areas between Canada and the French Republic (St. Pierre et Miquelon), 10 June 1992*
- *The Eritrea—Yemen Arbitration, Award of the Arbitral Tribunal in the First Stage of the Proceedings (Territorial Sovereignty and Scope of the Dispute), 3 October 1996*
- *The Eritrea—Yemen, Award of the Arbitral Tribunal in the Second Stage of the Proceedings (Maritime Delimitation), 17 December 1999*
- *Barbados/Trinidad and Tobago, Award of the Arbitral Tribunal, 11 April 2006*
- *Guyana/Suriname, Award of the Arbitral Tribunal, 17 September 2007.*
- *Bay of Bengal Maritime Boundary Arbitration between Bangladesh and India (Bangladesh v. India), P.C.A. Award 7 July 2014.*

(II)

- *Petroleum Development (Trucial Coast) Ltd. v. Sheik of Abu Dhabi Award of Lord Asquith of Bishopstone, 1951*
- *Saudi Arabia v. Arabian American Oil Company (Armaco)*, 1963
- *Sapphire International Petroleums Ltd. v National Iranian Oil Company*, 1967.

Agreements

(I)

- *Treaty between Great Britain and Northern Ireland and Venezuela relating to the submarine areas of the Gulf of Paria*, 1942
- *Bahrain-Saudi Arabia Frontier Agreement*, 1958
- *Agreement between the Government of the Czechoslovak Republic and the Austrian Federal Government concerning the Working of Common Deposits of Natural Gas and Petroleum*, 1960
- *Agreement between the Government of the Czechoslovak Republic and the Austrian Federal Government concerning the Principles of Geological Co-operation between the Czechoslovak Republic and the Republic of Austria*, 1960
- *Supplementary Agreement to the Treaty Concerning Arrangements for Co-operation in the Ems Estuary (Ems-Dollard Treaty)*, 1962
- *Treaty between the Kingdom of the Netherlands and the Federal Republic of Germany concerning the lateral delimitation of the continental shelf in the vicinity of the coast*, 1964
- *Agreement between the State of Kuwait and the Kingdom of Saudi Arabia relating to the partition of the Neutral Zone*, 1965
- *Agreement between Denmark and Norway Relating to the Delimitation of the Continental Shelf*, 1965
- *Agreement between the Government of the United Kingdom and Northern Ireland and the Government of the Kingdom of Norway relating to the*

Delimitation of the Continental Shelf between the Two Countries, 1965

- *Agreement concerning the Sovereignty over the Islands of Farsi and Al-Arabiyah and the Delimitation of the Boundary Line Separating the Submarine Areas between Iran and Saudi Arabia*, 1968
- *Agreement concerning the boundary line dividing the continental shelf between Iran and Qatar*, 1969
- *Agreement between the Government of the Republic of Indonesia and the Government of Malaysia relating to the delimitation of the continental shelves between the two Countries*, 1969
- *Agreement concerning settlement of offshore boundaries and ownership of islands between Qatar and the United Arab Emirates (with declarations and map)*, 1969
- *Treaty to Resolve Pending Boundary Differences and Maintain the Rio Grande and Colorado River as International Boundary*, 1970
- *Treaty between the Kingdom of Denmark and the Federal Republic of Germany concerning the Delimitation of the Continental Shelf under the North Sea*, 1971
- *Agreement between the Government of the Kingdom of Denmark and the Government of the United Kingdom of Great Britain and Northern Ireland relating to the Delimitation of the Continental Shelf between the Two Countries*, 1971
- *Treaty between the Kingdom of the Netherlands and the Federal Republic of Germany on Delimitation of the Continental Shelf under the North Sea*, 1971
- *Protocol to the Treaties of 28 January 1971 between the Federal Republic of Germany and Denmark and the Kingdom of the Netherlands, respectively, concerning the delimitation of the continental shelf under the North Sea*, 1971
- *Agreement between the United Kingdom of Great Britain and Northern Ireland and the Federal Republic of Germany relating to the Delimitation of the Continental Shelf under the North Sea between the Two Countries*, 1971
- *Protocol between the Government of the Kingdom of the Netherlands and the Government of the United Kingdom of Great Britain and Northern Ireland amending the Agreement of 6 October 1965 relating to the delimitation of the continental shelf under the North Sea between the two countries*, 1971
- *Agreement between the Government of the Commonwealth of Australia and the Government of the Republic of Indonesia Establishing Certain Seabed Boundaries*, 1971
- *Agreement between the Government of the Republic of Indonesia, the Government of Malaysia and the Government of the Kingdom of Thailand Relating to the Delimitation of the Continental Shelf Boundaries in the Northern Part of the Strait of Malacca*, 1971
- *Treaty concerning the Delimitation of the Continental shelf Under the North Sea*, 1971
- *Agreement between the Government of the Commonwealth of Australia and the Government of the Republic of Indonesia Establishing Certain Seabed Boundaries in the Area of the Timor and Arafura Seas, Supplementary to the Agreement of 18 May 1971*, 1972

- *Agreement between Australia and Indonesia Concerning Certain Boundaries between Papua New Guinea and Indonesia*, 1973
- *Agreement between Denmark and Norway relating to the delimitation of the continental shelf, 8 December 1965—Exchange of notes constituting an agreement amending the above-mentioned Agreement*, 1974
- *Agreement between Japan and the Republic of Korea Concerning Joint Development of the Southern Part of the Continental Shelf Adjacent to the Two Countries*, 1974
- *Agreement between the Government of the Democratic Republic of the Sudan and the Government of the Kingdom of Saudi Arabia relating to the Joint Exploitation of the Natural Resources of the Sea-bed and Subsoil of the Red Sea in the Common Zone*, 1974
- *Memorandum of Understanding between the Government of Australia and the Government of the Republic of Indonesia Regarding the Operations of Indonesian Traditional Fishermen in Areas of the Australian Exclusive Fishing Zone and Continental Shelf*, 1974
- *Agreement between Japan and the Republic of Korea Concerning the Establishment of Boundary in the Northern Part of the Continental Shelf Adjacent to the Two Countries*, 1974
- *Convention between the Government of the French Republic and the Government of the Spanish State on the delimitation of the continental shelves of the two states in the Bay of Biscay*, 1974
- *Agreement between the Government of the United Kingdom of Great Britain and Northern Ireland and the Government of the Kingdom of Norway relating to the exploitation of the Frigg Field Reservoir and the transmission of gas therefrom to the United Kingdom*, 1976
- *Protocol supplementary to the Agreement of 10 March 1965 between the Government of the United Kingdom of Great Britain and Northern Ireland and the Government of the Kingdom of Norway relating to the delimitation of the continental shelf between the two countries*, 1978
- *Agreed Minutes of the Malaysia-Thailand Official's Meeting on Delimitation of the Continental Shelf Boundary between Malaysia and Thailand in the Gulf of Thailand and in the South China Sea, 27 February–1 March 1978*, 1978
- *Treaty on Maritime Boundaries between the United States of America and the United Mexican States*, 1978
- *Treaty between Australia and the Independent State of Papua New Guinea Concerning Sovereignty and Maritime Boundaries in the area between the Two Countries, including the area known as Torres Strait, and Related Matters*, 1978
- *Memorandum of Understanding between the Kingdom of Thailand and Malaysia in the Establishment of a Joint Authority for the Exploitation of the Resources of the Sea-bed in a defined Area of the Continental Shelf of the Two Countries in the Gulf of Thailand*, 1979
- *Treaty between the Kingdom of Thailand and Malaysia relating to the Delimitation of the Territorial Sea of the Two Countries*, 1979

- *Memorandum of Understanding between the Kingdom of Thailand and Malaysia on the Delimitation of the Continental Shelf Boundary between the Two Countries in the Gulf of Thailand*, 1979
- *Agreement between the Government of the United Kingdom of Great Britain and Northern Ireland and the Government of the Kingdom of Norway relating to the exploitation of the Murchison Field Reservoir and the offtake of petroleum therefrom*, 1979
- *Agreement between the Government of the United Kingdom of Great Britain and Northern Ireland and the Government of Norway relating to the exploitation of the Statfjord Field Reservoirs and the offtake of petroleum therefrom*, 1979
- *Agreement between Iceland and Norway concerning Fishery and Continental Shelf questions between Iceland and Jan Mayen*, 1980
- *Agreement on the Continental Shelf between Iceland and Jan Mayen*, 1981
- *Agreement between Iceland and Norway on the Continental Shelf between Iceland and Jan Mayen*, 1981
- *Memorandum of Understanding between the Government of Australia concerning the Implementation of a Provisional Fisheries Surveillance and Enforcement Arrangement*, 1981
- *Agreement on Historic Waters of Vietnam and Kampuchea*, 1982
- *Agreement between the Great Libyan Arab Socialist People's Jamahariya and the Republic of Tunisia to Implement the Judgment of the International Court of Justice in the Tunisia/Libya Continental Shelf Case*, 1988
- *Treaty between Australia and the Republic of Indonesia on the Zone of Cooperation in an Area between the Indonesian Province of East Timor and Northern Australia*, 1989
- *Agreement between the Government of Malaysia and the Government of the Kingdom of Thailand on the constitution and Other Matters relating to the Establishment of the Malaysia-Thailand Joint Authority*, 1990
- *Treaty between the Republic of Trinidad and Tobago and the Republic of Venezuela on the Delimitation of Marine and Submarine Areas*, 1990
- *Agreement between the United Kingdom of Great Britain and Northern Ireland and the Netherlands relating to the exploitation of the Markham Field Reservoirs and the offtake of petroleum therefrom*, 1992
- *Agreement between the Government of Australia and the Government of the Republic of Indonesia relating to cooperation in fisheries*, 1992
- *Memorandum of Understanding between Malaysia and the Socialist Republic of Vietnam for the Exploration and Exploitation of Petroleum in a Defined Area of the Continental Shelf Involving the Two Countries*, 1992
- *Maritime Treaty between Jamaica and the Republic of Colombia*, 1993
- *Management and Cooperation Agreement between the Government of the Republic of Senegal and the Government of the Republic of Guinea-Bissau*, 1993
- *Joint Declaration on Cooperation over Offshore Activities in the Southwest Atlantic*, 1995

- *Protocol of Agreement Relating to the Organization and Operation of the Agency for Management and Cooperation between the Republic of Guinea-Bissau and the Republic of Senegal instituted by the agreement of 14 October 1993*, 1995
- *Agreement for the Implementation of the Provisions of the United Nations Convention on the Law of the Sea of 10 December 1982 relating to the Conservation and Management of Straddling Fish Stocks and Highly Migratory Fish Stocks*, 1995
- *Treaty between the Government of Australia and the Government of the Republic of Indonesia Establishing an Exclusive Economic Zone Boundary and Certain Seabed Boundaries*, 1997
- *Agreement between the Government of the Kingdom of Thailand and the Government of the Socialist Republic of Vietnam on the Delimitation of the Maritime Boundary between the Two Countries in the Gulf of Thailand*, 1997
- *Agreement between the State of Kuwait and the Kingdom of Saudi Arabia Regarding the Submerged Zone Contiguous to the Partitioned Zone*, 2000
- *Exchange of Notes Constituting an Agreement between the Government of Australia and the United Nations Transitional Administration in East Timor (UNTAET) concerning the Continued Operation of the Treaty between Australia and the Republic of Indonesia on the Zone of Cooperation in an Area between the Indonesian Province of East Timor and Northern Australia of 11 December 1989*, 2000
- *Agreement on the Delimitation of the Territorial Seas, Exclusive Economic Zones and Continental Shelves in the Beibu Gulf [Gulf of Tonkin] between the People's Republic of China and the Socialist Republic of Vietnam*, 2000
- *Treaty between the Government of the United States of America and the Government of the United Mexican States on the delimitation of the Continental Shelf in the Western Gulf of Mexico beyond 200 nautical miles*, 2000
- *Treaty between the Federal Republic of Nigeria and the Republic of Equatorial Guinea Concerning their Maritime Boundary*, 2000
- *Memorandum of Understanding between the Royal Government of Cambodia and the Royal Thai Government regarding the Area of their Overlapping Maritime Claims to the Continental Shelf*, 2001
- *Treaty between the Federal Republic of Nigeria and the Democratic Republic of Sao Tome and Principe on the Joint Development of Petroleum and Other Resources, in Respect of Areas of the Exclusive Economic Zone of the Two States*, 2001
- *Protocol in Implementation of Article 6.2 of the Treaty between the Federal Republic of Nigeria and the Republic of Equatorial Guinea Concerning their Maritime Boundary*, 2002
- *Treaty between Equatorial Guinea and Nigeria on Joint Exploration of Crude Oil, especially at the Zafiro-Ekanga Oil Field located at the maritime boundary of both countries*, 2002
- *Exchange of Notes Constituting an Agreement between the Government of Australia and the Government of the Democratic Republic of Timor-Leste*

Concerning Arrangements for Exploration and Exploitation of Petroleum in an Area of the Timor Sea between Australia and East Timor, 2002
- *Memorandum of Understanding between the Government of the Democratic Republic of East Timor and the Government of Australia Concerning an International Unitization Agreement for the Greater Sunrise field*, 2002
- *Memorandum of Understanding between the Government of the Democratic Republic of Timor-Leste and the Government of Australia relating to the Exploitation of the Sunrise and Troubadour Petroleum Fields in the Timor Sea*, 2003
- *Memorandum of Understanding for the Common Petroleum Development of the Lower Congo Basin between the Democratic Republic of Congo and the Republic of Angola*, 2003
- *Agreement between the Government of Australia and the Government of the Democratic Republic of Timor-Leste relating to the Unitization of the Sunrise and Troubadour fields*, 2003
- *Exclusive Economic Zone Co-operation Treaty signed between the Republic of Guyana and the State of Barbados*, 2003
- *Memorandum of Understanding Concerning the Procedure for Unitisation of Hydrocarbon Reservoirs that Extend Across the Delimitation Line between the Republic of Trinidad and Tobago and the Bolivarian Republic of Venezuela*, 2003
- *Letter of Intent—Enhanced bilateral cooperation between the Republic of Trinidad and Tobago and the Bolivarian Republic of Venezuela in the Energy Sector*, 2003
- *Framework Agreement between the Government of the United Kingdom of Great Britain and Northern Ireland and the Government of the Kingdom of Norway Concerning Cross-boundary Petroleum Co-operation*, 2005
- *Treaty between Australia and the Democratic Republic of Timor-Leste on Certain Maritime Arrangements in the Timor Sea*, 2006
- *Unitisation Agreement for the Exploitation and Development of Hydrocarbon Reservoirs of the Loran-Manatee Field that extends across the Delimitation Line between the Republic of Trinidad and Tobago and the Bolivarian Republic of Venezuela*, 2010
- *Agreement on the basic principles guiding the settlement of maritime issues between the Socialist Republic of Viet Nam and the People's Republic of China*, 2011
- *Agreement between the United States of America and the United Mexican States concerning Transboundary Hydrocarbon Reservoirs in the Gulf of Mexico*, 2012
- *Treaty Concerning the Joint Management of the Continental Shelf in the Mascarene Plateau Region between the Government of the Republic of Seychelles and the Government of the Republic of Mauritius*, 2012
- *Treaty Concerning the Joint Exercise of Sovereign Rights Over the Continental Shelf in the Mascarene Plateau Region*, 2012.

(II)

- *Convention for the Regulation of the Policing of the North Sea Fisheries*, 1882
- *Convention on the Fraser River System Sockeye Fisheries*, 1930
- *Charter of the United Nations*, 1945
- *Universal Declaration of Human Rights*, 1948
- *Convention for the Northwest Atlantic Fisheries*, 1949
- *Declaration of Santiago de Chile*, 1952
- *Northern Pacific Fisheries Convention*, 1952
- *Treaty on Preservation of North Pacific Halibut Fishery*, 1953
- *International Convention for the Prevention of Pollution of the Sea by Oil*, 1954
- *Geneva Convention on the Continental Shelf*, 1958
- *Geneva Convention on the High Seas*, 1958
- *Geneva Convention on the Territorial Sea and Contiguous Zone*, 1958
- *International Convention on Load Lines*, 1966
- *Vienna Convention on the Law of Treaties*, 1969
- *International Convention on Civil Liability for Oil Pollution Damage*, 1969
- *International Convention relating to Intervention on the High Seas in cases of Oil Pollution Casualties*, 1969
- *Convention on the International Regulations for Preventing Collisions at Sea*, 1972 (and 1987)
- *International Convention on the Establishment of an International Fund for Compensation for Oil Pollution Damage*, 1971 (and 1984, 1992 and 2003 Protocols)
- *Convention on the Prevention of Marine Pollution by Dumping of Waste and Other Matter*, 1972
- *Prevention of Pollution of the Mediterranean Sea by Dumping from Ships and Aircraft*, 1976 [renamed *Protocol for the Prevention and Elimination of Pollution of the Mediterranean Sea by Dumping from Ships and Aircraft or Incineration at Sea*, 1995]
- *International Convention for the Preservation of Pollution from Ships*, 1973 (as amended by 1978 Protocol)
- *International Convention for the Safety of Life At Sea*, 1974
- *International Convention on Standards of Training, Certification and Watchkeeping for Seafarers*, 1978
- *Vienna Convention on Succession of States in respect of Treaties*, 1978
- *United Nations Convention on the Law of the Sea*, 1982
- *Treaty Concerning Pacific Salmon*, 1985
- *Protocol for the Suppression of Unlawful Acts Against the Safety of Fixed Platforms Located on the Continental Shelf*, 1988
- *International Convention on Oil Pollution Preparedness, Response and Co-operation*, 1990
- *Convention on Biological Diversity*, 1992

- *International Convention on Liability and Compensation for Damage in connection with the Carriage of Hazardous and Noxious Substances by Sea*, 1996
- *Protocol to the Convention on the Prevention of Marine Pollution by Dumping of Wastes and Other Matter of 29 December 1972*, 1996
- *Protocol for the Suppression of Unlawful Acts against the Safety of Maritime Navigation Fixed Platforms Located on the Continental Shelf*, 2005.

International Law Commission

- *Yearbook of the International Law Commission, 1953, vol. 2*
- *Yearbook of the International Law Commission, 1956, vol. 2*
- *Yearbook of the International Law Commission, Second Report on State Responsibility, by Mr. Roberto Ago, Special Rapporteur, The Origin of International Responsibility*, UN Doc. A/CN.4/233 (1970)
- *Draft Articles on Responsibility of States for Internationally Wrongful Acts, with commentaries*, 2001
- *Shared natural resources: first report on outlines*, by Chusei Yamada, Doc. A/CN.4/533 (Fifty-fifth session: Geneva, 5 May–6 June and 7 July–8 August 2003)
- *Official Records of the General Assembly*, Fifty-fifth Session, Supplement No. 10 (Doc. A/55/10)
- *Fragmentation of International Law: Difficulties arising from the Diversification and Expansion of International Law*, by Martti Koskenniemi, Doc. A/CN.4/L.682, 13 April 2006 (Fifty-eighth session: Geneva, 1 May-9 June and 3 July-11 August 2006)
- *Fifth report on shared natural resources*, by Chusei Yamada, Doc. A/CN.4/591 (Sixtieth session: 5 May–6 June and 7 July–August 2008)
- *Report of the International Law Commission* (Sixty-second session: 3 May–4 June and 5 July–6 August 2010), Doc. A/65/10 (Chapter VII: Shared natural resources).

International Maritime Organization

- IMO resolution A.414(11), 15 November 1979, *Code for the Construction and Equipment of Mobile Offshore Drilling Units*
- IMO resolution A.572(14), 20 November 1985, *General Provisions on Ships' Routeing*
- IMO resolution A.671(16), 19 October 1989, *Safety Zones and Safety of Navigation around Offshore Installations and Structures*
- IMO resolution A.649(16), 19 October 1989, *Code for the Construction and Equipment of Mobile Offshore Drilling Units*

- IMO resolution A.672(16), 19 October 1989, *Guidelines and Standards for the Removal of Offshore Installations and Structures on the Continental Shelf and in the Exclusive Economic Zone*
- IMO resolution A.765(18), 4 November 1993, *Guidelines on the Safety of Towed Ships and Other Floating Objects, including Installations, Structures and Platforms at Sea*
- IMO resolution A.891(21), 25 November 1999, *Recommendations on Training of Personnel on Mobile Offshore Units (MOUs)*
- *Specific Guidelines for Assessment of Platforms or Other Man-made Structures at Sea*, adopted at the *Twenty-second Consultative Meeting of Contracting Parties to the 1972 London Convention, from 18–22 September 2000*
- IMO resolution MEPC/Circ.406, 10 November 2003, *Guidelines for Application of MARPOL Annex I Requirements to FPSOs and FSUs*
- IMO MSC/Circ. 1111, 7 June 2004, *Guidance Relating to the Implementation of SOLAS Chapter XI-2 and the ISPS Code*
- IMO resolution MEPC.139(53), 22 July 2005, *Guidelines for the Application of the Revised MARPOL Annex I Requirements to Floating Production, Storage and Offloading Facilities (FPSOs) and Floating Storage Unites (FSUs)*
- IMO resolution MEPC.142(54), 24 March 2006, *Amendments to the Guidelines for the Application of the Revised MARPOL Annex I Requirements to Floating Production, Storage and Offloading Facilities (FPSOs) and Floating Storage Units (FSUs) (Resolution MEPC.139(53)*
- IMO LEG/MISC.6, 10 September 2008, *Implications of the United Nations Convention on the Law of the Sea for the International Maritime Organization*
- IMO LEG/MISC.6, 10 September 2008, *Implications of the United Nations Convention on the Law of the Sea for the International Maritime Organization*
- *3rd Meeting of Contracting Parties to the 1996 Protocol to the 1972 Convention on the Prevention of Marine Pollution by Dumping of Wastes and Other Matter on 31 October 2008*
- IMO resolution A.1023(26), 2 December 2009, *Code for the Construction and Equipment of Mobile Offshore Drilling Units*
- IMO MSC-MEPC.2/Circ.9, 25 May 2010, *Guidance for the Application of Safety, Security and Environmental Protection provisions to FPSOs and FSUs.*

Resolutions of the United Nations

(I)

- Resolution (UNGA) n. 523 (VI), 12 January 1952
- Resolution (UNGA) n. 626 (VII), 21 December 1952
- Resolution (UNGA) n. 1514 (VX), 14 December 1960
- Resolution (UNGA) n. 1515 (XV), 15 December 1960
- Resolution (UNGA) n. 1803 (XVII), 14 December 1962

- Resolution (UNGA) n. 1803 (XVII), 14 December 1962
- Resolution (UNGA) n. 2158 (XXI), 25 November 1966
- Resolution (UNGA) n. 2340 (XXII), 18 December 1967
- Resolution (UNGA) n. 2398 (XXIII), 3 December 1968
- Resolution (UNGA) n. 2467 A (XXIII), 21 December 1968
- Resolution (UNGA) n. 2574 A (XXII), 15 December 1969
- Resolution (UNGA) n. 2574 (XXIV), 15 December 1969
- Resolution (UNGA) n. 2625 (XXXV), 24 October 1970
- Resolution (UNGA) n. 2625 (XXV), 24 October 1970
- Resolution (UNGA) n. 2749 (XXV), 17 December 1970
- Resolution (UNGA) n. 2997 (XXVII), 15 December 1972
- Resolution (UNGA) n. 2996 (XXVII), 15 December 1972
- Resolution (UNGA) n. 2995 (XXVII), 15 December 1972
- Resolution (UNGA) n. 2295 (XXVII), 15 December 1972
- Resolution (UNGA) n. 3129 (XXVIII), 13 December 1973
- Resolution (UNGA) n. 3201 (S-VI), 1 May 1974
- Resolution (UNGA) n. 3202 (S-VI), 1 May 1974
- Resolution (UNGA) n. 3201(S-VI), 1 May 1974
- Resolution (UNGA) n. 3281 (XXIX), 12 December 1974
- Resolution (UNGA) n. 3202 (S-VII), 16 September 1975
- Resolution (UNGA) n. 3362 (S-VII), 16 September 1975
- Resolution (UNGA) n. 3362 (S-VII), 16 September 1975
- Resolution (UNGA) n. 42/115, 7 December 1987
- Resolution (UNGA) n. 42/114, 7 December 1987
- Resolution (UNGA) n. 43/123, 8 December 1988
- Resolution (UNGA) n. 44/228, 22 December 1989
- Resolution (UNGA) n. 45/98, 14 December 1990
- Resolution (UNGA) n. 43/124, 14 December 1990
- Resolution (UNGA) n. 55/1999, 20 December 2000.

(II)

- Resolution (UNSC) n. 389, 22 April 1976
- Resolution (UNSC) n. 384, 22 December 1975
- Resolution (UNSC) n. 687, 3 April 1991
- Resolution (UNSC) n. 1272, 25 October 1999
- Resolution (UNSC) n. 1272, 25 October 1999
- Resolution (UNSC) n. 1264, 15 September 1999
- Resolution (UNSC) n. 1262, 27 August 1999
- Resolution (UNSC) n. 1246, 11 June 1999
- Resolution (UNSC) n. 1246, 11 June 1999
- Resolution (UNSC) n. 1236, 7 May 1999.

(III)

- Resolution (UNESC) n. 1346 (XLV), 30 July 1968

United Nations Conference on the Law of the Sea

- R.S.N.T., Document A/CONF.62/WP.8/Rev.1/PartII, UNCLOS III Off. Rec. Volume V
- *Official Records, Volume VI: Fourth Committee (Continental Shelf), Summary records of meetings and Annexes, 24 February-27 April 1958,* Doc: A/CONF.13/42
- I.C.N.T., Document A/CONF.62/WP.10, UNCLOS III Off. Rec. Volume VIII
- *Report by the Chairman of Negotiating Group 7 on the work of the Group at its 17–27th meetings*, NG7/24, 14 September 1978, Document A/CONF.62/RCNG/1 and Document A/CONF.62/RCNG/2, UNCLOS III Off. Rec. Volume X
- *Report of the Chairman of the Negotiating Group 7*, Document NG7/45, UNCLOS Off. Rec. Volume XII
- *Report of the Chairman of negotiating group 7*, Document A/CONF.62/L.47, UNCLOS III Off. Rec., Volume XVIII.

Other United Nations documents

- *Memorandum on the Regime of the High Seas*, 14 July 1950 (Doc. A/CN.4/32)
- United Nations Doc A/C.1/PV 1515 and Doc A/C.1/PV 1516, 1 November 1967
- *Economic Commission for Asia and the Far East, Annual Report (18 April 1967–30 April 1968), Official Records of the 45th Session of the Economic and Social Council, 2nd Supplement*
- *Memorandum by the President of the Conference on document A/CONF.62/WP.10* (Document A/CONF.62/WP.10/Add.1), 22 July 1977, p. 68.
- UNEP Doc. GC.6/CRP.2, 19 May 1978
- UN Security Council, *Decision taken by the Governing Council of the United Nations, Compensation Commission during its third session at the 18th meeting held on 28 November 1991, as revised at the 24th Meeting, 16 March 1992* (S/AC.26/1991/7/Rev.1, 17 March 1992), *Criteria for additional Categories of Claims: United Nations Compensation Commission, Governing Council, Fifth Session, Geneva, 16–20 March 1992*, para. 35.
- *Rio Declaration on Environment and Development* (A/CONF.151/26, Vol. I)
- *Statement of Forest Principles*, the *United Nations Framework Convention on Climate Change* (A/CONF.151/26, Vol. III)
- *Programme for the Further Implementation of Agenda 21*, (UNGA Doc. A/RES/S-19/2, 19 September 1997)
- Secretary-General Report S/1999/1024, para. 35, 4 October 1999
- *Johannesburg Declaration on Sustainable Development* adopted at the *2002 World Summit on Sustainable Development* and the *Plan of Implementation of the World Summit on Sustainable Development*

- *Report of the World Summit on Sustainable Development, Johannesburg, South Africa, 26 August–4 September 2002* (United Nations publication, Sales No. E.03.II.A1 and corrigendum), chap. I, resolution 1, annex
- UNEP (DEC)/MED IG.15/Inf. 13, 1 October 2003, *Guidelines Dumping of Platforms and Other Man-Made Structures at Sea.*
- *2009 London Convention and Protocol/UNEP Guidelines for the Placement of Artificial Reefs, 30th Consultative Meeting of Contracting Parties to the 1972 Convention on the Prevention of Marine Pollution by Dumping of Wastes and Other Matter, 27 to 31 October 2008*
- *Receipt of the Joint Submission made by the Republic of Mauritius and the Republic of Seychelles to the Commission on the Limits of the Continental Shelf, 12 December 2008*, (Doc. CLCS.14.2008.LOS)
- *Summary of the Recommendations of the Commission on the Limits of the Continental Shelf in regard to the Joint Submission made by Mauritius and Seychelles concerning the Mascarene Plateau region on 1 December 2008.*

Other Documents

(I)

- Council Directives 92/91/EEC, of 3 November 1992, and 92/104/EEC, of 3 December 1992, both amended by Directive 2007/30/EC, of the European Parliament and of the Council, of 20 June 2007

(II)

- OSPAR [*Convention for the Protection of the Marine Environment of the North-East Atlantic*] *Decision 98/3 on the Disposal of Disused Offshore Installations, 22–23 July 1998*
- OSPAR *Guidelines on Artificial Reefs in relation to Living Marine Resources*, 2012
- OSPAR *Assessment of Construction or Placement of Artificial Reefs*, 2009.

(III)

- *Report and Recommendations to the Governments of Iceland and Norway of the Conciliation Commission on the Continental Shelf Area between Iceland and Jan Mayen*, done on June 1981.

(IV)

- Constitution of the United States of America
- *Policy of the United States with respect to the Natural Resources of the Subsoil and Sea Bed of the Continental Shelf*, Proclamation n. 2667, 28 September 1945, 10 Fed. Reg. 12.303 (1945)

- *Ryan Consolidated Petroleum Corporation v. Pickens, Supreme Court of Texas*, 1955, 155 Tex. 221, 285 S.W.2d 201
- *Westmoreland and C. Nat Gas Co. v. De Witt*, 1899, 130 Pa.235
- *Brown v. Spilman*, 155 U.S. Supreme Court 665, 14 January 1895
- *Carter Oil Co. v. Dees*, Appellate Court of Illinois, Fourth District, 1950, 340Ill.App.449, 92 N.E.2d 519
- *Dobson v. Arkansas Oil & Gas Commission*, Supreme Court of Arkansas, 1951, 218 Ark. 160, 235 S.W. 2d 33
- *Millette v. Phillips Petroleum Co.*, Supreme Court of Mississippi, in Banc, 1950, 209 Miss. 687, 48 So.2d 344
- *Ohio Oil Co. v. Indiana*, Supreme Court of the United States of America, 1900, 177 U.S. 190,20 S.Ct.576, 44 L.Ed. 729
- *Japhet v. McRae*, Commission of Appeals of Texas, Section B, 1925, 276 S.W. 669
- *Barnard v. Monongahela Natural Gas Co.*, Supreme Court of Pennsylvania, 1906, 216 Pa. 362, 65 A. 801
- House Bill 25, 12 August 1931
- Acts 1977, 65th Leg. p. 2517, ch. 871, art. I, para. 1, eff. Sept. 1, 1977, amended by Acts 1979, 66th Leg., p. 673, ch. 300, para. 1, eff. May 29, 1979;
- Acts 1981, 67th Leg., p. 3166, ch. 830, para. 2, eff. June 17, 1981
- Acts 1995, 74th Leg., ch. 870, para. 1, eff. Sept. 1, 1995
- Acts 1995, 74th Leg., ch. 871, para. 1, eff. Sept. 1, 1995
- Indonesian Law 7, 17 July 1976 (revoked on 20 October 1999)
- Resolutions of the Angolan Government n. 3/08 of 21 January 2008
- Resolutions of the Angolan Government n. 8/08 of 21 January 2008
- Resolutions of the Angolan Government n. 33/08 of 23 April 2008.

Press Releases and Public Statements

- *China's Path of Peaceful Development and Its View of Regional Security*, 3 November 2008
- *China-Japan Joint Statement on All-round Promotion of Strategic Relationship of Mutual Benefit*, 22 May 2008
- *Joint press statement by leaders on the occasion of the working visit of Yab Dato' Seri Abdullah Haji Ahmad Badawi, Prime Minister of Malaysia to Brunei Darussalam on 15–16 March 2009*
- Communiqué by the Ministry of Foreign Affairs of the People's Republic of China: *Ambassador Liu Jianchao pays Courtesy call on Philippine Foreign Affairs Secretary Romulo, 13 March 2009*
- Ministry of Foreign Affairs of Malaysia: *Brunei Will Share Oil Revenue on Disputed Oil Block—Anifah*, 7 June 2010
- *M'sia-Brunei agreement will benefit both countries*, Borneo Bulletin: 3 May 2010

- Statement by the Ministry of Foreign Affairs of Viet Nam: *Viet Nam protests Tan Bao Hao's activities which violate sovereignty and jurisdictional rights of Viet Nam,* 7 November 2011
- *Joint Statement on the 15th Annual Leaders' Consulation between Brunei Darussalam and Malaysia, 12 September 2011*
- Statements by the Ministry of Foreign Affairs of the People's Republic of China: *Yang Jiechi Meets with Vietnamese Foreign Minister Pham Binh Minh, 11 July 2012*
- *Vice President Xi Jinping Meets with Vietnamese Army Chief, 13 April 2012*
- *Vice Premier Li Keqiang Meets with Vietnamese Deputy Prime Minister Hoang Trung Hai, 31 March 2012*
- Statement by the Ministry of Foreign Affairs of Viet Nam: *Viet Nam, China negotiate cooperation at sea, 30 May 2012*
- Statement by the Ministry of Foreign Affairs of Viet Nam: *Viet Nam, China discuss sea area off Tonkin Gulf, 24 May 2012*
- Statement by the Ministry of Foreign Affairs of the People's Republic of China: *Vice Foreign Minister Zhang Zhijun Lodges a Solemn Representation Concerning the Passing of the Maritime Law of Vietnam by the Vietnamese Parliament*, 21 June 2012
- Statement by the Ministry of Foreign Affairs of Viet Nam: *Foreign Minister Pham Binh Minh clarifies content of Viet Nam Maritime Law*, 26 June 2012
- Joint Communiqué 46th ASEAN Foreign Ministers' Meeting Badar Seri Begawan, Brunei Darussalam, 29–30 June 2013
- *China and the Philippines held 19th Foreign Ministry Consultations, 14 June 2013*
- *Notification and Statement of Claim on the West Philippine Sea, 22 January 2013*
- *Chinese Spokesperson Hong Lei's remarks on China returned the Philippines' Notification on the submission of South China Sea issue to international Arbitration, 19 February 2013*
- *Foreign Ministry Spokesperson Hua Chunying's Remarks on the Philippines' Efforts in the Pushing for the Establishment of the Arbitral Tribunal in Relation to the Disputes between China and the Philippines in the South China Sea, 26 April 2013*
- *The Department of Foreign Affairs' Statement on China's Response to the Philippines' Arbitration Case, 19 February 2013*
- I.T.L.O.S. Press Release n. 191, 25 April 2013, *Arbitrators appointed in the arbitral proceedings instituted by the Republic of the Philippines against the People's Republic of China.*
- I.T.L.O.S. Press Release n. 197, 24 June 2013, *New arbitrate and president appointed in the arbitral proceedings instituted by the Republic of the Philippines against the People's Republic of China.*

Bibliography

Akehurst, M. (1977). Custom as a source of international law. 47 *British Yearbook of International Law 1974*(1975), 31–42.
Albaharna, H. M. (1968). A note on the Kuwait-Saudi Arabia Neutral Zone Agreement of July 7th, 1965, relating to the partition of the zone. *International and Comparative Law Quarterly*, *17*, 730–735.
Albaharna, H. M. (1968). *The legal status of the arabian gulf states*. Manchester: Manchester University Press.
Alexander, L. M. (2001). Baseline delimitations and maritime boundaries. In *Law of the Sea* (pp. 3–26). Dartmouth, Aldershot, Burlington US, Singapore, Sydney: Ashgate.
Alexander, L. M. (1974). Regionalism and the Law of the Sea: The case of semi-enclosed seas. *Ocean Development and International Law Journal*, *2*(2), 151–186.
Alexander, L. M. (1982). *Marine regionalism in the Southeast Asian seas*. Honolulu: East-West Center.
Alexander, L. M. (1968). National jurisdiction and the use of the sea. *National Resources Journal*, *8*(3), 373–400.
Allott, P. (1992). Mare Nostrum: A new international Law of the Sea. *American Journal of International Law* 4, *86*, 764–787.
Amin, S. H. (1980). Law of the continental shelf delimitation: the Gulf example. *Netherlands International Law Review*, 27, 335–346.
Anand, R. P. (1996). South Asia and the Law of the Sea: Problems and prospects. In T. A. Mensah (Ed.), *Ocean governance: Strategies and approaches for the 21st century* (pp. 331–353). Honolulu: University of Hawaii.
Anderson, D. H. (1998). Strategies for dispute resolution: negotiating joint agreements. In G. Blake, M. Pratt, C. Schofield & J. A. Brown (Eds.), *Boundaries and energy: Problems and prospects* (pp. 473–484). London/The Hague/Boston.
Anderson, D. H. (2006). Negotiating maritime boundary agreements: A personal view. In R. Lagoni & D. Vignes (Eds.), *Maritime delimitation* (pp. 121–141). Leiden, Boston: Martinus Nijhoff Publishers.
Andrassy, J. (1951-II). Les relations internationales de voisinage. *Recueil des Cours de l'Académie de Droit International*, *79*, 77–180.
Andrassy, J. (1970). *International law and the resources of the sea*. New York, London: Columbia University Press.
Antunes, N. M. (2004a). O novo regime jus-internacionalista do mar: a consagração ex vi pacti de um mare nostrum. In *Estudos em Direito Internacional Público* (pp. 17–38). Coimbra: Almedina.
Antunes, N. M. (2004b). Spatial allocation of continental shelf rights in the Timor Sea. Reflections on maritime delimitation and joint development. In *Estudos em Direito Internacional Público* (pp. 273–320). Coimbra: Almedina.

V. Becker-Weinberg, *Joint Development of Hydrocarbon Deposits in the Law of the Sea*, Hamburg Studies on Maritime Affairs 30, DOI 10.1007/978-3-662-43570-0

Antunes, N. M. (2004c). The 1999 Eritrea-Yemen maritime delimitation award and the development of international law. *Estudos em Direito Internacional Público* (Almedina: Coimbra, 2004), pp. 225–268.

Antunes, N. M. (2003). *Towards the conceptualisation of maritime delimitation-legal and technical aspects of political process*. Leiven, Boston, Martinus Nijhoff Publishers.

Antunes, N. M. (2000). The pending maritime delimitation in the Cameroon v. Nigeria case: a piece in the jigsaw puzzle of the Gulf of Guinea. *International Journal of Marine and Coastal Law, 15*(2), 163–192.

Appolis, G. (1981). *L'Emprise Maritime de L'Etat Côtier*. Paris: Editions A. Pedone.

Ariffin, D. H. (1985). The Malaysian philosophy of joint development. In M. J. Valencia (Eds.), *Geology and hydrocarbon of the South China sea and possibilities of joint development, Proceedings of the Second EAPI/CCOP Workshop, East-West Center, Honolulu, Hawaii, 22–26 August 1983* (pp. 533–538). New York, Oxford, Toronto, Sydney, Paris, Frankfort: Pergamon Press.

Armanet, P. (1984). The economic interest underlying the first declaration on a maritime zone. In F. O. Vicuña (Eds.), *The exclusive economic zone. A latin American perspective* (pp. 27–30). Boulder Colorado: Westview Press.

Arsana, I. M. A., & Schofield, C. (2012). Adding further complexity? Extended continental shelf submissions in East and Southeast Asia. In C. Schofield (Eds.), *Maritime energy resources in Asia: Legal regimes and cooperation* (pp. 35–59). Seattle: The National Bureau of Asian Research.

Attard, D. J. (1987). *The exclusive economic zone in international law*. Oxford: Claredon Press.

Auburn, F. M., & Forbes, V. L. (1993). The timor gap treaty and the Law of the Sea Convention. *Ocean Yearbook, 10*, 40–53.

Aust, A. (2007). *Modern treaty law and practice*. Cambridge: Cambridge University Press.

Aust, A. (2012). Alternatives to treaty-making: MOUs as political commitments. In D. B. Hollis (Ed.), *The Oxford guide to treaties*. Oxford: Oxford University Press, pp. 46−72.

Azcárraga y. B., & de José L. (1949). Los derechos sobre la plataforma continental. *Revista Española de Derecho Internacional, 2*, 47–99.

Bain, J. S. (1949a). Rostow's proposals for petroleum policy. *Journal of Political Economy, 57*(1), 55–60.

Bain, J. S. (1949b) A rejoinder. *Journal of Political Economy, 57*(1), 68–69.

Bangert, K. (1999). The effective enforcement of high seas fishing regimes: The case of the convention for the regulation of the policing of the north sea fisheries of 6 May 1882. In: G. S. Goodwin-Gill & S. Talmon (Eds.), *The reality of international law. Essays in honour of Ian Brownlie* (pp. 1–20). Oxford: Clarendon Press.

Barberis, J. A. (1975). Los recursos minerales compartidos entre Estados y el derecho internacional. *Derecho de la Integraciónm, 18–19*(3), 45–48.

Bardonnet, D. (1989). Frontières terrestres et frontières maritimes. *Annuaire Français de Droit International, 35*, 1–64.

Bastida, A. E., Ifesi-Okoye, A., Mahmud, S., Ross, J., & Wälde, T. (2006–2007). Cross-border unitization and joint development agreements: An international law perspective. *Houston Journal of International Law, 29*, 355–422.

Beauchamp, K. P. (1986). The management function of ocean boundaries. *San Diego Law Review Association, 23*(3), 611–660.

Becker-Weinberg, V. (2013). Theory and practice of joint development in international law. In Z. Gao, J. Yu, H. Zhang, & J. Wu (Eds.), *Cooperation and development in the South China sea*. Beijing: China Democracy and Legal System Publishing House.

Becker-Weinberg, V. (2010). The internationalization of marine natural resources in UNCLOS. In R. Lagoni, P. Ehlers & M. Paschke (Eds.), *Recent developments in the Law of the Sea* (pp. 9–54). Berlin, Munster, Vienna, Zurich, London: LIT Verlag.

Becker-Weinberg, V. (2008). A nacionalização do petróleo e o princípio do aproveitamento conjunto entre Estados (pp. 369–418). M. Oliveira Martins (Eds.), Lisbon: AAFDL.

Becker-Weinberg, V. (2013a). Recalling the Falkland Islands (Malvinas) sovereignty formula. *Ocean Yearbook, 27,* 411–433.

Becker-Weinberg, V. (2013b). Joint development arrangements in Northeast Asia and the Gulf of Tonkin. In R. Beckman, I. Townsend-Gault, C. Schofield, T. Davenport & L. Bernard (Eds.), *Beyond territorial disputes in the South China sea: Legal frameworks for the joint development of hydrocarbon deposits* (pp. 219–248) Cheltenham-UK, Northampton-MA, USA: Edward Elgar.

Becker-Weinberg, V. (2013c). Seabed activities and the protection and preservation of the marine environment in disputed maritime areas of the Asia-Pacific region. In *Securing the Ocean for the Next Generation*, Papers from the Law of the Sea Institute, UC Berkeley-Korean Institute of Ocean Science and Technology Conference, held in Seoul, Korea, May 2012. http://www.law.berkeley.edu/files/Becker-Weinberg-final.pdf. Accessed Aug 2013. Also published by the Law of the Sea Institute-UC Berkley Publishers (2013), pp. 253–299.

Beckman, R. (2013). The UN convention on the Law of the Sea and the maritime disputes in the South China sea. *American Journal of International Law, 107,* 142–163.

Beckman, R., & Schofield, C. (2009). Moving beyond disputes over island sovereignty: ICJ decision sets stage for maritime boundary delimitation in the Singapore strait. *Ocean Development and International Law, 40,* 1–35.

Beekman, K. (1996–2002). Transboundary damage to the environment *per se*: remedial measures and standing. In *Revue Belge de Droit International* (pp. 453–492).

Bennouna, M. (1983). Les droits d'exploitation des ressources minérales des océans. In D. Bardonnet & G. Virally (Eds.), *Le Nouveau Droit international de la Mer* (pp. 117–139). Paris: A. Pedone.

Benvenisti, E. (2002). *Sharing transboundary resources. International law and optimal resource use* (pp. 21, 43). Cambridge: Cambridge University Press.

Benvenisti, E. (1996). Collective action in the utilization of shared freshwater: the challenges of international water resources law. *American Journal of International Law, 90,* (1996), pp. 384-415.

Bernard, T. (1996). *Co-operative agreements in the extractive petroleum industry.* London, The Hague, Boston: Kluwer Law International.

Birnie, P., Boyle, A., & Redgwell, C. (2009). *International law and the environment.* Oxford, New York: Oxford University Press.

Bjørlykke, K. (2011a). Introduction to petroleum geology. In *Petroleum geoscience: from sedimentary environments to rock physics* (pp. 10–11). Berlin: Springer.

Bjørlykke, K. (2011b). Petroleum migration. In: *Petroleum geoscience: from sedimentary environments to rock physics* (pp. 349–360). Berlin: Springer.

Blake, G. H., & Swarbrick, R. E. (1998). Hydrocarbon and international boundaries: a global overview. In G. Blake, M. Pratt, C. Schofield & J. Allison Brown (Eds.), *Boundaries and energy: Problems and prospects* (pp. 3–27). London, The Hague, Boston: Kluwer Law International.

Blake, G. (1995). Introduction: boundless resources? In G. H. Blake, W. J. Hildesley, M. A. Pratt, R. J. Ridley & C. H. Schofield (Eds.), *The peaceful management of transboundary resources* (pp. xiii–xviii). London, Dordrecht, Boston: Graham and Trotman/Martinus Nijhoff.

Blecher, M. D. (1979). Equitable delimitation of continental shelf. *American Journal of International Law, 73,* 60–88.

Blinn, K., Duval, C., Le Leuch, H., & Pertuzio, A. (1986). *International petroleum exploration and exploitation agreements. Legal, economic and policy aspects.* Bristol: Euromoney Publications.

Borgese, E. M. (1994). Pacem in Maribus XIX-background paper. In P. Bautista Payoyo (Ed.), *Ocean governance: sustainable development of the seas* (pp. 350–365). Tokyo, New York, Paris: United Nations University Press.

Borgese, E. (1972). *Mann Pacem in Maribus.* New York: Dodd, Mead and Company.

Borgese, E. M. (1977). The new international economic order and the Law of the Sea. In D. Walsh (Ed.), *The Law of the Sea. issues in ocean resource management* (pp. 82–111). New York: Praeger Publishers.

Borgese, E. M. (1982). The role of the Sea-bed Authority in the '80s and '90s. The common heritage of mankind. In R.-J. Dupuy (Ed.), *La Gestion des Ressources pour l'Humanité: Le Droit de la Mer/the management of humanity's resources: The Law of the Sea* (pp. 35–48). The Hague, Boston, London: Martinus Nijhoff Publishers.

Borgese, Elisabeth. (1986). *Mann the future of the oceans. A report of the club of Rome.* Montreal: Harvest House.

Bos, M. (1977). The recognized manifestations of international law. A new theory of "sources". *German Yearbook of International Law, 20*, 9–76.

Bowett, D. W. (1978). The arbitration between the United Kingdom and France concerning the continental shelf boundary in the english channel and South-Western approaches. *British Yearbook of International Law, 49*, 1–29.

Boyle, A. (2007). Relationship between international environmental law and other branches of international law. In D. Bodansky, J. Brunnée, & E. Hey (Eds.), *The Oxford handbook of international environmental law*. Oxford: Oxford University Press.

Brown, E. D. (1994). *The international Law of the Sea, introductory manual* (Vol. 1). Aldershot, Broofield USA, Singapore, Sydney: Dartmouth.

Brown, E. D. (1992a). *Sea-bed energy and minerals: The international legal regime. The continental shelf* (Vol. 1). Dordrecht, Boston, London: Martinus Nijhoff Publishers.

Brown, E. D. (1992b). The significance of a possible EC EEZ for the law relating to artificial islands, installations, and structures, and to cables and pipelines, in the exclusive economic zone. *Ocean Development and International Law, 23*, 115–144.

Brown, E. D. (1983). The Tunisia-Libya continental shelf case. *Marine Policy, 7*(3), 142–162.

Brown, E. D. (1979). The Anglo-French continental shelf case. *San Diego Law Review, 19*, 461–530.

Brown, E. D. (1986). *Sea-bed energy and mineral resources and the Law of the Sea—volume II: The area beyond the limits of national jurisdiction.* : London: Graham and Trotman.

Brownlie, I. (1979-I). Legal status of natural resources in international law (some aspects). *Recueil des Cours de l'Académie de Droit International, 162*, 245–318.

Buchholz, H. J. (2001). The Baltic Sea: Lessons learned. In M. J. Valencia (Ed.), *Maritime regime building. Lessons learned and their relevance for Northeast Asia* (pp. 17–25). The Hague, Boston, London: Martinus Nijhoff Publishers.

Bundy, R. R. (1995). Natural resource development (oil and gas) and boundary disputes. In G. H. Blake, W. J. Hildesley, M. A. Pratt, R. J. Ridley & C. H. Schofield (Eds.), *The peaceful management of transboundary resources* (pp. 23–40). London, Dordrecht, Boston: Graham and Trotman/Martinus Nijhoff.

Bundy, R. R. (1994). State practice in maritime delimitation. In G. H. Blake (Ed.), *World boundaries. Maritime boundaries* (Vol. 5, pp. 18–40). London, New York: Routledge.

Burmester, H. Australia and the Law of the Sea. In J. Crawford & D. R. Rothwell (Eds.), *The Law of the Sea in the asian pacific region* (pp. 51–64). Dordrecht, Boston, London: Martinus Nijhoff Publishers.

Burmester, H. (1990). The zone of co-operation between Australia and Indonesia: a preliminary outline with particular reference to applicable law. In H. Fox (Ed.), *Joint development of offshore oil and gas. The institute's revised model agreement. Conference papers. The Australia/Indonesia zone of co-operation treaty 1989* (Vol. II, pp. 128–139). London: The British Institute of International and Comparative Law.

Burney, L. H. (1996). A pragmatic approach to decision making in the next era of oil and gas jurisprudence. Journal of *Energy, Natural Resources and Environmental Law, 16*, 1–90.

Byers, M. (1999). *Custom, power and the power of rules: international relations and customary International Law.* Cambridge: Cambridge University Press.

Cahier, P. (1991). Les sources du droit relatif à la délimitation du plateau continental. In M. Virally (Ed.), *Le Droit International au service de la Paix, de la Justice et du Développement*, mélanges (pp. 175–182). Paris: Pedone.

Cameron, P. D. (2006). The rules of engagement: developing cross-border petroleum deposits in the north sea and the caribbean. *International and Comparative Law Quarterly*, *55*, 559–585.

Caminos, H. (1984). The regime of fisheries on the exclusive economic zone. In F. O. Vicuña (Ed.), *The exclusive economic zone. A latin American perspective* (pp. 143–158). (Westview Press: Boulder Colorado, 1984), pp. 143-158.

Caminos, H. (2012). Enforcement jurisdiction under the United Nations convention on the Law of the Sea. An overview. In H. P. Hestermeyer, D. König, N. Matz-Lück, V. Röben, A. Seibert-Fohr, P.-T. Stoll & S. Vöneky (Eds.), *Coexistence, cooperation and solidarity. liber amicorum rüdiger wolfrum* (Vol. 1, pp. 737–776). Leiden, Boston: Martinus Nijhoff Publishers.

Campbell, H. W. (1959). Oil and gas conservation in Illinois. *University of Illinois Law Forum*, *1959*, 570–583.

Carleton, C. (2006.) Maritime delimitation in complex island situations: a case study on the Caribbean Sea. In R. Lagoni & D. Vignes (Eds.), *Maritime delimitation* (pp. 153–188). Leiden, Boston: Martinus Nijhoff Publishers.

Carreau, D. (2004). *Droit international* (8th ed.). Paris: Pedone.

Carrillo-Salcedo, J.-A. (1985). *El Derecho Internacional en un Mundo en Cambio*. Madrid: Tecnos.

Centre for Oceans Law and Policy, University of Virginia School of Law. (1993). In S. N. Nandan & S. Rosenne (Eds.), *United nations convention on the Law of the Sea 1982: A commentary* (Vol. 2). Dordrecht, Boston, London: Martinus Nijhoff Publishers.

Charney, J. I. (1995). Central East Asian maritime boundaries and the Law of the Sea. In *American Journal of International Law*, *89*(4), 724–749.

Charney, J. I. (1996). Progress in international maritime boundary delimitation law. *American Journal of International Law*, *90*(1), 227–256.

Charney, J. I. (1984). Ocean boundaries between nations: a theory for progress. *American Journal of International Law 78*(3), 582–606.

Charney, J. I. (2002). International maritime boundaries for the continental shelf: the relevance of natural prolongation. In N. Ando, E. McWhinney & R. Wolfrum (Eds.), *Liber amicorum Judge Shigeru Oda*, (Vol. 2, pp. 1011–1029). The Hague, London: Kluwer Law International.

Charney, J. I. (1989). The delimitation of ocean boundaries. In D. G. Dallmeyer & L. Devorsey Jr (Eds.), *Rights to oceanic resources. Deciding and drawing maritime boundaries* (pp. 25–49). Dordrecht, Boston, London: Martinus Nijhoff Publishers.

Charney, J. I. (1982.) The offshore jurisdiction of the States of the United States and the Provinces of Canada-a comparison. *Ocean Development and International Law*, *12*(2), 301–335.

Charney, J. I. (1975). The equitable sharing of revenues from seabed mining. In Herbert G. Knight et al. (Eds.), *Policy issues in ocean law: Studies of the working group on technical issues of the* Law of the Sea (pp. 53–120). St. Paul Minnesota.

Chinkin, C. M. (1993). East timor moves into the world court. *European Journal of International Law*, *4*(2), 206–222.

Churchill, R. R. (1990). Joint development zones: international legal issues. In H. Fox (Ed.), *Joint development of offshore oil and gas. The institute's revised model agreement. Conference papers. The Australia/Indonesia zone of co-operation treaty 1989* (Vol. II, pp. 55–67). London: The British Institute of International and Comparative Law.

Churchill, R. R. (1994). The Greenland—Jan Mayen case and its significance for the international law of maritime boundary delimitation. *International Journal of Marine and Coastal Law*, *9*(1) 1–29.

Churchill, R. R. & Lowe, A. V. (1999). *The Law of the Sea* (3rd edn). Manchester: Manchester University Press.

Clain, L. E. (1984–1985). Gulf of Maine—a disappointing first in the delimitation of a single maritime boundary. *Virginia Journal of International Law, 25*, 521–620.

Colmenares, N. A., & Escobar, J. J. (2002). Ocean and coastal issues and policy responses in the Caribbean. In *Ocean and Coastal Management, 45*, 905–924.

Colombos, J. C. (1967). *The international Law of the Sea* (6th edn). New York: David McKay Company Inc.

Colson, D. (2003). The delimitation of the outer continental shelf between neighboring status. in *American Journal of International Law, 97*, 91–107.

Combacau, J. (1985). *Le Droit de la Mer*. Paris: Presses Universitaires de France.

Conforti, B. La délimitation et les méthodes pratiques. Equité, délimitation et critères de délimitation. In D. Pharand & U. Leanza (Eds.), *The continental shelf and the exclusive economic zone/Le Plateau Continental et la Zone Économique Exclusive* (pp. 55–56). Dordrecht, Boston, London: Martinus Nijhoff Publishers.

Cook, C. (1981–1983). Filling the gap-delimiting the Australia-Indonesia maritime boundary. *Australian Yearbook of International Law, 10*, 131–175.

Crawford, J. (2012). *Brownlie's principles of public international law*. Oxford Oxford: University Press.

d'Amato, A. A. (1971). Trashing customary international law. *American Journal of International Law, 81*, 101–108.

d'Amato, A. A. (1961–1962). Treaties as a source of general rules of international law. *Harvard International Law Club Bulletin, 3*, 1–43.

Davenport, T., Townsend-Gault, I., Beckman, R., Schofield, C., Ong, D., & Becker-Weinberg, V. et al. (2013). *Conference on joint development and the South China Sea—conference report.* http://cil.nus.edu.sg/wp/wp-content/uploads/2011/06/Report-of-CIL-Conference-on-Joint-Development-and-the-South-China-Sea-2011-04.08.2011.pdf. Accessed Aug 2013.

Decaux, E. (1982). L'arrêt de la Cour Internationale de Justice dans l'affaire du plateau continental (Tunisie/Libye). *Annuaire Français de Droit International, 28*, 357–391.

Decaux, E. (1981). L'arrêt de la C.I.J. sur la requête a fin d'intervention de Malte, dans l'affaire du plateau continental entre la Tunisie et la Libye (14 avril 1981). *Annuaire Français de Droit International, 27*, pp. 177–202.

Dellapenna, J. W. (2003). The customary international law of transboundary fresh waters. In M. Fitzmaurice & M. Szuniewicz (Eds.), *Exploitation of natural resources in the 21st century* (pp. 153–158). The Hague, London, New York: Kluwer Law International.

Deng, X. (1984). Speech at the third plenary session of the Central Advisory Commission of the Communist Party of China, October 22, 1984. *Selected Works of Deng Xiaoping* (Vol. 3). Beijing: Foreign Languages Press.

Devaux-Charbonnel, J. (1956-2). Le régime juridique de la recherche et de l'exploitation du pétrole dans le plateau continental. *Annuaire Français de Droit International*, 320–333.

Dinh, N. Q., Dailler, P., & Pellet, A. (2002). *Droit international public* (7th edn). Paris: Librairie Générale de Doit et de Jurisprudence/E.J.A.

Djalal, H. (1998). South China Sea island disputes. In M. H. Nordquist & J. N. Moore (Eds.), *Security flashpoints. Oil, Islands, sea access and military confrontation* (pp. 109–133). The Hague, Boston, London: Martinus Nijhoff Publishers.

Doggett, W. J. (1963). Practical legal problems encountered in the formation, operation and dissolution of fieldwide oil and gas units. *Oklahoma Law Review, 16*, 1–58.

Drummond, D. O., Sherman, L. R., & McCarthy, E. R. Jr (2004–2005). The rule of capture in Texas—still so misunderstood after all these years. *Texas Tech Law Review, 37*, 1–97.

Dupuy, P.-M. (2002). A general stocktaking of the connections between the multilateral dimension of obligations and codification of the law of responsibility. *European Journal of International Law, 13*(5), 1053–1081.

Dupuy, P.-M. (2007). Formation of customary international law and general principles. In D. Bodansky, J. Brunnée, & E. Hey (Eds.), *The Oxford handbook of international environmental law*. Oxford: Oxford University Press.

Dupuy, R.-J. (1979). *L'Océan Partagé*. Paris: A. Pedone.

Dzurek, D. J. (1996). The Spratly Islands dispute: who's on first? *Maritime Boundaries*, *2*(1), 1–67.

Dzurek, Daniel J. Southeast Asian offshore oil disputes. 11 *Ocean Yearbook* (1994), pp. 157–178.

Dzurek, D. J. (1998). Maritime agreements and oil exploration in the Gulf of Thailand. In G. Blake, M. Prat, C. Schofield, J. Allison Brown (Eds.), *Boundaries and energy: Problems and prospects* (pp. 117–135). London, The Hague, Boston: Kluwer Law International.

Eckman, D. W. (1973). Statutory fieldwide oil and gas units: a review for future agreements. *Natural Resources Law*, *6*, 339–387.

Elisa M. (2009). *Corporate accountability in international environmental law*. Oxford: Oxford University Press.

Ely, Northcutt "The conservation of oil. 51 *Harvard Law Review* (1937–1938), pp. 1209–1244.

Esmaeili, H. (2001). *The legal regime of offshore oil rigs in international law*. Aldershot, Burlington USA, Singapore, Sydney: Dartmouth/Ashgate.

Evans, M. (1989). *Relevant circumstances and maritime delimitation*. : Oxford: Claredon Press.

Evans, M. D. (2006). Maritime boundary delimitation: where do we go from here? In D. Freestone, R. Barnes, & D. Ong (Eds.), *The Law of the Sea: Progress and prospects*. Oxford: Oxford University Press.

Feldman, M. B. (1983). The Tunisian-Libya continental shelf case: geographic justice or judicial compromise?. *American Journal of International Law*, *77*(2), 219–238.

Fesharaki, F. (1981). Joint development of offshore petroleum resources: the Persian Gulf experience. In M. J. Valencia (Ed.), *The South China Sea. Hydrocarbon potential and possibilities of joint development*. (pp. 1325–1334). Oxford, New York Toronto, Sydney, Paris, Frankfurt: Pergamon Press.

Fitzmaurice, M. (2001). International protection of the environment. *Recueil de Cours de l'Académie de Droit Internationale de La Haye*, *293*, 9–488.

Fitzmaurice, M. (2007). International responsibility and liability. In D. Bodansky, J. Brunnée, & E. Hey (Eds.), *The Oxford Handbook of International Environmental Law*. Oxford: Oxford University Press.

Flormann-Pfaff, A. (1994). *Lagerstätten im Völkerrecht: Joint Development: Zusammenarbeit bei anerkannten und streitigen Grenzen*. Berlin: Duncker und Humblot.

Fox, H., McDade, P., Rankin, D. R., Strati, A., & Huey, P. (1989). In M. Bathurst (Ed.), *Joint development of offshore oil and gas. A model agreement for states for joint development with explanatory commentary*. London: BIICL.

Franckx, E. (2007). The 200-mile: between creeping jurisdiction and creeping common heritage?. *George Washington International Law Review*, *39*, 467–498.

Franckx, E. (2004). Baltic Sea update. In J. I. Charney & Lewis M. A. (Eds.), *International maritime boundaries* (Vol. 3, pp. 2557–2571). The Hague, Boston, London: Martinus Nijhoff Publishers.

Franckx, E. (1993). Baltic Sea maritime boundaries. In In J. I. Charney & Lewis M. A. (Eds.), *International maritime boundaries* (Vol. 1, pp. 345–368). Dordrecht, Boston, London: Martinus Nijhoff Publishers.

Franckx, E. (2010). The International Seabed Authority and the common heritage of mankind: the need for states to establish the outer limits of their continental shelf. *International Journal of Marine and Coastal Law*, *25*, 543–567.

Fu, K. C. (2013). Safeguarding China's national interests in the South China sea: Rectification, services, leadership, and maritime delimitation. *17-1 China Oceans Law Review*, pp. 12–27

Gao, J. (2009). A note on the 2008 cooperation consensus between China and Japan in the East China Sea. *Ocean Development and International Law Journal*, *40*, 291–296.

Gao, Z. (1983). Legal aspects of joint development in international law. In In M. Kusuma-Atmadja, T. A. Mensah & B. H. Oxman (Eds.), *Sustainable development and preservation of the oceans: The challenges of UNCLOS and agenda 21* (pp. 629–644). Hawaii: The Law of the Sea Institute, University of Hawaii.

Gao, Z. (1998). The legal concept and aspects of joint development in international law. *Ocean Yearbook, 13*, 107–124.

Gao, Z. (1994). *International petroleum contracts. Current trends and new directions*. London: Grahm and Trotman/Martinus Nijhoff.

Garcia A. F. V. (1993). In D. Pharand & U. Leanza (Eds.), *Continental shelf and the exclusive economic zone/Le Plateau Continental et la Zone Économique Exclusive*. Dordrecht, Boston, London: Martinus Nijhoff Publishers.

Gardiner, R. K. (2010). *Treaty interpretation*. Oxford: Oxford University Press.

Gardiner, R. K. (2012). The Vienna Convention rules on treaty interpretation. In D. B. Hollis (Ed.), *The oxford guide to treaties*. Oxford: Oxford University Press.

German, W. P. Z. (1932). Compulsory unit operation of oil pools. *California Law Review, 20*, 111–131.

Getting the Deal Through. (2012). In B. Palmer (Ed.), *Oil regulation 2012*. London: Law Business Research Ltd.

Giardina, A. (1973). Nationalisation et indemnisation en Droit International. In *Le Droit Pétrolier et la Souveraineté des Pays Producteurs* (pp. 39–49). Paris: Librairie Générale de Droit et de Jurisprudence.

Gidel, G. (1952). *Le Plateau Continental*. Madrid: International Bar Association.

Gold, E., & Mitchell, C. L. (1988). The new *Law of the Sea* in the Eastern Caribbean: towards a new ocean development action plan. In E. Gold (Ed.), *A new law of the Sea for the Caribbean. An examination of marine law and policy. Issues in the lesser antilles* (pp. 265–276). Berlin, Heidelberg, New York, London, Paris, Tokyo: Springer.

Goldie, L. F. E. (1970). The North Sea continental shelf case. A ray of hope for the International Court? *New York Law Forum, 6*(2), 327–377.

Greenfield, J. (1992). *China's practice in the Law of the Sea*. Oxford: Claredon Press.

Groves, H. (2005). Offshore oil and gas resources: economics, politics and the rule of law in the Nigeria-Sao Tome e Principe joint development zone. *Journal of International Affairs, 59*(1), 81–96.

Haglund, C. (1933–1934). The new conservation movement with respect to petroleum and natural gas. *Kentucky Law Journal, 22*, 543–581.

Haller-Trost, R. (1998). *The contested maritime and territorial boundaries of Malaysia. An International Law Perspective*. London, The Hague, Boston: Kluwer Law International.

Handlan, A. L., & Sykes, K. L. (1983–1984). Pooling and unitization: legal and ethical considerations. *Tulsa Law Journal, 19*, 309–336.

Hardwicke, R. E. (1935). The rule of capture and its implications as applied to oil and gas. *Texas Law Review, 13*(4), 391–422.

Hardwicke, R. E. (1951–1952). Unitization statutes: voluntary action or compulsion. *Rocky Mountain Law Review, 24*, 29–43.

Hardwicke, R. E. (1941). Oil conservation: statutes, administration, and court review. *Mississippi Law Journal, 13*, 381–416.

Hardy, G. W. (1966). Drainage of oil and gas from adjoining tracts—a further development. *Natural Resources Journal, 6*, pp. 47–51.

Harrison, J. (2011). *Making the Law of the Sea: A study in the development of international law*. Cambridge: Cambridge University Press.

Hayashi, M. (1993). The management of transboundary fish stocks under the LOS Convention. *International Journal of Marine and Coastal Law, 8*(2), 245–261.

Hemingway, R. W. (1991). *The Law of Oil and Gas* (3rd edn). St. Paul, Minnesota: West Publishing Co.

Highet, K. (1989). Whatever became of natural prolongation?. In D. G. Dallmeyer & L. Devorsey Jr. (Eds.), *Rights to oceanic resources. Deciding and drawing maritime boundaries* (pp. 87–100). Dordrecht, Boston, London: Martinus Nijhoff Publishers.

Hilmar, R., & Schmidt-Thomé, M. (2004). Hydrocarbon potential of the Baltic Sea region. In E. M. Emelyanou, M. Schmidt-Thomé & M. Spiridonov (Eds.), *Mineral resources of the Baltic Sea-exploration, exploitation and sustainable development, Zeitschrift für Angewandte Geologie* (Vol. 2, pp. 17–28).

Hong Thao, N. (2005). Maritime delimitation and fishery cooperation in the Tonkin Golf. *Ocean Development and International Law Journal, 36*, 35–44.

Hong Thao, N., & Amer, R. (2009). Regional conflict management: challenges of the border disputes of Cambodia, Laos, and Vietnam. *Austrian Journal of South-East Asian Studies, 2*(2), 53–79.

Hong Thao, N., & Amer, R. (2006). Managing Vietnam's maritime boundary disputes. *Ocean Development and International Law, 38*, 305–324.

Hossain, K. (1990). Choice of petroleum development regime on joint development of offshore oil and gas. In H. Fox (Ed.), *Joint development of offshore oil and gas. The institute's revised model agreement. Conference papers. The Australia/Indonesia zone of co-operation treaty 1989* (Vol. II, pp. 72–76). London: The British Institute of International and Comparative Law.

Huneidi, I. (1990). The Saudi/Kuwait joint development areas of the neutral zone, onshore and offshore. In H. Fox (Ed.), *Joint development of offshore oil and gas. The institute's revised model agreement. Conference papers. The Australia/Indonesia zone of co-operation treaty 1989* (Vol. II, pp. 77–87). London: The British Institute of International and Comparative Law.

Hurst, C. J. B. (1923–1924). Whose is the bed of the Sea? *British Yearbook of International Law, 4*, 34–43.

International Law Association. (1978). In D. Rauschning (Ed.), *Report of the international committee on the legal aspects of the conservation of the environment*. Manila: Fifty-eighth Conference.

International Law Association. (1980). In D. Rauschning (Ed.), *Report of the international committee on the legal aspects of the conservation of the environment*. Manila: Fifty-ninth Conference.

International Law Association. (1988). In R. Lagoni (Ed.), *Report of the international committee on the EEZ on joint development of non-living resources in the exclusive economic zone*. Warsaw: Sixty-third Conference.

International Law Association. (2000). In R. Mullerson (Ed.), *Report of the international committee on formation of customary (General) international law*. London Conference.

International Law Association. (2004). In A. O. Elferink (Ed.), *Report of the international committee on legal issues of the outer continental shelf to the biennial international law association conference in Berlin, Germany, August 2004*. London: Seventy-First Conference.

International Law Association. (2006). In In A. O. Elferink (Ed.), *Second report of the international committee on legal issues of the outer continental shelf to the biennial international law association conference in Toronto, Canada, June 2006*. London: Seventy-Second Conference.

International Law Association. (2008). In D. M. Ong (Ed.), *Third report of the international committee on legal issues of the outer continental shelf to the biennial international law association conference in Rio de Janeiro, Brazil, 17–21 August 2008*. Rio de Janeiro: Seventy-Third Conference.

Jacobs, J. C. (1947–1948). Unit operation of oil and gas fields. *Yale Law Journal, 57*, 1207–1228.

Jacobson, J. L. (1975). The United States and the *Law of the Sea* conference: interim resource options. In H. G. Knight et al. (Eds.), *Policy Issues in ocean law: Studies of the working group on technical issues of the Law of the Sea* (pp. 121–189). St. Paul Minnesota.

Jessup, P. C. (1929). L'exploitation des richesses de la mer. *Recueil des Cours de l'Académie de Droit International, 29*, 405–508.

Ji, G. (2009). Sino-Japanese jurisdictional delimitation in East China Sea: approaches to dispute settlement. In S.-Y. Hong & J. V. Dyke (Eds.), *Maritime boundary disputes, settlement processes, and the Law of the Sea* (pp. 77–105). Martinus Nijhoff Publishers.

Johnston, D. M. (1988). *The Theory and history of ocean boundary-making*. Kingston, Montreal: McGill-Queen's University Press.

Johnston, D. M., & Valencia, M. J. (1991). In S. Oda (Ed.), *Pacific ocean boundary problems. Status and solutions*. Martinus Nijhoff Publishers: Dordrecht, Boston, London.

Joyner, C. C. (1998a). The Spratly Islands dispute: rethinking the interplay of law, diplomacy, and geopolitics in the South China Sea. *International Journal of Marine and Coastal Law, 13*(2), 193–236.

Joyner, C. C. (1998b). Toward a Spratly resource development authority: precursor agreements and confidence-building measures. In M. H. Nordquist & J. N. Moore (Eds.), *Security flashpoints. oil, islands, sea access and military confrontation* (pp. 219–242). The Hague, Boston, London: Martinus Nijhoff Publishers.

Junger, P. D. (1958–1959). The Wyoming oil and gas conservation act—private rights and public policy. *Wyoming Law Journal, 13*, 1–42.

Karaman, I. V. (2012). *Dispute resolution in the Law of the Sea*. Leiden: Martinus Nijhoff Publishers.

Kaye, S. (2004). Negotiation and dispute resolution: a case study in international boundary making—the Australia-Indonesia boundary. In A. G. Oude Elferink & D. R. Rothwell (Eds.), *Oceans management in the 21st century: Institutional frameworks and responses* (pp. 143–166). Leiden, Boston: Martinus Nijhoff Publishers.

Kenworthy, W. E. (1960). Joint development of international waters. *American Journal of International Law, 54*(3), 592–602.

Kim, S. P. (2004). In V. Lowe (Ed.), *Maritime delimitation and interim arrangements in North East Asia*. The Hague, London, New York: Martinus Nijhoff Publishers.

King, A. A. (1947–1948). Pooling and unitization of oil and gas leases. *Michigan Law Review, 46*, 311–340.

Kiss, A.-C. (1985). The common heritage of mankind: utopia or reality? *International Journal, 40*, 423–441.

Kiss, A.-C. (1982). La notion de patrimoine commun de l'humanité. *Recueil des Cours de l'Académie de Droit International, 175*(2), 99–256.

Kittichaisaree, K. (1987). *The Law of the Sea and maritime boundary delimitation in South-East Asia*. Oxford, New York: Oxford University Press.

Klein, N. (2006). Provisional measures and provisional arrangements in maritime boundary disputes. *International Journal of Marine and Coastal Law, 21*, 423–460.

Knauss, J. A. (1985). Creeping jurisdiction and customary international law. *Ocean Development and International Law, 15*(2), 209–216.

Koers, A. W. (1974). Implications of alternative regimes for the Northeast Atlantic treaty system. In G. Pontecorvo & N. H. Hagist (Eds.), *Fisheries conflicts in the Northeast Atlantic: Problems of management and jurisdiction*. Cambridge Massachusetts: Ballinger Publishing Company.

Kolb, R. (2003). In V. Lowe (Ed.), *Case law on equitable maritime delimitation/jurisprudence sur les Délimitations Maritimes Selon L'Équeté. Digest and commentaries/Répertoire et commentaries* The Hague, London, New York: Martinus Nijhoff Publishers.

Kramer, B. M., & Anderson, O. L. (2005). The rule of capture—an oil and gas perspective. *Environmental Law, 35*, 899–954.

Kusuma-Atmadja, M. (1983). Joint development of oil and as by neighboring countries. In M. Kusuma-Atmadja, T. A. Mensah & B. Oxman (Eds.), *Sustainable development and preservation of the oceans: the challenges of UNCLOS and agenda* (Vol. 21, pp. 592–609). The Law of the Sea Institute, University of Hawaii.

Kusumowidagdo, J. U. In Almqvist and Wiksell International (Ed.), *Consultations clauses as means of providing for treaty obedience. A study in the law of treaties*. Stockholm.

Kwiatkowska, B. (1993). Economic and environmental considerations in maritime boundary delimitations. In J. I. Charney & L. M. Alexander (Eds.), *International maritime boundaries* (Vol. 1, pp. 75–113). Dordrecht, Boston, London: Martinus Nijhoff Publishers.

Kwiatkowska, B. (1991). Creeping jurisdiction beyond 200 miles in the light of the 1982 *Law of the Sea* convention and state practice. *Ocean Development and International Law*, 22(2), 159–187.

Kwiatkowska, B. (2001a). Equitable maritime boundary delimitation: a legal perspective. In *Law of the sea*. (pp. 241–258). Dartmouth, Aldershot, Burlington US, Singapore, Sydney: Ashgate.

Kwiatkowska, B. (2001b). The Eritrea-Yemen Arbitration: landmark progress in the acquisition of territorial sovereignty and equitable maritime boundary delimitation. *Ocean Development and International Law*, *32*(1), 1–25.

Kwiatkowska, B. (1989). *The 200 mile exclusive economic zone in the new Law of the Sea*. Dordrecht, Boston, London: Martinus Nijhoff Publishers.

Kwiatkowska, B. (2002). The Qatar v. Bahrain maritime delimitation and territorial questions case. *Ocean Development and International Law Journal*, *33*(3,4), 227–262.

Kwiatkowska, B. (2003) The Qatar v. Bahrain maritime delimitation and territorial questions case. *Maritime Boundaries*, *3*(6), 1–44.

Kwiatkowska, B. (2004). The contribution of the 2002 (Cameroon v. Nigeria: Equatorial Guinea intervening) land and maritime boundary between Cameroon and Nigeria judgment of the International Court of justice to equitable maritime boundary delimitation. *The Hague Yearbook of International Law*, *17*, 95–135.

Kwiatkowska, B., & Soons, A. H. A. (1990). Entitlement to maritime areas of rocks which cannot sustain human habitation or economic life of their own. *Netherlands Yearbook of International Law*, *21*, 139–181.

Kwon, P. H. (2000). *The Law of the Sea and Northeast Asia. A challenge for cooperation* (pp. 89–97). The Hague, London, Boston: Kluwer Law International.

Lagoni, R. (1979). Oil and gas deposits across national frontiers. *American Journal of International Law*, *73*(1), 215–243.

Lagoni, R. (1984). Interim measures pending maritime delimitation agreements. *American Journal of International Law*, *78*(2), 345–368.

Lagoni, R. (2006). Festlandsockel. In W. G. Vitzthum (Ed.), *Handbuch des Seerechts* (pp. 161–221). Munich: C.H. Beck.

Lagoni, R. (2012). Cables and pipeline survey at sea. In H. P. Hestermeyer, D. König, N. Matz-Lück, V. Röben, A. Seibert-Fohr & P.-T. Stoll et al. (Eds.), *Coexistence, cooperation and solidarity, liber amicorum rüdiger wolfrum* (Vol. 1, pp. 933–955). Leiden, Boston: Martinus Nijhoff Publishers.

Lagoni, R. (1989). Overlapping claims to continental shelf areas (preconditions and possible solutions). *The aegean issues: problems and prospects.* Foreign Policy Institute.

Lauterpracht, H. (1950). Sovereignty over submarine areas. *British Yearbook of International Law*, *27*, 376–433.

Leanza, U. La délimitation du plateau continental et de la zone économique exclusive: une introduction. In D. Pharand & U. Leanza (Eds.), *The continental shelf and the exclusive economic zone/Le Plateau Continental et la Zone Économique Exclusive* (pp. 37–54). Dordrecht, Boston, London: Martinus Nijhoff Publishers.

Leanza, U. The delimitation of the continental shelf of the Mediterranean Sea. *International Journal of Marine and Coastal Law*, 8(3), 373–395.

Lee, S.-H. (1990). Korea's claims to maritime jurisdiction. *Korean Journal of Comparative Law,* 62–89.

Lee, S., & Park, Y. K. (2013). Maritime delimitation and joint resource development in the East China Sea. In *Securing the Ocean for the Next Generation*, Papers from the Law of the Sea Institute, UC Berkeley-Korean Institute of Ocean Science and Technology Conference, held in Seoul, Korea, May 2012. http://www.law.berkeley.edu/files/Lee-Park-final.pdf. Accessed Aug 2013.

Lévy, J.-P. Ocean management and a new Law of the Sea. *Ocean Management*, *1*, 129–143.

Lévy, J.-P. (1995-1). Les Nations Unies et la Convention de 1982 sur le Droit de la Mer. *Revue Belge de Droit International*, 11–31.

Lewis, V. A., & Challenger, B. (1988). Regional cooperation and ocean development—the OECS experience. in: E. Gold (Ed.), *A new Law of the Sea for the caribbean. An examination of marine law and policy. Issues in the lesser antilles* (pp. 249–265). Berlin, Heidelberg, New York, London, Paris, Tokyo: Springer.

Li, J., & Amer, R. (2012). Recent practices in dispute management in the South China Sea. In C. Schofield (Ed.), *Maritime energy resources in asia: legal regimes and cooperation* (pp. 79–103). Seattle: The National Bureau of Asian Research.

Lucchini, L. (2006). La délimitation des frontières maritimes dans la jurisprudence internationale: vue d'ensemble. In R. Lagoni & D. Vignes (Eds.), *Maritime délimitation* (pp. 1–18). Leiden, Boston: Martinus Nijhoff Publishers.

Lucchini, L., & Voelckel, M. (1990). *Droit de la Mer: La Mer et son Droit, Les espaces maritimes* (Vol. 1). Paris: A. Pedone.

Lucchini, L., & Voelckel, M. (1996). *Droit de la Mer: Délimitation, Navigation et Pêche* (Vol. 2). Paris: A. Pédone.

Lupinacci, J. C. (1984). The legal status of the exclusive economic zone in the 1982 convention on the *Law of the Sea*. In F. O. Vicuña (Ed.), *The exclusive economic zone. A latin american perspective* (pp. 75–121). Boulder Colorado: Westview Press.

MacDonald, J. M. (1994). Artificial reef debate: habitat enhancement or waste disposal?. *Ocean Development and International Law*, *25*(1), 87–118.

MacLeod, N. (2007). Unitisation. In G. Gordon & J. Paterson (Eds.), *Oil and gas law. Current practice and emerging trends* (pp. 303–333). Dundee: Dundee University Press.

Maffei, M. C. (1993). The case of East Timor before the International Court of Justice—some tentative comments. *European Journal of International Law*. (pp. 223–238).

Maljean-Dubois, S., & Rajamani, L. (2011). *La mise en oeuvre du droit international de l'environnement/Implementation of international environmental law* Leiden, Boston: Martinus Nijhoff Publishers.

Manin, P. (1978). Le Traité de Frigg. *Annuaire Français de Droit International*, *24*, 792–809.

Marffy, A. de (1996). Recent practice of states in the Baltic Sea. In R. Platzöder & P. Verlaan (Eds.), *The Baltic Sea: new developments in national policies and international cooperation* (pp. 91–100). The Hague, London, Boston: Martinus Nijhoff Publishers.

Marshall, J. H., & Meyers, N. L. (1931). Legal planning of petroleum production. *Yale Law Journal*, *41*, 33–68.

Masterson, W. D. Jr. (1952). A 1952 survey of gas and oil law. *Southwestern Law Journal*, *6*, 1–46.

Matz-Lück, N. (2010). The concept of the common heritage of mankind: its viability as a management tool for deep-sea genetic resources. In A. G. Oude Elferink & E. J. Molenaar (Eds.), *The international legal regime of areas beyond national jurisdiction: current and future developments* (pp. 61–75). Leiden, Boston: Martinus Nijhoff Publishers.

McDonald, S. L. (1973–1974). Unit operation of oil reservoirs as an instrument of conservation. *Notre Dame Law*, *49*, 305–316.

McDorman, T. L. (2002). The role of the commission on the limits of the continental shelf: A technical body in a political world. *The International Journal of Marine and Coastal Law*, *17*(3), 301–324.

McDorman, T. L. (1993). The South China Sea islands dispute in the 1990s - a new multilateral process and continuing friction. *International Journal of Marine and Coastal Law*, *8*(2), 263–282.

McDorman, T. L. (2004). Malaysia-Vietnam. In J. I. Charney & L. M. Alexander (Eds.), *International maritime boundaries* (Vol. 3, pp. 2335–2344). The Hague, Boston, London: Martinus Nijhoff Publishers.

McDorman, T. L. (2005). People's Republic of China-Vietnam. In D. A. Colson & R. W. Smith (Eds.), *International maritime boundaries* (Vol. 5, pp. 3745–3753). Leiden, Boston: Martinus Nijhoff Publishers.

McDorman, T. L. (1990). The Canada-France maritime boundary case: drawing a line around St. Pierre and Miquelon. *American Journal of International Law*, *84*, 157–189.

McDorman, T. L. (2002). The role of the Commission on the Limits of the Continental Shelf: a technical body in a political world. *International Journal of Marine and Coastal Law*, *17*(3), 301–324.

McDougal, M. S., & Burke, W. T. (1962). *The public order of the oceans. A contemporary international Law of the Sea*. New Haven, London: Yale University Press.

McLaughlin, R. J. (2009). Maritime boundary delimitation and cooperative management of transboundary hydrocarbons in the ultra-deepwaters of the Gulf of Mexico. In S.-Y. Hong & J. V. Dyke (Eds.), *Maritime boundary disputes, settlement processes, and the Law of the Sea* (pp. 199–230). Martinus Nijhoff Publishers.

McRae, D. M. Delimitation of the continental shelf between the United Kingdom and France: the channel arbitration. *Canadian Yearbook of International Law*, *15*, 173–197.

Mehdiyoun, K. (2000). Ownership of oil and gas resources in the Caspian Sea. *American Journal of International Law*, *94*(1), 179–189.

Mendelson, M. The formation of customary international law. *Recueil des Cours de l'Académie de Droit International*, *272*, 159–410.

Mendelson, M. (1995). The subjective element in customary international law. *British Yearbook of International Law*, *66*, 177–208.

Mendelson, M. (2001). The curious case of Qatar v. Bahrain in the International Court of Justice. *British Yearbook of International Law*, *72*, 183–211.

Mendelson, M. (2002). On the quasi normative effect of maritime boundary agreements. In N. Ando, E. McWhinney & R. Wolfrum (Eds.), *Liber amicorum Judge Shigeru Oda* (Vol. 2, pp. 1069–1086). The Hague, London, New York: Kluwer Law International.

Mendelson, M. (2004). The Cameroon-Nigeria case in the ICJ: some territorial sovereignty and boundary delimitation issues. *British Yearbook of International Law*, *75*, 223–247.

Merrill, M. H. (1953). Recent unitization cases. *Oklahoma Law Review*, *6* 169–170.

Merrill, M. H. (1940–1941). The evolution of oil and gas law. 13 *Mississippi Law Journal* (1940-1941), pp. 281-291.

Mito, L. A. (1998). The Timor Gap treaty as a model for joint development in the Spratly islands. *American University Law Review*, *13*(3), 727–764.

Miyoshi, M. (1999). The joint development of offshore oil and gas in relation to maritime boundary delimitation. *Maritime Briefing/International Boundaries Research Unit*, *2*(5), 1–53.

Miyoshi, M. (1993). Considerations of Equity in the settlement of territorial and boundary disputes. In *International law in japanese perspective* (Vol. 2). Dordrecht, Boston, London: Martinus Nijhoff Publishers.

Miyoshi, M. (2002). Considerations of equity in maritime boundary cases before the International Court of Justice. In N. Ando, E. McWhinney & R. Wolfrum (Eds.), *Liber Amicorum Judge Shigeru Oda*, (Vol. 2, pp. 1087–1101). The Hague, London, New York: Kluwer Law International.

Miyoshi, M. (1988). The basic concept of joint development of hydrocarbon resources on the continental shelf. *International Journal of Estuarine and Costal Law*, *3*(1), 1–16.

Miyoshi, M. (1985). The Japan-South Korea agreement on joint development of the continental shelf. In M. J. Valencia (Ed.), *Geology and hydrocarbon of the South China Sea and possibilities of joint development, Proceedings of the Second EAPI/CCOP Workshop. East-West Center, Honolulu, Hawaii, 22–26 August 1983* (pp. 545–553). New York, Oxford, Toronto, Sydney, Paris, Frankfurt: Pergamon Press.

Miyoshi, M. (1983). Is joint development possible in the South China Sea? In Mochtar Kusuma-Atmadja, Thomas A. Mensah and Bernard H. Oxman (Eds.), *Sustainable development and preservation of the oceans: The challenges of UNCLOS and agenda* (Vol. 21, pp. 610–624). The Law of the Sea Institute/University of Hawaii.

Miyoshi, M. (1990). The Japan/South Korea joint development agreement 1974. In Hazel Fox (Ed.), *Joint development of offshore oil and gas. The institute's revised model agreement. Conference papers. The Australia/Indonesia zone of co-operation treaty 1989.* (pp. 89–97). London: The British Institute of International and Comparative Law.

Miyoshi, M. (2009). Some thoughts on maritime boundary delimitation. In S.-Y. Hong & J. V. Dyke (Eds.), *Maritime boundary disputes, settlement processes, and the Law of the Sea.* (pp. 107–118). Martinus Nijhoff Publishers.

Moses, L. (1942–1943). Some legal and economic aspects of unit operations of oil fields. *Texas Law Review, 21*, 748–772.

Mouton, M.W. (1954-I). The continental shelf. *Recueil des Cours de l'Académie de Droit International, 85*, 347–463.

Murphy, B. M. (1946–1947). The legislative and administrative concept of oil and gas conservation in Arkansas, 1917-1947. *Arkansas Law Review 1*(236), 236–247.

Murphy, B. K. (1994). Dangerous ground: the Spratly islands and international law. *Ocean and Coastal L. J, 1*(1), 187–212.

Nakauchi, K. (1979). Problems of delimitation in the East China Sea and the Sea of Japan. *Ocean Development and International Law Journal, 6*, 305–316.

Nelson, L. D. M. (1990). The roles of equity in the delimitation of maritime boundaries. *American Journal of International Law, 84*(4), 837–858.

Nelson, L. D. M. (1988). The delimitation of maritime boundaries in the Caribbean. In D. M. Johnston & Phillip M. Saunders (Eds.), *Ocean boundary making: Regional issues and developments* (pp. 142–203). London, New York, Sydney: Croom Helm.

Nong, H., & Shicun, W. (2006). The energy security of China and oil and gas exploitation in the South China Sea. In M. H. Nordquist, J. N. Moore & K.-c. Fu (Eds.), *Recent developments in the Law of the Sea and China* (pp. 145–154). Leiden, Boston: Martinus Nijhoff Publishers.

Nordholt, H. S. (1985). Delimitation of the continental shelf in the East China Sea. *Netherlands International Law Review, 32*, 123–160.

Nordquist, M. H. (2012). Textual interpretation of article 121 in the UN Convention on the Law of the Sea. In H. P. Hestermeyer, D. König, N. Matz-Lück, V. Röben, A. Seibert-Fohr & P.-T. Stoll et al. (Eds.), *Coexistence, cooperation and solidarity. Liber Amicorum Rüdiger Wolfrum*, (Vol. 1, pp. 991–1035). Leiden, Boston: Martinus Nijhoff Publishers.

Nweihed, K. G. (1980). EZ(uneasy) delimitation in the semi-enclosed Caribbean Sea: recent agreements between Venezuela and her neighbours. *Ocean Development and International Law, 8*(1), 1–33.

Oda, S. (1989). *International control of sea resources.* Dordrecht, Boston, London: Martinus Nijohff Publishers.

Oda, S. (1977). *The Law of the Sea in our time—I. New developments 1966–1975.* Leyden: Sijthoff.

Oda, S. (1982). Sharing of ocean resources-unresolved issues in the Law of the Sea. In R.-J. Dupuy (Ed.), *La Gestion des Ressources pour l'Humanité: Le Droit de la Mer/The management of humanity's resources: The Law of the Sea.* The Hague, Boston, London: Martinus Nijhoff Publishers.

Oda, S. (2001). Fisheries under the United Nations Convention on the Law of the Sea. In H. Caminos (Ed.), *Law of the Sea.* (pp. 283–299). Aldershot, Burlington US, Singapore, Sydney: Ashgate/Dartmouth.

Odell, P. R. (1978). Oil and gas exploration and exploitation in the North Sea. *Ocean Yearbook, 1*, 139–159.

Oduntan, G. (2008). The emergent legal regime for exploration of hydrocarbons in the Gulf of Guinea: imperative considerations for participating States and multinationals. *International Comparative Law Quarterly*, *57*, 253–302.

Ogley, R. C. (1981). The law of the sea draft convention and the new international economic order. *Marine Policy*, *5*(3), 240–251.

Ong, D. (2000). The legal status of the 1989 Australia-Indonesia Timor Gap Treaty following the end of Indonesian rule in East Timor. *Netherlands Yearbook of International Law*, *31*, 67–129.

Ong, D. (2002). The new Timor Sea Arrangement 2001: is joint development of common offshore oil and gas deposits mandated under International Law? *International Journal of Marine and Coastal Law*, *17*(1), 79–122.

Ong, D. (1995). South-east Asian state practice on the joint development of offshore oil and gas deposits. G. H. Blake, W. J. Hildesley, M. A. Pratt, R. J. Ridley & C. H. Schofield (Eds.), *The peaceful management of transboundary resources* (pp. 77–96). London, Dordrecht, Boston: Graham and Trotman/Martinus Nijhoff.

Ong, D. (2003). The progressive integration of environmental protection within offshore joint development agreements. In M. Fitzmaurice & M. Szuniewicz (Eds.), *Exploitation of natural resources in the 21st century* (pp. 113–141). The Hague, London, New York: Kluwer Law International.

Ong D. (1999a). Joint development of common offshore oil and gas deposits: "mere" state practice or customary International Law? *American Journal of International Law*, *93*(4), 771–804.

Ong D. (1999b). The 1979 and 1990 Malaysia-Thailand joint development agreements: a model for international legal co-operation in common offshore petroleum deposits? *International Journal of Marine and Coastal Law*, *14*(2), 207–246.

Ong D. (1991). Thailand/Malaysia. Joint development agreement 1990. *International Journal of Estuarine and Coastal Law*, *6*(1), 57–60.

Onorato, W. T. (1985). Joint development of seabed hydrocarbon resources: an overview of precedents in the North Sea. In M. J. Valencia (Ed.), *The south china sea. hydrocarbon potential and possibilities of joint development* (pp. 1311–1324). New York, Oxford, Toronto, Sydney, Paris, Frankfurt: Pergamon Press.

Onorato, W. T. (1990). Joint development in the international petroleum sector: the Yemeni variant. *International and Comparative Law Quarterly*, *39*, 653–662.

Onorato, W. T. (1968). Apportionment of an international common petroleum deposit. *International and Comparative Law Quarterly*, *17*, 85–102.

Onorato, W. T. (1977). Apportionment of an international common petroleum deposits. *International and Comparative Law Quarterly*, *26*, 324–337.

Onorato, W. T. (1985). A case study in joint development: the Saudi Arabia-Kuwait portioned neutral zone. In M. J. Valencia (Ed.), *Geology and hydrocarbon potential of the South China Sea and possibilities of joint development* (pp. 539–544). New York, Oxford, Toronto, Sydney, Paris, Frankfurt: Pergamon Press.

Onorato, W. T. (2001). The Timor Gap Treaty revisted ten years on. In S. Schlemmer-Schulte & K.-Y. Tung (Eds.), *International finance and development law. Liber Amicorum Ibrahim F.I. Shihata* , (pp. 527–547). The Hague, London, Boston, New York: Sabine Kluwer Law International.

Onorato, W. T. (1986). Promoting foreign investment through international petroleum joint development regimes. *International Centre for Settlement of Investiment Disputes Review—Foreign Investment Law Journal*, *1*(1), 81–88.

Onorato, W. T., & Valencia, M. J. (2000). The new Timor Gap Treaty: legal and political implications. *International Centre for Settlement of Investiment Disputes Review—Foreign Investment Law Journal*, *28*, 59–85.

Onorato, W. T., & Valencia, M. J. (1990). International cooperation for petroleum development: the Timor Gap Treaty. *International Centre for Settlement of Investiment Disputes Review—Foreign Investment Law Journal, 5*(1), 1–29.

Organization of the Petroleum Exporting Countries. (2011). In J. Griffin (Ed.), *World oil outlook 2011*. OPEC Secretariat.

Orrego Vicuña, F. (1976). *Francisco Los Fundos Marinos y Oceanicos. Jurisdicción nacional y Régimen internacional*. Santiago: Editorial Andres Bello.

Orrego Vicuña, F. (1989). *Francisco the exclusive economic zone. Regime and legal nature under international law*. Cambridge, New York, Port Chester, Melbourne, Sydney: Cambridge University Press.

Orrego Vicuña, F. (1999). *Francisco the changing international law of high seas fisheries*. Cambridge: Cambridge University Press.

Orrego Vicuña, F. (1990). La zone économique exclusive dans la législation et la pratique des États. In J. Combacau & P.-M. Dupuy (Eds.), *Droit de La Mer* (Vol. 2, pp. 5–56). Paris: Pedone.

Orrego Vicuña, F. (1994). Regional cooperation in non-living resources: joint management zones. In P. B. Payoyo (Ed.), *Ocean governance: sustainable development of the seas* (pp. 171–182). Tokyo, New York, Paris: United Nations University Press.

Østreng, W. (1985). Reaching agreement on international exploitation of ocean mineral resources (with special reference to the joint development area between Jan Mayen and Iceland). In Mark J. Valencia (Ed.), *Geology and hydrocarbon potential of the South China Sea and possibilities of joint development* (pp. 555–571). New York, Oxford, Toronto, Sydney, Paris, Frankfurt: Pergamon Press.

Oude Elferink, A. G. (2007). The regime of the area: delineating the scope of application of the common heritage principle and freedom of the high seas. *International Journal of Marine and Coastal Law, 22*(1), 143–176.

Oude Elferink, A. G. (2001). The Islands in the South China Sea: how does their presence limit the extent of the high seas and the Area and the maritime zones of the mainland coasts? *Ocean Development and International Law, 32*(2), 169–191.

Oude Elferink, A. G. (1994). In Shigeru Oda (Ed.), *The law of maritime boundary delimitation: A case study of the Russian Federation*. Dordrecht, Boston, LondonMartinus Nijhoff Publishers.

Oude Elferink, G. (2013). *The delimitation of the continental shelf between Denmark, Germany and the Netherlands. Arguing law, practicing politics?*. Cambridge: Cambridge University Press.

Oxman, B. (1995). Law of the sea forum: the 1994 Agreement on implementation of the seabed provisions of the Convention on the Law of the Sea. The 1994 Agreement and the Convention. *American Journal of International Law, 88*(4), 687–696.

Oxman, B. H. (1993). Political, strategic, and historical considerations. In J. I. Charney & L. M. Alexander (Eds.), *International maritime boundaries*, (Vol. 1, pp. 3–40). Dordrecht, Boston, London: Martinus Nijhoff Publishers.

Park, C.-H. (1981). Joint development of mineral resources in disputed waters: the case of Japan and South Korea in the East China Sea. In Mark J. Valencia (Ed.), *The South China Sea. hydrocarbon potential and possibilities of joint development* (pp. 1335–1354). Oxford, New York, Toronto, Sydney, Paris, Frankfurt: Pergamon Press.

Park, C.-H. (1988). *East Asia and the Law of the Sea*. Seoul: Seoul National University Press.

Park, C.-H. (1979). China and maritime boundary issue. *German Yearbook of International Law, 22*, 119–141.

Park, C. -H. (1973). Oil under troubled waters: the Northeast Asia sea-bed controversy. *Harvard International Law Journal, 14*, 212–260.

Payoyo, P. B. (1997). In Shigeru Oda (Ed.), *Cries of the Sea. World inequality, sustainable development and the common heritage of humanity*. The Hague, London, Boston: Martins Nijhoff Publishers.

Peter, H. S. (2007). The evolution of international environmental law. In D. Bodansky, J. Brunnée, & E. Hey (Eds.), *The Oxford handbook of international environmental law* (pp. 30–43). Oxford: Oxford University Press.

Peters, P., Soons, A. H. A., & Zima, L. A. (1984). Removal of installations in the Exclusive Economic Zone. *Netherlands Yearbook of International Law*, *15*, 167–207.

Pharan, D. (1993). International decisions. Affaire du plateau continental de la Mer du Nord (1969). In D. Pharand & U. Leanza (Eds.), *The continental shelf and the exclusive economic zone/Le Plateau Continental et la Zone Économique Exclusive* (pp. 95–104). Dordrecht, Boston, London: Martinus Nijhoff Publishers.

Pierce, D. E. (1983). Coordinated reservoir development—an alternative to the rule of capture for the ownership and development of oil and gas—part I. *Journal of Energy Law and Policy*, *4*, 50–55.

Pierce, R. J. (1987–1988). State regulation of natural gas in a federally deregulated market: the tragedy of the commons revisited. *Cornell Law Review*, *73*, 15–53.

Pinto, M. C. W. (2002). Maritime boundary issues and their resolution an overview. In N. Ando, E. McWhinney & R. Wolfrum (Eds.), *Liber Amicorum Judge Shigeru Oda*, (Vol. 2, pp. 1115–1142). The Hague, London, New York: Kluwer Law International.

Post, A. M. (1983). In S. Oda (Ed.), *Deepsea mining and the Law of the Sea*. The Hague, Boston, Lancaster: Martins Nijhoff Publishers.

Prescott, J. R. V. (1975). *The political geography of the oceans. Problems in modern geography*. London, Vancouver: David and Charles Newton Abbot.

Prescott, J. R. V. (2004). Guinea-Bissau-Senegal. In J. I. Charney & L. M. Alexander (Eds.), *International maritime boundaries*, (Vol. 3, pp. 2250–2255). The Hague, Boston, London: Martinus Nijhoff Publishers.

Prescott, V., & Schofield, C. (2005). *The maritime political boundaries of the world*. Leiden, Boston: Martinus Nijhoff Publishers.

Prescott, V., & Schofield, C. (2001). Undelimited maritime boundaries of the Asian Rim in the Pacific Ocean. *Maritime Boundaries*, *3*(1), 1–68.

Pressler, H. P. Jr. (1945–1946). Legal problems involved in cycling gas in gas fields. *Texas Law Review*, *24*, 29–32.

Prewett, B. (1950–1951). Act 134 compulsory unitization of oil and gas pools. *Arkansas Law Review*, *5*, 392–394.

Proel*β*, A. (2006). Ausschlie*β*liche Wirtschaftszone. In W. G. Vitzthum (Ed.), *Handbuch des Seerechts* (pp. 222–263). Munich: C.H. Beck.),

Queneudec, J.-P. (1989). L'arbitrage relatif à la détermination de la frontière maritime entre la Guinée-Bissau et le Sénégal 31 Juillet 1989. *Annuaire Français de Droit International*, *35*, 323–339.

Queneudec, J.-P. (1981). Note sur l'arret de la cour internationale de justice relatif a la délimitation du plateau continental entre la Tunisie et la Libye. *Annuaire Français de Droit International*, *27*, 201–212.

Queneudec, J.-P. (1979). L'affaire de la délimitation du plateau continental entre la France et le Royaume-Uni. *Revue Générale de Droit International Public*, *83*, 53–103.

Razavi, A. (1997). In S. Oda (Ed.), *Continental shelf delimitation and related maritime issues in the Persian Gulf*. The Hague, Boston, London: Martinus Nijhoff Publishers.

Redgwell, C. (2010). International environmental law. In M. D. Evans (Ed.), *International law*. Oxford: Oxford University Press.

Redgwell, C. (2007). National implementation. In D. Bodansky, J. Brunnée, & E. Hey (Eds.), *The Oxford handbook of international environmental law*. Oxford: Oxford University Press.

Redgwell, C. (2006). From permission to prohibition: The 1982 Convention on the Law of the Sea and Protection of the Marine Environment. In D. Freestone, R. Barnes, & D. Ong (Eds.), *The Law of the Sea: Progress and prospects*. Oxford: Oxford University Press.

Rhee, S.-M. (1981). Equitable solutions to the maritime boundary dispute between the United States and Canada in the Gulf of Maine. *American Journal of International Law*, *75*, 590–628.

Richardson, E. L. (1988). Jan Mayen in perspective. *American Journal of International Law, 82*, 443–458.

Robinson, D. R., Colson, D. A., & Rashkow, B. C. (1985). Some perspectives on adjudicating before the world court: the Gulf of Maine case. *American Journal of International Law, 79*(3), 578–597.

Robson, C. (1995). Transboundary petroleum reservoirs: Legal issues and solutions. In G. H. Blake, W. J. Hildesley, M. A. Pratt, R. J. Ridley & C. H. Schofield (Eds.), *The peaceful management of transboundary resources* (pp. 3–21). London, Dordrecht, Boston: Graham and Trotman/Martinus Nijhoff.

Rodgers, A. B., & Utton, A. E. (1985). The Ixtapa draft agreement relating to the use of transboundary groundwaters. *Natural Resources Journal, 25*, 713–722.

Rogers, R. G., & Gault, E. S. (1973). Mississippi compulsory field-wide unitization. *Mississippi Law Journal 44*, pp. 185–207.

Rogoff, M. A. (1994). The obligation to negotiate in international law: rules and realities. *Michigan Journal of International Law, 16*, 141–185.

Rosenne, S. (2001). Geography in international maritime boundary-making. In H. Caminos (Ed.), *Law of the Sea* (pp. 225–258). Dartmouth, Aldershot, Burlington US, Singapore, Sydney: Ashgate.

Rosenne, S. (2007). Dispute management and resolution. In G. Gordon & J. Paterson (Eds.), *Oil and gas law. Current practice and emerging trends* (pp. 431–460). Dundee: Dundee University Press.

Rosseau, C. (1980). *Droit international public. Les relations internationales* (Vol. 4). Paris: Sirey.

Rostow, E. V. (1949). A reply. *Journal of Political Economy, 57*(1), 60–68.

Rothwell, D. R. (1996). The law of the sea in the Asian-Pacific region: An overview of trends and developments. *Proceedings of the International Law Association (ILA) First Asian-Pacific Regional Conference* (pp. 52–81). Chinese Society of International Law.

Rothwell, D. R., & Stephens, T. (2010). *The international Law of the Sea*. Oxford and Portland, Oregon: Hart Publishing.

Sands, P. (2003). *Principles of international environmental law*. Cambridge: Cambridge University Press.

Scelle, G. (1955). *Plateau Continental et Droit International*. Paris: A. Pedone.

Schicun, W. (2013). Solving disputes for regional cooperation and development in the South China sea: A Chinese perspective. Oxford, Cambridge, Philadelphia, New Delhi: Chandos Publishing.

Schneider, J. (1985). The Gulf of Maine Case: The nature of an equitable result. *American Journal of International Law, 79*(3), 539–577.

Schofield, C. (2009). The trouble with islands: the definition and role of islands and rocks in maritime delimitation. In S.-Y. Hong & J. V. Dyke (Eds.), *Maritime Boundary disputes, settlement processes, and the Law of the Sea* (pp. 19–38). Martinus Nijhoff Publishers.

Schofield, C., & Tan-Mullins, M. (2008). Maritime claims, conflicts and cooperation in the Gulf of Thailand. *Ocean Yearbook, 22*, 75–116.

Schofield, C., & Wang, D. K.-H. (2012). The regime of islands under UNCLOS: implications for the South China Sea. In C. Schofield (Ed.), *Maritime energy resources in Asia: Legal regimes and cooperation* (pp. 61–77). SeattleThe National Bureau of Asian Research.

Schrijver, N. (1997). *Sovereignty over natural resources: Balancing rights and duties*. Cambridge: Cambridge University Press.

Scovazzi, T. (2001). The evolution of international law of the sea: new issues, new challenges. *Recueil des Cours de l'Académie de Droit International, 286*, 39–244.

Shaw, M. (2003). *International law*. Cambridge: Cambridge University Press.

Sheppard, J., & Chakalall, B. (1988). Regional planning to protect and develop the Caribbean marine environment. In E. Gold (Ed.), *A new Law of the Sea for the Caribbean. An examination of marine law and policy. Issues in the Lesser Antilles* (pp. 163–178). Berlin, Heidelberg, New York, London, Paris, Tokyo: Springer.

Shicun, W. (2007). "Joint development": An ad hoc solution to the South China Sea Dispute. *China Oceans Law Review, 2*, 1–10.

Shihata, I. F. I., & Onorato, W. T. (1988). Joint development of international petroleum resources in undefined and disputed areas. In G. Blake, M. Pratt, C. Schofield & J. A. Brown (Eds.), *Boundaries and energy: Problems and prospects* (pp. 433–451). London/The Hague/Boston.

Smith, B. D. (1988). *State responsibility and the marine environment: The rules of decision.* Oxford: Clarendon Press.

Smith, E. (1965). The Texas compulsory pooling act. *Texas Law Review, 43*, 1003–1034.

Smith, E. (1966). The Texas compulsory pooling act. *Texas Law Review, 44*, 387–423.

Smith, R. W. (1983). Joint (development) zones: a review of past practice and thoughts on the future. In M. Kusuma-Atmadja, T. A. Mensah & B. H. Oxman (Eds.), *Sustainable development and preservation of the oceans: The challenges of UNCLOS and agenda* (Vol. 21, pp. 654–662). The Law of the Sea Institute/University of Hawaii.

Song, Y.-h. (2000). One China, but two sets of maritime legislation: Developments, implications, and challenges for the United States. In H. N. Scheiber (Ed.), *Law of the Sea. common heritage and emerging challenges* (pp. 209–247). The Hague, London, Boston: Martinus Nijhoff Publishers.

Song, Y.-h. (2009). "Okinotorishima: A "rock" or and "island"? Recent maritime controversy between Japan and Taiwan/China. In S.-Y. Hong & J. V. Dyke (Eds.), *Maritime boundary disputes, settlement processes, and the Law of the Sea* (pp. 145–176). Martinus Nijhoff Publishers.

Song, Y.-h. (2013). Article 121(3) of the law of the sea convention and the disputed offshore islands in East Asia: A tribute to Judge Choon-Ho Park. In J. V. Dyke, S. P. Broder, S. Lee & J.-H. Paik (Eds.), *Governing ocean resources. New challenges and emerging regimes. A tribute to Judge Choon-Ho Park* (pp. 61–97). Leiden, Boston: Martinus Nijhoff Publishers.

Song, Y.-h., & Zou, K. (2000) Maritime legislation of mainland China and Taiwan: Developments, comparison, implications, and potential challenges for the United States. *Ocean Development and International Law Journal, 31*(4), 303–345.

Strati, A. (1990). Potential areas for joint development. In In H. Fox (Ed.), *Joint development of offshore oil and gas. The institute's revised model agreement. conference papers. The Australia/Indonesia zone of co-operation treaty 1989* (Vol. II, pp. 109–122). London: The British Institute of International and Comparative Law.

Styles, S. (2007). Joint operating agreements. In G. Gordon & J. Paterson (Eds.), *Oil and gas law. Current practice and emerging trends* (pp. 265–301). Dundee: Dundee University Press.

Suarez, S. (2008). *The outer limits of the continental shelf. Legal aspects of their establishment.* Berlin, Heidelberg, New York: Springer.

Suter, K. (1993). Timor Gap treaty. The continuing controversy. *Marine Policy, 17*(4), 294–302.

Symonides, J. (1984). Status of the enclosed and semi-enclosed seas. *German Yearbook of International Law, 27*, 315–333.

Székely, A. (1986). The international law of submarine transboundary hydrocarbon resources: legal limits to behaviour and experiences for the Gulf of Mexico. *Natural Resource Journal, 26*, 733–768.

Tanaka, Y. (2008). *A dual approach to ocean governance: The cases of zonal and integrated management in international Law of the Sea.* Surrey, Burlington: The Ashgate International Law Series.

Tanaka, Y. (2004a). Zonal and integrated management approaches to ocean governance: reflections on a dual approach in international Law of the Sea. *International Journal of Marine and Coastal Law, 19*(4), 483–514.

Tanaka, Y. (2004b). Reflections on maritime delimitation in the Cameroon/Nigeria case. *International and Comparative Law Quarterly, 53*, 369–406.

Tanaka, Y. (2012). *The international Law of the Sea*. Cambridge: Cambridge University Press.
Tanaka, Y. (2003). Reflections on maritime delimitation in the Qatar/Bahrain case. *International and Comparative Law Quarterly, 52*, 53–80.
Tanaka, Y. (2004). Reflections on the Eritrea/Yemen Arbitration of 17 December 1999 (second phase: maritime delimitation). *Netherlands International Law Review, 48*, 197–225.
Tanja, G. J. (1990). *The legal determination of international maritime boundaries. The progressive development of continental shelf, EFZ and EEZ law*. Deventer, Boston: Kluwer Law and Taxation Publishers.
Taverne, B. (1994). *An introduction to the regulation of the petroleum industry. Laws, contracts and conventions, international energy and resources law and policy series*. London, Dordrecht, Boston: Graham and Trotman/Martinus Nijhoff.
Thomas, P. L. (1965–1966). Prospects for compulsory fieldwide unitization in Texas. *Texas Law Review, 44*, 510–532.
Townsend-Gault, I. (1990a). The impact of a joint development zone on previously granted interests. In In H. Fox (Ed.), *Joint development of offshore oil and gas. The institute's revised model agreement. conference papers. The Australia/Indonesia zone of co-operation treaty 1989* (pp. 174–184). London: The British Institute of International and Comparative Law.
Townsend-Gault, I. (1990b). The Malaysia/Thailand joint development arrangement. In H. Fox (Ed.), *Joint development of offshore oil and gas. The institute's revised model agreement. Conference papers. The Australia/Indonesia zone of co-operation treaty 1989* (Vol. II, pp. 102–107). London: The British Institute of International and Comparative Law.
Townsend-Gault, I. (1988a). Joint development of offshore mineral resources-progress and prospects for the future. *Natural Resources Forum, 12*(3), 275–285.
Townsend-Gault, I. (1988b). Offshore boundary delimitation in the Arabian/Persian Gulf. In D. M. Johnston & P. M. Saunders (Eds.), *Ocean boundary making: regional issues and developments* (pp. 204–228). London, New York, Sydney: Croom Helm.
Townsend-Gault, I. (1988c). Petroleum development offshore: legal and contractual issues. In N. Beredjick & T. Wälde (Eds.), *Petroleum investment policies in developing countries* (pp. 101–163). London: Graham and Trotman.
Townsend-Gault, I. (2012). Maritime cooperation in a functional perspective. In C. Schofield (Ed.), *Maritime energy resources in Asia: Legal regimes and cooperation* (pp. 7–21). Seattle: The National Bureau of Asian Research.
Townsend-Gault, I. (2002). Maritime boundaries in the Arabian Gulf. In C. Schofield, D. Newman, A. Drysdale & J. A. Brown (Eds.), *The Razor's edge: International boundaries and political geography* (pp. 223–236). London, The Hague, New YorkKluwer Law International.
Townsend-Gault, I., & Stormont, W. G. (1995). Offshore petroleum joint development arrangements: functional instrument? Compromise? Obligation? In G. H. Blake, W. J. Hildesley, M. A. Pratt, R. J. Ridley & C. H. Schofield (Eds.), *The peaceful management of transboundary resources* (pp. 51–76). London, Dordrecht, Boston: Graham and Trotman/ Martinus Nijhoff.
Treves, T. (1989). La limite extérieure du plateau continental: évolution récente de la pratique. *Annuaire Français de Droit International, 35*, 724–735.
Treves, T. (1989). Codification du droit international et pratique des États dans le droit de la mer. *Recueil de Cours de l'Académie de Droit Internationale de La Haye, 223*, 9–302.
Treviranus, H.-D. (1963). Der deutsch-niederländische Ems-Dollart-Vertrag. *Zeitschrift für Ausländisches Offentliches Recht und Völkerrecht, 23*, 536–553.
Triggs, G. (2004). The Timor Sea Treaty and the international agreement for the Greater Sunrise: practical solutions in the Timor Sea. *Australian Yearbook of International Law, 23*, 161–176.
United Nations. (1994). *The Law of the Sea. Practice of States at the time of entry into force of the United Nations Convention on the Law of the Sea*. New York: UN Publications.
United Nations. (1950). La doctrine du plateau continental et la pratique. *Yearbook of the International Law Commission, 2*, 108–113.

Utton, A. E. (1968). Institutional arrangements for developing North Sea oil and gas. *Virginia Journal of International Law, 9*(1), 66–81.

Utton, A. E., & McHugh, P. D. (1986). On an institutional arrangement for developing oil and gas in the Gulf of Mexico. *Natural Resource Journal, 26*, 717–732.

Valencia, M. J. (1985a). Joint jurisdiction and development in southeast Asia seas: factors and candidate areas. In M. J. Valencia (Ed.), *Geology and hydrocarbon potential of the South China Sea and possibilities of joint development* (pp. 573–579). New York, Oxford, Toronto, Sydney, Paris, Frankfurt: Pergamon Press.

Valencia, M. J. (1985b). *South-East Asian Seas: Oil under trouble waters. hydrocarbon potential, jurisdictional issues, and international relations*. Oxford, New York: Oxford University Press.

Valencia, M. J. (2000). Regional maritime regime building: prospects in northeast and southeast Asia. *Ocean Development and International Law Journal, 31*(3), 223–247.

Valencia, M. J. (1986). Taming troubled waters: Joint development of oil and mineral resources in overlapping claim areas. *San Diego Law Review Association, 23*(3), 661–684.

Valencia, M. J. (2001). Relevance of lessons learned to Northeast Asia. In M. J. Valencia (Ed.), *Maritime regime building. Lessons learned and their relevance for Northeast Asia* (pp. 131–148). The Hague, Boston, London: Martinus Nijhoff Publishers.

Valencia, M. J. (2005). Conclusions and the way forward. *Marine Policy, 29*(2), 185–187.

Valencia, Mark J. Conclusions, regime building and the way forward. 28-1 *Marine Policy* (January 2004), pp. 89–96.

Valencia, M. J. (2003). Regime building in the East China Sea. *Ocean Development and International Law Journal, 34*(1), 189–208.

Valencia, M. J. (1988). The Yellow Sea. Transnational marine resource management issues. *Marine Policy, 12*(4), 382–395.

Valencia, M. J. (1990). International conflict over maritime resources in South-East Asia: trends in politicization and militarization. In L. T. Gee & M. J. Valencia (Eds.), *Conflict over natural resources in South-East Asia and the Pacific, natural resources of South-east Asia* (pp. 94–144). Singapore, Oxford, New York: Oxford University Press.

Valencia, M. J. (1989). Northeast Asia: petroleum potential, jurisdictional claims, and international relations. *Ocean Development and International Law Journal, 20*(1), 35–61.

Valencia, M. J., & Miyoshi, M. (1986) Southeast Asia seas: joint development of hydrocarbons in overlapping claim areas. *Ocean Development and International Law Journal, 16*(3), pp. 211–254.

Valencia, M. J., & Van Dyke, J. (1998). Comprehensive solutions to the South China Sea disputes: some options. In G. Blake, M. Pratt, C. Schofield & J. A. Brown (Eds.), *Boundaries and energy: Problems and prospects* (pp. 85–115). London, The Hague, Boston: Kluwer Law International.

Valero, G. M. C. (1994). Spratly archipelago dispute. Is the question of sovereignty still relevant? *Marine Policy, 18*(4), 314–344.

Vallée, C. (1971). *Le Plateau Continental dans le Droit Positif Actuel*. Paris: A. Pedone.

Van Dyke, J. (2004). Giving teeth to the environmental obligations in the LOS Convention. In A. G. Oude Elferink & D. R. Rothwell (Eds.), *Oceans management in the 21st century: Institutional frameworks and responses*. (pp. 167–176). Leiden/Boston: Martinus Nijhoff Publishers.

Van Dyke, J. M. (2000). *Law of the Sea. The common heritage and emerging challenges*. H. N. Scheiber (Ed.), The Hague, London, Boston: Martinus Nijhoff Publishers.

Van Dyke, J. M. (2009). Disputes over islands and maritime boundaries in East Asia. In S.-Y. Hong & J. V. Dyke (Eds.), *Maritime boundary disputes, settlement processes, and the Law of the Sea* (pp. 39–75). Martinus Nijhoff Publishers.

Van Dyke, J. M. (2003). The Republic of Korea's maritime boundaries. *International Journal of Marine and Coastal Law, 18*(4), 509–540.

Van Dyke, J. M., & Bennett, D. L. (1993). Islands and the delimitation of the Ocean Space in the South China Sea. *Ocean Yearbook, 10*, 54–89.

Vanderzwagg, D. (2000). Transboundary challenges and cooperation in the Gulf of Maine region: Riding a restless sea toward misty shores. In H. N. Scheiber (Ed.), *Law of the Sea. Common heritage and emerging challenges* (pp. 265–285). The Hague, London, Boston: Martinus Nijhoff Publishers.
Vargas, J.A. (1979). The legal nature of the patrimonial sea: a first step towards the definition of the exclusive economic zone. *German Yearbook of International Law*, *22*, 142–177.
Veasey, J. A. (1919–1920). The law of oil and gas—part I. *Michigan Law Review*, *18*(445), 445–469.
Victor P. (1998). National rights to hydrocarbon resources of the continental margin beyond 200 nautical miles. In G. Blake, M. Pratt, C. Schofield & J. A. Brown (Eds.), *Boundaries and energy: Problems and prospects. International boundary studies series* (pp. 51–82). London, The Hague, Boston: Kluwer Law International.
von Heinegg, W. H. Treaties, fundamental change of circumstances. *Max Planck Encyclopedia of Public International Law* [online].
von Walter, A. Arbitration on oil concession disputes. *Max Planck Encyclopedia of Public International Law* [online].
Vuilleimier, J. (1957-2). Vers un régime juridique propre à la recherché et à l'exploitation des gisements de pétrole. *Zeitschrift für Schweizerisches Recht/Revue de Droit Suisse/Revista di Diritto Svizzero*, *76*, 185a–293a.
Wälde, T. (1988). Investment policies in the international petroleum industry—responses to the current crisis. In N. Beredjick & T. Wälde (Eds.), *Petroleum investment policies in developing countries* (pp. 7–27). London: Graham and Trotman.
Wälde, T. (1990). Financial and contractual perspectives in negotiating joint petroleum development agreements. In H. Fox (Ed.), *Joint development of offshore oil and gas. The institute's revised model agreement. Conference papers. The Australia/Indonesia zone of co-operation treaty 1989* (Vol. II, pp. 156–169). London: The British Institute of International and Comparative Law.
Walker A. W. Jr. (1937–1938). Property rights in oil and gas and their effect upon police regulation of production. *Texas Law Review*, *16*, 370–381.
Weaver, J. L., Asmus, D. F. (2006). Unitizing oil and gas fields around the world: a comparative analysis of national laws and private contracts. *Houston Journal of International Law*, *28*(3), 3–198.
Weil, P. (1911). Des espaces maritimes aux territoires maritimes: vers une conception territorialiste de la délimitation maritime. In M. M. Virally (Ed.), *Le Droit International au service de la Paix, de la Justice et du Développement* (pp. 501–511). Paris: Pedone.
Weil, P. (1988). *Perspectives du Droit de la délimitation maritime*. Paris: Pedone.
Wenk, E. (1973). International institutions for rational management of ocean space. *Ocean Management*, *1*, 171–200.
Wenk, E. (1981). Global principles for national marine policies: A challenge for the future. *Comparative marine policy: Perspectives from Europe, Scandinavia, Canada, and the United States* (pp. 3–20). New York: Praeger Publishers.
Wenk, E. (1972). *The politics of the ocean*. University of Washington Press.
Willheim, E. (1989). Australia-Indonesia sea-bed boundary negotiations: proposals for a joint development zone in the. Timor Gap. *Natural Resource Journal*, *29*, 821–842.
Williams, A. (2007). The Pacific Salmon Treaty: a historical analysis and prescription for the future. *Journal of Environmental Law and Litigation*, *22*, 153–195.
Williams, H. R. (1951–1952). Conservation of oil and gas. *Harvard Law Review 65*, 1155–1183.
Wolfrum, R. (2004). Legitimacy of international law and the exercise of administrative functions: the example of the International Seabed Authority, the International Maritime Organization (IMO) and the International Fisheries Organizations. *German Law Journal*, *9*(11), 2039–2059.
Wolfrum, R. (1978). Die Fischerei auf Hoher See. *eitschrift für Ausländisches Offentliches Recht und Völkerrecht*, *38*, 659–709.

Wolfrum, R. (1998). Means of ensuring compliance with and enforcement of international environmental law. *Recueil de Cours de L'Académie de Droit Internationale de La Haye*, 272, 9–154.

Wolfrum, R. (2006). Hohe See und Tiefseeboden (Gebiet). In W. G. Vitzthum (Ed.), *Handbuch des Seerechts* (pp. 287–345). Munich: Verlag C.H. Beck.

Wolfrum, R. (1984). *Die Internationalisierung staatsfreier Räume: die Entwicklung einer internationalen Verwaltung für Antarktis, Weltraum, Hohe See und Meeresboden*. Berlin: Springer.

Wolfrum, R. (1983). The principle of the common heritage of mankind. *Zeitschrift für Ausländisches Öffentliches Recht und Völkerrecht*, *43*, 312–337.

Wolfrum, R. (2004). Legitimacy of international law and the exercise of administrative functions: the example of the International Seabed Authority, the International Maritime Organization (IMO) and the International Fisheries Organizations. *German Law Journal*, *9*(11), 2039–2059.

Wood, M. C. (2007). The International Seabed Authority: fifth to twelfth sessions (1999–2006). *Max Planck Yearbook of United Nations Law*, *11*, 47–98.

Wright, S. W. (1986–1987). The Arkansas law of oil and gas. *University of Arkansas at Little Rock Law Journal*, *9*(2), 223–253.

Xue, G. (2011). How much can a rock get? A reflection from the Okinotorishima Rocks. *13-1 China Oceans Law Review*, 1–34.

Yacouba, C. (2003–2004). Les gisements en mer des hydrocarbures transfrontalières: régime juridique en droit de la délimitation maritime. *Ottawa Law Review 35*, 64–73.

Yankov, A. (2004). The concept of protection and sustainable development in the marine environment. *Ocean Yearbook*, *18*, 267–283.

Yannis, A. (2002). The concept of suspended sovereignty in International Law and its implications in international politics. *European Journal of International Law*, *13*(5), 1037–1052.

Young, R. (1965). Offshore claims and problems in the North Sea. *American Journal of International Law*, *59*(3), 505–522.

Young, R. (1948). Recent developments with respect to the continental shelf. *American Journal of International Law*, *42*, 849–857.

Young, R. (1965). The Persian Golf. In R. Churchill, M. Nordquist, S. Houston Lay (Eds.), 3 *New directions in the Law of the Sea* (pp. 231–240). New York: Oceana Publications Inc.

Young, R. (1970). Equitable solutions for offshore boundaries: the 1968 Saudi Arabia-Iran Agreement. *American Journal of International Law*, *64*, 152–157.

Yturriaga, J. A. de. (1997). In S. Oda (Ed.), *The international regime of fisheries. From UNCLOS 1982 to the Presential Sea*. The Hague, Boston, London: Martinus Nijhoff Publishers.

Yu, S. K.-T. (1996). The law of EEZ/Shelf boundary delimitation: the practice of States in the South China Sea. *Proceedings of the International Law Association (ILA) First Asian-Pacific Regional Conference* (pp. 31–51). Chinese Society of International Law.

Yuan, P. C. (1982). China's jurisdiction over its offshore petroleum resources. *Ocean Development and International Law*, *12*(2), 191–208.

Yuan, P. C. (1988). China's offshore oil development policy and legislation: an overall analysis. *International Journal of Estuarine and Coastal Law*, *3*(2), 101–137.

Yuan, P. C. (1982). China's sovereignty over its offshore oil. In M. J. Valencia (Ed.), *Geology and hydrocarbon potential of the South China Sea and possibilities of joint development* (pp. 520–532). New York, Oxford, Toronto, Sydney, Paris, Frankfurt: Pergamon Press.

Zacher, M. W., & McConnell, J. G. (1990). Down to the sea with stakes: the evolving law of the sea and the future of the deep seabed regime. *Ocean Development and International Law*, *21*(1), 71–103.

Zhang, H., & Mi, C. (2014). Jurisprudential tenability of the Philippines v China arbitration on South China sea disputes? In S. Talmon, B. B. Jia (Eds.) *The South China sea arbitration: A Chinese perspective* (pp. 137–158). Oxford and Portland, Oregon: Hart Publishing.

Zimmermann, A. State Succession in other matters than treaties. *Max Planck Encyclopedia of Public International Law* [online].

Zou, K. (1999a). The Chinese traditional maritime boundary line in the South China Sea and its legal consequences for the resolution of the dispute over the Spratly Islands. *International Journal of Marine and Coastal Law*, 14(1), 27–55.

Zou, K. (1999b). Maritime boundary delimitation in the Gulf of Tonkin. *Ocean Development and International Law, 30*, 235–254.

Zou, K. (2005a). In V. Lowe (Ed.), *China's Marine Legal System and the Law of the Sea.* Leiden, Boston: Martinus Nijhoff Publishers.

Zou, K. (2005b). A new model of joint development for the South China Sea. In M. H. Nordquist, J. N. Moore & K.-c. Fu (Eds.), *Recent developments in the law of the Sean and China* (pp. 155–168). Leiden, Boston: Martinus Nijhoff Publishers.

Zou, K. (2005c). Implementing the United Nations Convention on the Law of the Sea in East Asia: issues and trends. *Singapore Year Book of International Law*, *9*, 37–53.

Zou, K. (2005d). The Sino-Vietnamese agreement on maritime boundary delimitation in the Gulf of Tonkin. *Ocean Development and International Law Journal, 36*, 13–24.

About the International Max Planck Research School of Maritime Affairs at the University of Hamburg

The International Max Planck Research School for Maritime Affairs at the University of Hamburg was established by the Max Planck Society for the Advancement of Science, in co-operation with the Max Planck Institute for Foreign Private Law and Private International Law (Hamburg), the Max Planck Institute for Comparative Foreign Public Law and International Law (Heidelberg), the Max Planck Institute for Meteorology (Hamburg) and the University of Hamburg. The School's research is focused on the legal, economic, and geophysical aspects of the use, protection, and organization of the oceans. Its researchers work in the fields of law, economics, and natural sciences. The School provides extensive research capacities as well as its own teaching curriculum. Currently, the School has 22 Directors who determine the general work of the School, act as supervisors for dissertations, elect applicants for the School's PhD-grants, and are the editors of this book series.

Prof. Dr. Dr. h.c. mult. Jürgen Basedow is Director of the Max Planck Institute for Foreign Private Law and Private International Law; *President and Professor Monika Breuch-Moritz* is the President of the German Federal Maritime and Hydrographic Agency; *Prof. Dr. Dr. h.c. Peter Ehlers* is the Director ret. of the German Federal Maritime and Hydrographic Agency; *Prof. Dr. Dr. h.c. Hartmut Graßl* is Director emeritus of the Max Planck Institute for Meteorology; *Dr. Tatiana Ilyina* is the Leader of the Research Group. Ocean Biogeochemistry" at the Max Planck Institute for Meteorology in Hamburg; *Prof. Dr. Florian Jeßberger* is Head of the International and Comparative Criminal Law Division at the University of Hamburg; *Prof. Dr. Lars Kaleschke* is Junior Professor at the Institute of Oceanography of the University of Hamburg; *Prof. Dr. Hans-Joachim Koch* is Director emeritus of the Seminar of Environmental Law at the University of Hamburg; *Prof. Dr. Robert Koch* is Director of the Institute of Insurance Law at the University of Hamburg; *Prof. Dr. Doris König* is the President of the Bucerius Law School; *Prof. Dr. Rainer Lagoni* is Director emeritus of the Institute of Maritime Law and the Law of the Sea at the University of Hamburg; *Prof. Dr. Gerhard Lammel* is Senior Scientist and Lecturer at the Max Planck Institute for Chemistry, Mainz; *Prof. Dr. Ulrich Magnus* is Managing Director of the Seminar of Foreign Law and Private International Law at the University of Hamburg; *Prof. Dr. Peter Mankowski* is Director of the Seminar of Foreign and Private

V. Becker-Weinberg, *Joint Development of Hydrocarbon Deposits in the Law of the Sea*, Hamburg Studies on Maritime Affairs 30, DOI 10.1007/978-3-662-43570-0

International Law at the University of Hamburg; *Prof. Stefan Oeter* is Managing Director of the Institute for International Affairs at the University of Hamburg; *Prof. Dr. Marian Paschke* is Managing Director of the Institute of Maritime Law and the Law of the Sea at the University of Hamburg; *PD Dr. Thomas Pohlmann* is Senior Scientist at the Centre for Marine and Climate Research and Member of the Institute of Oceanography at the University of Hamburg; *Dr. Uwe A. Schneider* is Assistant Professor at the Research Unit Sustainability and Global Change of the University of Hamburg; *Prof. Dr. Detlef Stammer* is Professor in Physical Oceanography and Remote Sensing at the Institute of Oceanography of the University of Hamburg; *Prof. Dr. Jürgen Sündermann* is Director emeritus of the Centre for Marine and Climate Research at the University of Hamburg; *Prof. Dr. Rüdiger Wolfrum* is Director emeritus at the Max Planck Institute for Comparative Foreign Public Law and International Law and a judge at the International Tribunal for the Law of the Sea; *Prof. Dr. Wilfried Zahel* is Professor emeritus at the Centre for Marine and Climate Research of the University of Hamburg.

At present, *Prof. Dr. Dr. h.c. Jürgen Basedow* and *Prof. Dr. Ulrich Magnus* serve as speakers of the Research School.

MIX
Papier aus verantwortungsvollen Quellen
Paper from responsible sources
FSC® C105338

If you have any concerns about our products,
you can contact us on
ProductSafety@springernature.com

In case Publisher is established outside the EU,
the EU authorized representative is:
Springer Nature Customer Service Center GmbH
Europaplatz 3, 69115 Heidelberg, Germany

Printed by Libri Plureos GmbH
in Hamburg, Germany